T0199798

Digital Protective Relays

Problems and Solutions

Digital Protective Relays

Problems and Solutions

Vladimir Gurevich

CRC Press
Taylor & Francis Group
Boca Raton London New York

CRC Press is an imprint of the
Taylor & Francis Group, an **informa** business

CRC Press
Taylor & Francis Group
6000 Broken Sound Parkway NW, Suite 300
Boca Raton, FL 33487-2742

First issued in paperback 2019

ISBN-13: 978-1-4398-3785-6 (hbk)
ISBN-13: 978-0-367-38338-1 (pbk)

Library of Congress Cataloging-in-Publication Data

Gurevich, Vladimir, 1956-
    Digital protective relays : problems and solutions / author, Vladimir Gurevich.
        p. cm.
    Includes bibliographical references and index.
    ISBN 978-1-4398-3785-6 (hardcover : alk. paper)
    1. Protective relays. 2. Digital electronics. I. Title.

    TK2861.G87 2011
    621.31'7--dc22   2010030228

Visit the Taylor & Francis Web site at
http://www.taylorandfrancis.com

and the CRC Press Web site at
http://www.crcpress.com

Contents

Preface

> The Electronic World is reality, the game occurs in the physical world.
>
> **—Sidey Myoo**

Today it is quite common to say that we are living in the "atomic age," but this is wrong; the fact is that we are, and have been for some time, living in the "electronics age." There is nothing on Earth (and in space) that in one way or another does not depend on electricity. Industry, manufacturing, transportation, communication, banks, health care—whatever the endeavor, it is driven by electricity.

We are so used to electricity that we take it for granted. And we are paying a price for this insouciance. Unfortunately, the electronics age lets us know what happens when electronic systems fail. In the last 20–30 years, we have witnessed several electronic disasters stemming from human error, for example massive power grid failures (blackouts), leading to huge losses and often death (in the United States, 1965, 1977, and 2003; France, 1978; Canada, 1982 and 2003; Italy, 2003; and Sweden, 1983 and 2003), aircraft crashes (the most recent being the crash of flight AF-447, an Airbus A330-200 from Rio de Janeiro to Paris, on June 1, 2009), and so on.

Integral microchips and microprocessors have come into our lives so swiftly and completely that sometimes it seems that modern equipment simply cannot exist without them, which is true. However, the dependence of modern equipment on microelectronics and microprocessors does not mean that there are no problems in this area. The integrity of many functions distributed earlier among separate devices of a complex system in a single microprocessor leads to the reduction of system reliability because damage to the microprocessor or to any number of peripheral elements serving the microprocessor leads to failure of the whole system but not of its separate functions as it was in pre-microprocessor time. Added to this is the extra sensitivity of microelectronic and microprocessor-based equipment to electromagnetic interferences (EMIs) and the possibility of intentional remote actions breaking the normal operation of the microprocessor-based devices (e.g., electromagnetic weapons and electromagnetic terrorism). Intensive investigations into the electromagnetic weapons field are being carried out in Russia, the United States, England, Germany, China, and India. Many world-leading companies work intensively in this sphere creating new devices of these weapon systems functioning at a distance from several dozens of meters to several kilometers, which while specialized in their use are still available to everybody (as they are freely sold on the market).

Relay protection of power units plays an important role in the hierarchy of the electronics age in preventing many disasters.

On the other hand, malfunction protective relays comprise one of the main causes of the heavy failures that periodically occur in power systems all over the world. According to the North American Electric Reliability Council, in 74% of the cases the reason for heavy failures in power systems was the incorrect actions of relay protection in trying to avoid the failure. Thus, the reliability of a power system depends on the reliability of relay protection in many respects.

The possibility of using computers for protecting elements of power systems was first suggested in 1965. George D. Rockefeller was the first to outline the details of using a computer for protecting all the equipment in a high-voltage substation and the lines emanating from the substation.[1]

Digital protective relays (DPRs) started to replace static relays in the beginning of the 1980s. At first these were "hybrid" solutions, where the time-critical filtering was performed with analog electronics. Typical examples are REZ1 (universal phase and ground distance relay for permissive and blocking schemes), RACID (universal phase and ground overcurrent relay for lines and cables), REG 100 (multifunction generator protection relay, with differential, underimpedance, overexcitation, overvoltage, and other protection functions), REB 100 (busbar protection), and others.

Today protective-relay usage patterns in the world's electric power business continue to grow, with the annual market exceeding $1.5 billion (all figures this volume given in U.S. dollars unless otherwise indicated), according to a recent study by the Newton-Evans Research Co.[2] In total, global demand for protective relays will approach $2 billion by 2009, estimates Newton-Evans. The percentage of digital relays in the mix of the millions of protective relays used by the world's utilities continues to increase. Nearly 60% of the installed generator relays and more than 50% of transmission line relays in North America are now digital units.

There are currently at least ten large suppliers of protective relays: ABB, Areva, GE, SEL, Siemens, NARI-Relays, Basler, Beckwith, Cooper Power, and Schneider.

On a global basis, electric utilities currently purchase only about $850 million in protective relays directly from manufacturers each year. As much as $120 million of this amount is electromechanical, which is still prevalent in Russia, Eastern Europe, and Central Asia and continues to account for another 10 to 20% or so of demand in most other world regions. North American utilities continue to account for about $35 million in annual purchases of electromechanical relays.

Despite some clear and very well-known advantages of digital relays (which are much discussed in technical journals), digital relays also have serious problems, about which researchers usually prefer to not mention. Why? But are the DPR ideal devices? If one is to trust numerous publications in the technical literature, yes! Then it is possible to explain the full absence of critical publications considering problems and disadvantages of the DPR. However, it seems rather strange that such complex technical systems as DPRs should not have disadvantages, like any other complex engineering systems. Alas, in a real world, as we well know, ideal devices do not exist.

This is the first book on the market that is not devoted to the well-known advantages of DPRs, as all other books on the subject are, but to its poorly known problems and disadvantages. It is thus unique in this sense.

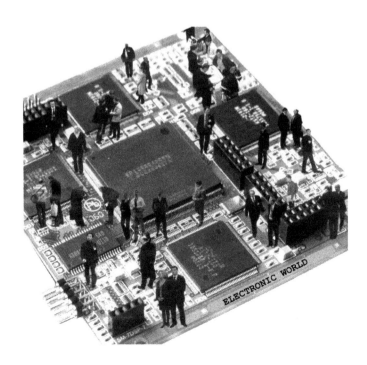

References

1. Rockefeller, G. D. Fault Protection with a Digital Computer. *IEEE Transactions*, 1969, PAS-88, pp. 438–464.
2. Newton-Evans Research Co. The World Market for Protective Relays in Electric Utilities: 2006–2008. Ellicott City, MD: Newton-Evans Research Co.

Acknowledgments

Acknowledgment is made to the following firms and organizations for their kind permission for allowing me to use various information and illustrations:

Alliance Semiconductor

Analog Devices, Inc.

Dionics, Inc.

Fairchild Semiconductors, Inc.

NxtPhase T&D Corp.

STMicroelectronics

Finally and most importantly, my utmost thanks and appreciation go to my wife, Tatiana, who has endured the writing of this (my fifth!) book, for her never-ending patience.

About the Author

Vladimir I. Gurevich was born in Kharkov, Ukraine in 1956. He received an M.S.E.E. degree (1978) at the Kharkov Technical University and a Ph.D. degree (1986) at Kharkov National Polytechnic University.

He has held the positions of assistant and associate professor at Kharkov Technical University and chief engineer and director of Inventor, Ltd.

He arrived in Israel in 1994 and works today at the Israel Electric Corporation as an engineer specialist and head of a section of the Central Electric Laboratory.

He is the author of more than 130 professional papers and 5 books and the holder of nearly 120 patents in the fields of electrical engineering and power electronics. In 2006, he became an honored professor with the Kharkov Technical University. Since 2007, he has served as an expert with the TC-94 Committee of the International Electrotechnical Commission (IEC) and in 2010, he became a member of the Israel National Committee of the International Council on Large Electric Systems (CIGRE).

Gurevich's books, which have been published by Taylor & Francis, include the following:

- *Protection Devices and Systems for High-Voltage Applications*
- *Electrical Relays: Principles and Applications*
- *Electronic Devices on Discrete Components for Industrial and Power Engineering*

His personal website is www.gurevich-publications.com

1

Basic Components of Digital Protective Relays

1.1 Semiconducting Materials and the *p-n* Junction

As is known, all substances depending on their electroconductivity are divided into three groups: conductors (usually metals) with a resistance of 10^{-6}–10^{-3} Ohm × cm, dielectrics with a resistance of 10^9–10^{20} Ohm × cm, and semiconductors (many native-grown and artificial crystals) covering an enormous intermediate range of values of specific electrical resistance.

The main peculiarity of crystal substances is typical, well-ordered atomic packing into peculiar blocks—crystals. Each crystal has several flat symmetric surfaces, and its internal structure is determined by the regular positional relationship of its atoms, which is called the *lattice*. Both in appearance and in structure, any crystal is like any other crystal of the same given substance. Crystals of various substances are different. For example, a crystal of table salt has the form of a cube. A single crystal may be quite large in size or so small that it can only be seen with the help of a microscope. Substances having no crystal structure are called *amorphous*. For example, glass is amorphous in contrast to quartz, which has a crystal structure.

Among the semiconductors that are now used in electronics, one should point out germanium, silicon, selenium, copper-oxide, copper sulfide, cadmium sulfide, gallium arsenide, and carborundum. To produce semiconductors, two elements are mostly used: germanium and silicon.

In order to understand the processes taking place in semiconductors, it is necessary to consider phenomena in the crystal structure of semiconductor materials, which occur when their atoms are held in a strictly determined relative position to each other due to weakly bound electrons on their external shells. Such electrons, together with electrons of neighboring atoms, form *valence bonds* between the atoms. Electrons taking part in such bonds are called *valence electrons*. In absolutely pure germanium or silicon at very low temperatures, there are no free electrons capable of creating electric current, because under such circumstances all four valence electrons of the external shells of each atom that can take part in the process of charge transfer are too strongly held by the valence bounds. That's why that substance is an insulator (dielectric) in the full sense of the word—it does not let electric current pass at all.

When the temperature is increased, due to the thermal motion some valence electrons detach from their bonds and can move along the crystal lattice. Such electrons are called *free electrons*. The valence bond from which the electron is detached is called a *hole*. It possesses properties of a positive electric charge, in contrast to the electron, which has a negative electric charge. The higher the temperature is, the greater the number of free electrons capable of moving along the lattice, and the higher the conductivity of the substance.

Moving along the crystal lattice, free electrons may run across holes—valence bonds missing some electrons—and fill up these bonds. Such a phenomenon is called *recombination*. At normal temperatures in the semiconductor material, free electrons occur constantly, and recombination of electrons and holes takes place.

If a piece of semiconductor material is put into an electric field by applying a positive or negative terminal to its ends, for instance, electrons will move through the lattice toward the positive electrode and holes—to the negative one. The conductivity of a semiconductor can be enhanced considerably by applying specially selected admixtures to it—metal or nonmetal ones. In the lattice, the atoms of these admixtures will replace some of the atoms of the semiconductors. Let us remind ourselves that external shells of atoms of germanium and silicon contain four valence electrons, and that electrons can only be taken from the external shell of the atom. In their turn the electrons can be added only to the external shell, and the maximum number of electrons on the external shell is eight.

When an atom of the admixture that has more valence electrons than required for valence bonds with neighboring atoms of the semiconductor, additional free electrons capable of moving along the lattice occur on it. As a result, the electroconductivity of the semiconductor increases. As germanium and silicon belong to the fourth group of the periodic table of chemical elements, donors for them may be elements of the fifth group, which have five electrons on the external shell of atoms. Phosphorus, arsenic, and stibium belong to such donors (also called *donor admixtures*).

If admixture atoms have fewer electrons than needed for valence bonds with surrounding semiconductor atoms, some of these bonds turn out to be vacant and holes will occur in them. Admixtures of this kind are called *p-type admixtures* because they absorb (accept) free electrons. For germanium and silicon, *p*-type admixtures are elements from the third group of the periodic table of chemical elements, and the external shells of their atoms contain three valence electrons. Boron, aluminum, gallium, and indium can be considered *p*-type admixtures (also called *accepter admixtures*).

In the crystal structure of a pure semiconductor, all valence bonds of neighboring atoms turn out to be fully filled, and occurrence of free electrons and holes can be caused only by deformation of the lattice, arising from thermal or other radiation. Because of this, the conductivity of a pure semiconductor is quite low under normal conditions.

If some donor admixture is injected, the four electrons of the admixture, together with the same number in the filled valence, bond with the latter. The fifth electron of each admixture atom appears to be "excessive" or "redundant," and therefore can freely move along the lattice.

When an accepter admixture is injected, only three filled valence bonds are formed between each atom of the admixture and neighboring atoms of the semiconductor. To fill up the fourth, one electron is lacking. This valence bond appears to be vacant. As a result, a hole occurs. Holes can move along the lattice like positive charges, but instead of an admixture atom, which has a fixed and permanent position in the crystal structure, the vacant valence bond moves.

It goes like this: an electron is known to be an elementary carrier of an electric charge. Affected by different causes, the electron can escape from the filled valence bond, having left a hole which is a vacant valence bond and which *behaves like a positive charge equaling numerically the negative charge of the electron*. Affected by the attracting force of its positive charge, the electron of another atom near the hole may "jump" to the hole. At that point, recombination of the hole and the electron occurs, their charges are mutually neutralized, and the valence bond is filled. The hole in this place of the lattice of the semiconductor disappears. In its turn a new hole, which has arisen in the valence bond from which the electron has escaped, may be filled with some other electron which has left a hole. Thus, *moving of electrons in the lattice of the semiconductor with a p-type admixture and recombination of them with holes can be regarded as the moving of holes*. For better understanding, one may imagine a concert hall in which for some reason some seats in the first row turn out to be vacant. As spectators from the second row move to the vacant seats in the first row, their seats are taken by spectators of the third row, and so on. One can say that in some sense, vacant seats "move" to the last rows of the concert halls, although in fact all the seats remain screwed to the floor. The "moving" of holes in the crystal is very much like the "moving" of such vacant seats.

Semiconductors with electroconductivity enhanced, due to an excess of free electrons caused by admixture injection, are called *semiconductors with electroconductivity* or, in short, *n-type semiconductors*. Semiconductors with electroconductivity influenced mostly by the moving of holes are called *semiconductors with p-type conductivity* or just *p-type semiconductors*.

There are practically no semiconductors with only electronic or only p-type conductivity. In a semiconductor of *n*-type, electric current is partially caused by the moving of holes arising in its lattice because of an escaping of electrons from some valence bonds, and in semiconductors of *p*-type, current is partially created by the moving of electrons. Because of this, it is better to define semiconductors of the *n*-type as semiconductors in which *the main current carriers are electrons* and semiconductors of the *p*-type as semiconductors in which *holes are the main current carriers.* Thus a semiconductor belongs to this or that type depending on what type of current carrier predominates in it. According to this, the other opposite charge carrier for any semiconductor of a given type is a *minor carrier.*

One should take into account that any semiconductor can be made a semiconductor of *n*- or *p*-type by putting certain admixtures into it. In order to obtain the required conductivity, it is enough to add a very small amount of the admixture, about one atom of the admixture for tens of millions of atoms of the semiconductor. All of this imposes special requirements for the purification of the original semiconductor material, and for accuracy in the dosage of admixture injection. One should also take into consideration that the speed of current carriers in a semiconductor is lower than in a metal conductor or in a vacuum. The moving of electrons is slowed down by obstacles in the form of inhomogeneities in the crystal. The moving of holes is half as slow because they move due to the jumping of electrons to vacant valence bounds. The mobility of electrons and holes in a semiconductor is increased when the temperature goes up. This leads to an increase of conductivity of the semiconductor.

The functioning of most semiconductors is based on the processes taking place in an intermediate layer formed in the semiconductor, at the boundary of the two zones with the conductivities of the two different types: *p* and *n*. The boundary is usually called the *p-n junction* or the *electron-hole junction*, in accordance with the main characteristics of the type of main charge carriers in the two adjoining zones of the semiconductor.

There are two types of *p-n* junctions, *planar* and *point junctions*, which are illustrated schematically in Figure 1.1. A planar junction is formed by

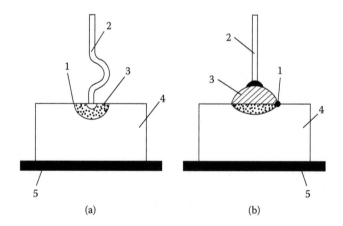

FIGURE 1.1
Construction of (a) point and (b) planar *p-n* junctions of the diode. 1: *p-n* junction; 2: wire terminal; 3: *p*-area; 4: crystal of *n*-type; and 5: metal heel piece.

moving a piece of the admixture—for instance, indium, to the surface of the germanium—of *n*-type, and further heating until the admixture is melted.

When a certain temperature is maintained for a certain period of time, there is diffusion of some admixture atoms to the plate of the semiconductor, to a small depth, and a zone with conductivity opposite of that of the original semiconductor is formed. In the above case, it is *p*-type for *n*-germanium.

Point junction results from tight electric contact of the thin metal conductor (wire), which is known to have electric conductivity, with the surface of the *p*-type semiconductor. This was the basic principle on which the first crystal detectors operated. To decrease dependence of diode properties on the position of the pointed end of the wire on the surface of the semiconductor, and the clearance of its momentary surface point, junctions are formed by fusing the end of the thin metal wire to the surface of a semiconductor of the *n*-type. Fusion is carried the moment a short-term powerful pulse of electric current is applied. Affected by the heat formed for this short period of time, some electrons escape from atoms of the semiconductor, which are near the contact point, and leave holes. As a result of this, some small part of the *n*-type semiconductor in the immediate vicinity of the contact turns into a semiconductor of the *p*-type (area 3 on Figure 1.1a).

Each part of semiconductor material, taken separately (that is, before contacting), was neutral, since there was a balance of free and bound charges (Figure 1.2a). In the *n*-type area, concentration of free electrons is quite high and that of holes is quite low. In the *p*-type area, on the contrary, the concentration of holes is high, and that of electrons is low. Joining of semiconductors with different concentrations of main current carriers causes diffusion of these carriers through the junction layer of these materials: the main carriers of the *p*-type semiconductor—holes—diffuse to the *n*-type area because the concentration of holes in it is very low. And vice versa: electrons from an *n*-type semiconductor with a high concentration of them diffuse to the *p*-type area, where there are few of them (Figure 1.2b).

On the boundary of the division of the two semiconductors, from each side a thin zone with conductivity opposite of that of the original semiconductor is formed. As a result, on the boundary (which is called a *p-n* junction) a space charge arises (the so-called potential barrier), which creates a diffusive electric field and prevents the main current carriers from flowing after balance has been achieved (Figure 1.2c).

Strongly pronounced dependence of electric conductivity of a *p-n* junction, from the polarity of external voltage applied to it, is typical of the *p-n* junction. This can never be noticed in a semiconductor with the same conductivity. If voltage applied from the outside creates an electric field coinciding with a diffusive electric field, the junction will be blocked and current will not pass through it (Figure 1.3).

Moreover, moving of minor carriers becomes more intense, which causes enlargement of the blocking layer and lifting of the barrier for main carriers.

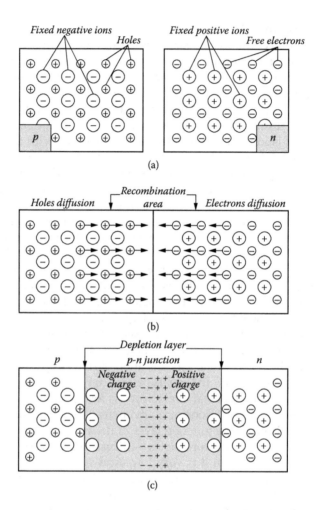

FIGURE 1.2
Formation of a blocking layer when semiconductors of different conductivity are connected.

In this case, it is usually said that the junction is *reversely bias*. Moving of minor carriers causes a small current to pass through the blocked junction. This is the so-called *reverse current* of the diode, or *leakage current*. The smaller it is, the better the diode is.

When the polarity of the voltage applied to the junction is changed, the number of main charge carriers in the junction zone increases. They neutralize the space charge of the blocking layer by reducing its width and lowering the potential barrier that prevented the main carriers from mobbing through the junction. It is usually said that the junction is *forward biased*. The voltage required for overcoming the potential barrier in the forward direction is about 0.2 V for germanium diodes, and 0.6–0.7 V for silicon ones. To

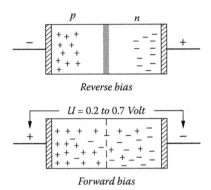

FIGURE 1.3
p-n junction with reverse and forward bias.

overcome the potential barrier in the reverse direction, tens and sometimes even thousands of volts are required.

If the barrier is overpassed, irreversible destruction of the junction and its breakdown take place, which is why threshold values of reverse voltage and forward current are indicated for junctions of different appliances.

Figure 1.4 illustrates an approximate volt-ampere characteristic of a single junction, which is dependent upon current passing through it on the polarity and external voltage applied to the junction. Currents of forward and reverse direction (up to the breakdown area) may differ by tens and hundreds of times. As a rule, planar junctions withstand higher voltages and currents than point ones, but do not work properly with high-frequency currents.

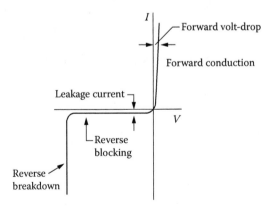

FIGURE 1.4
Volt-ampere characteristic of a single *p-n* junction (diode).

1.2 The Principle behind Transistors

The idea of somehow using semiconductors had been tossed about before World War II, but knowledge about how they worked was scant, and manufacturing semiconductors was difficult. In 1945, however, the vice president for research at Bell Laboratories established a research group to look into the problem. The group was led by William Shockley and included Walter Brattain, John Bardeen, and others, physicists who had worked with quantum theory, especially in solids. The team was talented and worked well together.

In 1947 John Bardeen and Walter Brattain, with colleagues, created the first successful amplifying semiconductor device. They called it a transistor (from *transfer* and *resistor*). In 1950 Shockley made improvements to it that made it easier to manufacture. His original idea eventually led to the development of the silicon chip. Shockley, Bardeen, and Brattain won the 1956 Nobel Prize for the development of the transistor. It allowed electronic devices to be built smaller, lighter, and even cheaper.

It can be seen in Figure 1.5 that a transistor contains two semiconductor diodes, connected together and having a common area. Two utmost layers of the semiconductor (one of them is called an *emitter*, and the other a *collector*) have *p*-type conductivity with a high concentration of holes, and the intermediate layer (called a *base*) has *n*-type conductivity with a low concentration of electrons. In electric circuits, low voltage is applied to the first (the emitter) *p-n* junction because the junction is connected in the forward (carrying) direction, and much higher voltage is applied to the second (the collector) junction in the reverse (cutoff) direction. In other words, the emitter junction is forward biased and the collector junction is reverse biased. The collector junction remains blocked until there is no current in the emitter-base circuit. The resistance of the whole crystal (from the emitter to the collector) is very high. As soon as the input circuit (Figure 1.5) is closed, holes from the emitter seem to be injected (emitted) to the base and quickly saturate it (including the area adjacent to the collector). As the concentration of holes in the emitter is much higher than the concentration of electrons in the base, after recombination

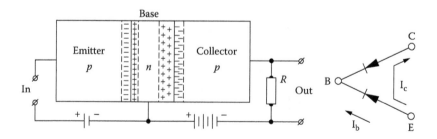

FIGURE 1.5
Circuit and the principle of operation of a transistor.

there are still many vacant holes in the base area, which is affected by the high voltage (a few or tens of volts) applied between the base and the collector, easily overpassing the barrier layer between the base and the collector.

Increased concentration of holes in the cutoff collector junction causes the resistance of this junction to fall rapidly, and it begins to *conduct current in the reverse direction*. The high strength of the electric field in the "base-collector" junction results in a very high sensitivity of the resistance of this junction in the reverse (cutoff) state to a concentration of the holes in it. That's why even a small number of holes injected from the emitter under the effect of weak input current can lead to sharp changes of conductivity of the whole structure, and considerable current in the collector circuit.

The ratio of collector current to base current is called the *current amplification factor*. In low-power transistors, this amplification factor has values of tens and hundreds, and in power transistors it has values of tens.

1.3 Some Transistor Kinds

In the 1970s, transistor engineering developed very rapidly. Hundreds of types of transistors and new variants of them appeared (Figure 1.6). Among them appeared transistors with reverse conductivity or *n-p-n* transistors, and also unijunction transistors (so called because these contain only one

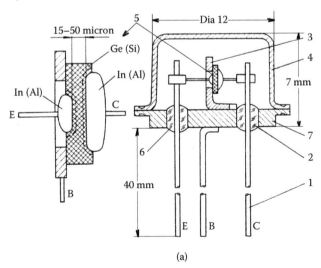

(a)

FIGURE 1.6
Transistors produced in the 1970s: (a) low-power transistor. 1: outlets; 2 and 6: glass insulators; 3: crystal holder; 4: protection cover; 5: silicon (germanium) crystal; 7: flange; 8: copper heat sink; 9: Kovar bushing; and 10: hole for gas removal after case welding and disk for sealing in.

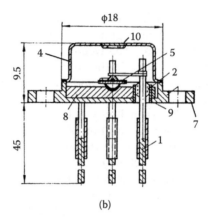

(b)

FIGURE 1.6 (continued)
(b) power transistor.

junction; such a transistor is sometimes also called a *two-base diode*; see Figure 1.7). This transistor contains one junction formed by welding a core made from *p*-material to a single-crystal wafer made from *n*-type material (silicon). The two outlets, serving as bases, are attached to the wafer. The core, placed asymmetrically with regard to the base, is called an *emitter*. Resistance between the bases is about a few thousand Ohms. Usually,

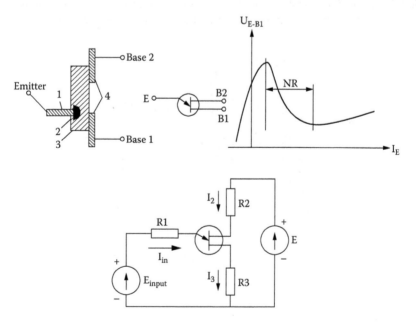

FIGURE 1.7
A unijunction transistor (or two-base diode) and its circuit. 1: *p*-type core; 2: *p-n* junction; 3: *n*-type plate; 4: Ohmic contacts; and NR: negative resistance area.

the base B_2 is biased in a positive direction from the base B_1. Application of positive voltage to the emitter causes a strong current of the emitter (with an insignificant voltage drop between the emitter E and the base B_1). One can observe the area of negative resistance (NR; see Figure 1.7) on the emitter characteristic of the transistor where the transistor is very rapidly enabled, operating like a switch.

In fact, modern transistors (Figure 1.8) are characterized by such a diversity of types that it is simply impossible to describe all of them in this book. Therefore, only a brief description of the most popular types of modern semiconductor devices, and the relays based on them, are presented here.

Besides the transistors described above, which are called *bipolar junction transistors* or just *bipolar transistors* (Figure 1.9), so-called field effect transistors (FETs; see Figure 1.10) have become very popular recently. The first person to attempt to construct a FET in 1948 was, again, William Shockley. But it took many years of additional experiments to create a working FET with a control *p-n* junction called a *unitron* (unipolar transistor); this was accomplished in 1952. Such a transistor was a semiconductor three-electrode device in which control of the current caused by the ordered motion of charge carriers of the same sign between two electrodes was carried out with the help of an electric field (that is why it is called a *field*) applied to the ford electrode.

Electrodes between which working currents pass are called *source* and *drain* electrodes. The source electrode is the one through which carriers flow

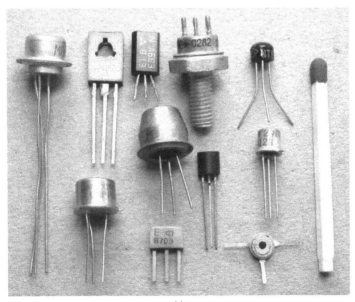

(a)

FIGURE 1.8
(a) low-signal transistors.

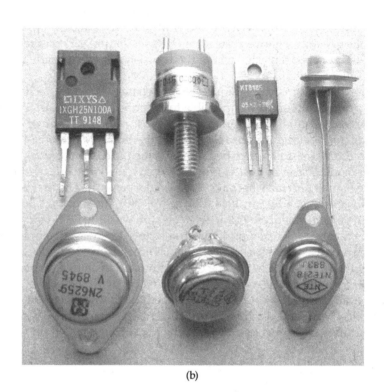

(b)

(c)

FIGURE 1.8 (continued).
(b) power transistors. (c) high-power transistors.

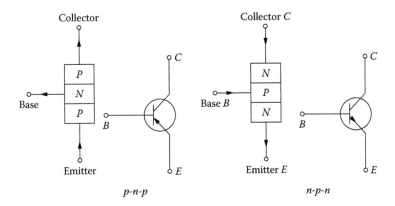

FIGURE 1.9
Structure and symbolic notation on the schemes of bipolar transistors of *p-n-p* and *n-p-n* types.

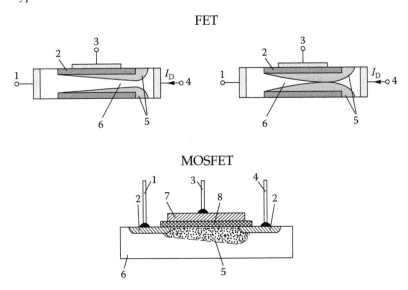

FIGURE 1.10
Simplified structure of field effect transistors (FETs) and metal-oxide semiconductor FETs (MOSFETs). 1: source; 2: *n*-type admixture; 3: gate; 4: drain; 5: area consolidated by current carriers (depletion layer); 6: conductive channel in silicon of *p*-type; 7: metal; and 8: silicon dioxide.

into the device. The third electrode is called a *gate*. Change of value of the working current in a unipolar transistor is carried out by changing the effective resistance of the current conducting area, the semiconductor material between the source and the drain called the *channel*. That change is made by increasing or decreasing area 5 (Figure 1.10). Increase of voltage of the initial junction bias leads to expansion of the depletion layer. As a result, the rest area of the section of the conductive channel in the silicon decreases and the transistor is blocked, and vice versa: when the value of the blocking voltage on the gate decreases, the area depleted by current carriers (area 5) contracts and turns into a pointed wedge. At the same time, the section of the conductive channel increases and the transistor is enabled.

Depending on the type of conductivity of semiconductor material of the channels, there are unipolar transistors with *p* and *n* channels. Because of the fact that control of the working current of unipolar transistors is carried out with the help of a channel, they are also called *channel transistors*. The third name of the same semiconductor device—a *field transistor* or *field effect transistor* (FET)—points out that working current control is carried out by an electric field (voltage) instead of electric current as in a bipolar transistor. The latter peculiarity of unipolar transistors, which allows them to obtain very high input resistances estimated in tens and hundreds of megohms, determined their most popular name: *field transistors*.

It should be noted that apart from field transistors with *p-n* junctions between the gate and the channel (FETs), there are also field transistors with an insulated gate: metal-oxide semiconductor FETs (MOSFETs). The latter were suggested by S. Hofstein and F. Heiman in 1963.

Field transistors with an insulated gate appeared as a result of searching for methods to further increase input resistance and frequency range extensions of field transistors with *p-n* junctions. The distinguishing feature of such field transistors is that the junction biased in a reverse direction is replaced with a control structure of "metal-oxide semiconductor," or a MOSFET structure in abbreviated form. As shown in Figure 1.10, this device is based on a silicon mono crystal, in this case of the *p*-type. The source and drain areas have conductivity opposite to the rest of the crystal, that is, of the *n*-type. The distance between the source and the drain is very small, usually about 1 micron. The semiconductor area between the source and the drain, which is capable of conducting current under certain conditions, is called a *channel*, as in the previous case.

In fact, the channel is an *n*-type area formed by diffusion of a small amount of the donor admixture to the crystal with *p*-type conductivity. The gate is a metal plate covering source and drain zones. It is isolated from the mono crystal by a dielectric layer only 0.1 micron thick. The film of silicon dioxide formed at this high temperature is used as a dielectric. Such film allows us to adjust the concentration of the main carriers in the channel area by changing both value and polarity of the gate voltage. This is the major difference between field transistors with *p-n* junctions (Figure 1.11), which can only operate well *with blocking voltage of the gate*, as opposed to the MOSFET transistors (Figure 1.12). The change of polarity of the bias voltage leads to junction unblocking and to a sharp reduction of the input resistance of the transistor.

The basic advantages of MOSFET transistors are as follows: first, there is an insulated gate allowing an increase in input resistance by at least 1000 times

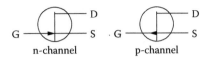

FIGURE 1.11
Symbolic notation of FETs with *n* and *p* channels. G: gate; S: source; and D: drain.

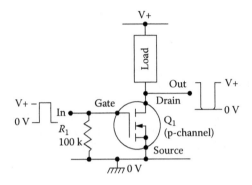

FIGURE 1.12
Symbolic notation and the circuit of a MOSFET.

in comparison with the input resistance of a field transistor with a *p-n* junction. In fact, it can reach a billion megohms. Second, gate and drain capacities become considerably lower and usually do not exceed 1–2 pF. Third, the limiting frequency of MOSFETs can reach 700–1000 MHz, which is at least ten times higher than that of standard field transistors.

Attempts to combine in one switching device the advantages of bipolar and field transistors led to the invention of a compound structure in 1978, which was called a *pobistor* (Figure 1.13). The idea of a modular junction of crystals of bipolar and field transistors in the same case was employed by Mitsubishi Electric to create a powerful switching semiconductor module (Figure 1.14).

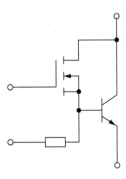

FIGURE 1.13
Compound structure *pobistor*.

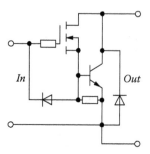

FIGURE 1.14
Scheme of a power-switching module CASCADE-CD, with a working voltage of 1000 V and currents more than 100 A (Mitsubishi Electric).

Further development of production technology of semiconductor devices allowed the development of a single-crystal device with a complex structure with properties of a pobistor: an *insulated gate bipolar transistor* (IGBT). This is a device which combines the fast-acting features and high-power capabilities of the bipolar transistor with the voltage control features of the MOSFET gate. In simple terms, the collector-emitter characteristics are similar to those of the bipolar transistor, but the control features are those of the MOSFET. The equivalent circuit and the circuit symbol are illustrated in Figure 1.15. Such a transistor (Figure 1.16) has a higher switching power than FETs and bipolar transistors, and its operation speed is between that of FETs and bipolar transistors. Unlike bipolar transistors, the IGBT transistor doesn't operate well in the amplification mode, and it is designed for use in the switching (relay) mode as a powerful high-speed switch.

The IGBT transistor is enabled by a signal of positive (with regard to the emitter) polarity, with voltage not more than 20 V. It can be blocked with zero potential on the gate; however, with some types of loads a signal of negative polarity on the gate may be required for reliable blocking (Figure 1.17). Many companies produce special devices for IGBT transistor control. They are made as separate integrated circuits or ready-to-use printed circuit cards,

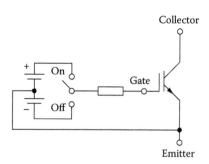

FIGURE 1.15
Insulated gate bipolar transistor (IGBT).

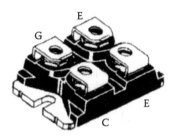

FIGURE 1.16
An IGBT-transistor IXDN75N120A produced by IXYS with a switched current up to 120 A, and maximum voltage of up to 1200 V (dissipated power is 630 W). With such high parameters, the device is quite small in size: 38 × 25 × 12 mm.

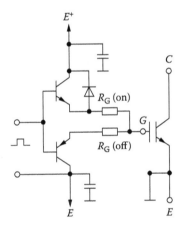

FIGURE 1.17
Model scheme of the IGBT transistor control, providing pulses of opposite polarity on the gate required for reliable blocking and unblocking of the transistor.

so-called *drivers* (Figure 1.18). Such drivers are universal as a rule and can be applied to any type of power IGBT transistors. Apart from forming control signals of the required level and form, such devices often protect the IGBT from short circuits.

1.4 General Modes of the Bipolar Transistor

As an element of an electric circuit, the transistor is usually used such that one of its electrodes is input, one is output, and the third is the common with respect to the input and output. A transistor is commonly connected to external circuits in a four-terminal configuration, referred to as a *quadrupole*.

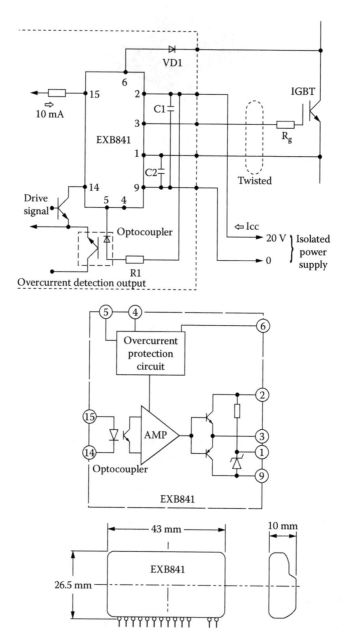

FIGURE 1.18
IGBT-driving hybrid integral circuit EXB841 type (Fuji Electric).

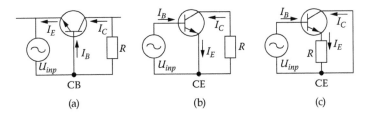

FIGURE 1.19
Transistor basic circuit configurations: (a) common base (CB), (b) common emitter (CE), and (c) common collector (CC). I_E: emitter current; I_B: base current; I_C: collector current; and R: load.

The source of an input signal requiring amplification is connected to the circuit of the input electrode, and the load on which the amplified signal is dissipated is connected to the circuit of the output electrode. Depending on which electrode is the common for the input and output circuits, transistor connections fall into three basic circuits, as shown in Figure 1.19.

In the common base (CB) circuit, I_E is the input signal and I_C is the output signal. The current amplification coefficient (also called the *current gain*, which is the ratio of amplifier output current to input current) of a transistor in this configuration is equal to $\alpha = I_C/I_E \approx 1$. A device may have low internal input resistance and high internal output resistance, and for this reason, a change of the load resistance exerts only a minimal influence on the output current (the functional scheme of this mode relates to current source). CB configurations of transistor connections are not commonly used in practice.

The common emitter (CE) circuit is used most often as an amplification stage. Current gain for this circuit is close to the transistor's gain and is equal to $\beta = I_C/I_B \approx 10-200$ and more, depending on the type of transistor used. For direct current, $\beta = h_{FE}$. (h_{FE} is the DC current gain, a parameter specified by the manufacturer of transistors). This circuit has rather high input resistance (i.e., it does not shunt and weaken the input signal) and low output resistance.

The common collector (CC) circuit is used quite often in cases where it is needed to stage with very high input resistance. Its circuit current amplification coefficient is close to that of the CE circuit; however, the main application of this circuit relates to the functional mode of the voltage, not current amplification, as precisely in this mode one manages to realize the scheme most fully: its very high input resistance. In the voltage amplification mode, the scheme has a gain of close to one.

Regardless of the transistor used in the circuit, it may function in four main modes determined by the polarity of voltage on the emitter and collector junctions. All possible bias modes are illustrated in Figure 1.20. They are the forward active mode of operation, the reverse active mode of operation, the saturation mode, and the cutoff mode.

As the base current I_B increases or decreases, the operating point moves up or down the load line (see Figure 1.21). If I_B increases too much, the operating point moves into the saturation region. In the saturation region, the

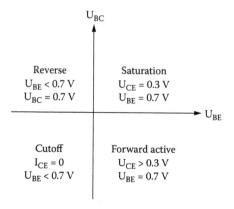

FIGURE 1.20
Possible bias modes of operation of a bipolar junction transistor.

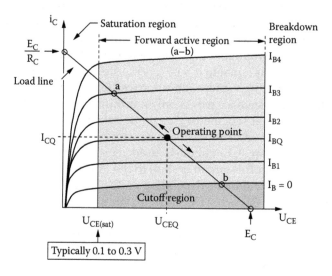

FIGURE 1.21
Output dynamic characteristic of a bipolar junction transistor.

transistor is fully turned ON and the value of collector current I_C is determined by the value of the load resistance R_L. The voltage drop across the transistor V_{CE} is near zero.

In the cutoff region, the transistor is fully turned OFF and the value of the collector current I_C is near zero. Full-power supply voltage appears across the transistor. Because there is no current flow through the transistor, there is no voltage drop across the load resistor R_L.

However, bipolar transistors do not have to be restricted to these two extreme modes of operation. As described above, base current "opens a gate" for a limited amount of current through the collector. If this limit for the

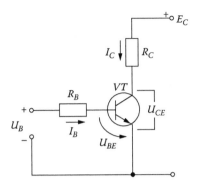

FIGURE 1.22
Transistor stage in CE mode.

controlled current is greater than zero but less than the maximum allowed by the power supply and load circuit, the transistor will "throttle" the collector current in a mode somewhere between cutoff and saturation. This mode of operation is called the *active mode*. A load line is a plot of collector-to-emitter voltage over a range of base currents. The dots marking where the load line intersects the various transistor curves represent the realistic operating conditions for those base currents given.

Let us examine the transistor stage circuit of CE type in detail (see Figure 1.22). It begins to get clearer as to what the load line (Figure 1.21) is. This line is built on the series of the static volt-ampere characteristics of transistors (these characteristics are parameters set by the manufacturer) on two cross-points with axes corresponding to the idle mode and short circuit.

Main ratios for this circuit:

$$I_B = \frac{U_B - U_{BE}}{R_B}, \ I_C = \beta I_B, \ U_{CE} = E_C - I_C R_C$$

For the first of them, $I_C = 0$, $U_{CE} = E_C$ (point on the abscissa axis); and for the second, $U_{CE} = 0$, $I_C = E_C/R_C$ (point on the ordinate axis). The intersection points of the load line with any of the static characteristics are called the *working points* corresponding to definite values of output current and output voltage. The transistor functions in the active mode (amplification) when the working point lies within the limits of the a-b interval. The functioning of the transistor stage in the amplification mode is characterized by the so-called *quiescent point*, or *Q-point*. The quiescent operating conditions may be shown on the graph in the form of a single point along the load line (see Figure 1.23). For a class A amplifier (widely used in simple automatic devices), the operating point for the quiescent mode (quiescent point) will be in the middle of the load line.

Pragmatically, the working Q-point of a transistor chosen according to its characteristic (Figure 1.21) should be set with the help of the so-called biasing

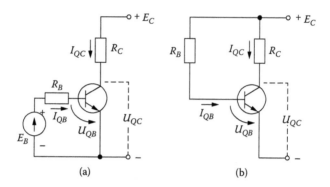

FIGURE 1.23
Circuit diagram of setting of a working quiescent point of transistors by a fixed current.

circuit. There are two methods for setting the transistor working point: by fixed current or by fixed voltage. The first one is implemented with the help of two circuit diagrams (see Figure 1.23). In the circuit of Figure 1.23a, the biasing circuit is formed by resistor R_B, which is calculated as follows:

$$R_B = \frac{E_B - U_{QB}}{I_{QB}}$$

where $U_{BE} \approx 0.7$–0.9 V when the base-emitter junction is forward biased; and I_{QB} is the quiescent base current from the output dynamic characteristic (Figure 1.21).

With only one biasing source, as is shown in Figure 1.23b, the quiescent mode is ensured by the power supply voltage E_C and by resistor R_B:

$$R_B = \frac{E_C - U_{QB}}{I_{QB}}$$

The method of setting biasing by fixed voltage is shown in Figure 1.24. It is the most widely used method of setting a transistor quiescent point by means of two resistors $R1$ and $R2$.

For this circuit, the following ratios are appropriate:

$$R_1 = \frac{E_C - U_{QB}}{I_{QB} + I_d}, \quad R_2 = \frac{U_{QB}}{I_d}, \quad I_d = (2-5)I_{QB}$$

As the environment temperature changes, the transistor current gain (β) changes (it increases as the temperature increases and decreases with a decrease in temperature), and the quiescent point position will change (Figure 1.25).

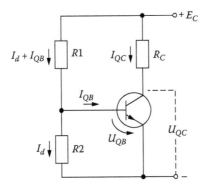

FIGURE 1.24
Circuit diagram of transistor stage with biasing by fixed voltage.

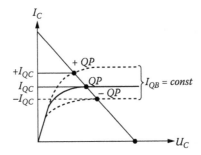

FIGURE 1.25
Changing of a transistor quiescent point with a displacement direct current I_{QB} that is temperature influenced.

In this circuit (Figure 1.26), for the resistance of the additional resistor R_E in the emitter circuit, one chooses based on the equation $R_E = (0.1–0.2) R_C$; and for the capacitance, C, of capacitor C, one chooses from the following equation:

$$\frac{1}{\omega C_E} \ll R_E$$

where ω is the minimum frequency of the enhanced signal.

Thanks to the biasing capability of this capacitor, in AC applications we obtain the amplification stage with a CE, and in DC applications we obtain the amplification stage with negative feedback.

By the biasing of working points higher than point *a* or lower than point *b*, the transistor goes into the saturation or cutoff mode, correspondingly. In the saturation mode, the transistor is fully turned on and the current $I_{SAT} = E_C/R_C$ flows through it; it does not increase more with an increase of the

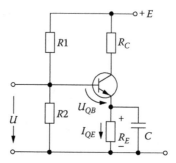

FIGURE 1.26
Amplifying stage with thermostabilization of a quiescent point.

input signal (that is why this mode is called *saturation*). To force the transistor into the saturation mode, one should make its base current not less than $I_B = I_{C(SAT)}/\beta$. In many automation devices, transistors function in the switch mode, that is, in two ultimate modes: saturation and cutoff. Conditions for safe cutoff or complete saturation of transistor are ensured, like in the case examined above, by the choice of biasing resistances R_B and R_C.

Equations for voltage in a base circuit:

$$U_B = -I_{C0}R_B + U_{BE} \text{ or } U_{BE} = U_B + I_{C0}R_B$$

The safe transistor cutoff is ensured under the condition that $U_{BE} \le 0$, where $R_B \le U_B/I_{C0}$. In this case, the value of the base resistance may be calculated as follows:

$$R_B \le \frac{U_B}{I_{C0max}}$$

where I_{C0max} is the maximum value of the collector's reverse current (transistor certified value).

In the circuit diagram of Figure 1.22, the positive input signal of a definite value turns the transistor on (the saturation mode). At this point, currents flowing in the transistor are equal:

$$I_B = \frac{U_B}{R_B}, \quad I_{C(SAT)} = \frac{E_C}{R_C} \le \beta I_B$$

Thus one may calculate the value of base resistance for transistor saturation mode:

$$R_B \le \beta_{min} R_C \frac{U_B}{E_C}$$

1.5 Transistor Devices in Switching Mode

One of the often-used operation modes of a transistor is the switching mode; even a single transistor can work as a high-speed switch (Figure 1.27). For switching current from one circuit to another, a two-transistor circuit (Figure 1.28) is used. In this circuit, stable offset voltage is applied to the base of the transistor (T_2), and control voltage to base T_1.

When $u_{inp} = u_{offset}$, the currents and voltages in the arms of the circuit are the same. If the input voltage (u_{inp}) exceeds the offset voltage (u_{offset}), transistor T_2 is gradually blocked and the whole current flows only through transistor T_1 and load resistor R_{C1}, and vice versa. When input voltage decreases below the level of the offset voltage ($u_{inp} < u_{offset}$), transistor T_1 is blocked and T_2 is unblocked, switching the sole current to the circuit of the resistor R_{C2}.

As is known, contacts of several electromagnetic relays, connected with each other in a certain way, are widely used in automation systems for carrying out the simplest logical operations with electric signals (Figure 1.29).

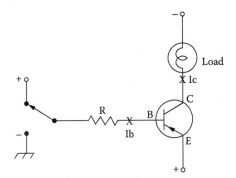

FIGURE 1.27
Electronic switch on a single transistor.

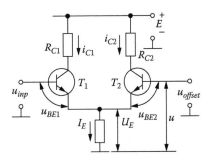

FIGURE 1.28
Transistor switch of two circuits.

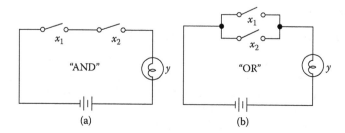

FIGURE 1.29
Implementation of the simplest logical operations, with the help of electromagnetic relay contacts.

For example, the logical operation AND is implemented with the help of several contacts connected in series, switched to the load circuit (Figure 1.29a). The signal Y will be the output of this circuit (that is, the bulb will be alight) only if signals on the first input X1 and on the second input X2 operate simultaneously (that is, when both contacts are closed). Another simple logical operation, OR (Figure 1.29b), is implemented with the help of several contacts connected in parallel. In this circuit, in order to obtain the signal Y on output (that is, for switching on the bulb), input of signal or on the first input (X1), on the second input (X2), or on both of the inputs simultaneously is required. Implementation of logical operations with electric circuits is one of the most important functions of relays. Transistor circuits successfully carry out this task. For example, the function NOT can be implemented on any type of single transistor (Figure 1.30).

In the circuit in Figure 1.30, when an input signal is missing, the transistor is blocked; that is, the whole voltage of the power source E is applied between the emitter and the collector (the drain and the source) of the transistor, and since the output signal is voltage on the collector (the source) of the transistor, that means that if there is no signal at the input, there will be a signal at the output of this circuit. And vice versa: when the signal is applied

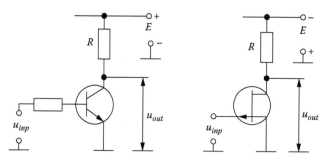

FIGURE 1.30
Logical element NOT implemented on bipolar and field transistors.

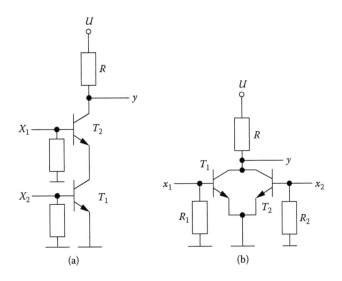

FIGURE 1.31
Transistor logical elements (a) AND-NOT and (b) OR-NOT.

at the input, the transistor is unblocked and voltage drops to a very small value (fractions of a volt), and therefore the signal disappears at the output.

The logical element AND-NOT can be implemented by different circuit methods. In the simplest case, this is a circuit from transistors connected in series (Figure 1.31a). When control signals are applied to both inputs X1 and X2 simultaneously, both transistors will be enabled and the voltage drop in the circuit with two transistors connected in series will decrease to a very small value. This means no output signal Y. In the second circuit diagram (Figure 1.31b), even one signal on any input (X1 or X2) is enough for the voltage on output Y to disappear.

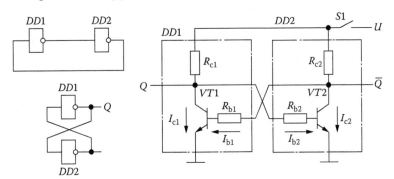

FIGURE 1.32
Bistable relay circuit with two logical elements NOT.

TABLE 1.1

Basic Logical Elements (According to Certain Standards
Logical Elements Are Also Indicated as Rectangles)

Logical Function	Conventional Symbols	Boolean Identities	Truth Table		
			Inputs		Output
			B	A	Y
AND		$A \times B = Y$	0	0	0
			0	1	0
			1	0	0
			1	1	1
OR		$A + B = Y$	0	0	0
			0	1	1
			1	0	1
			1	1	1
NOT		$A = \bar{A}$		0	1
				1	0
AND-NOT (NAND)		$\overline{A \times B} = Y$	0	0	1
			0	1	1
			1	0	1
			1	1	0
OR-NOT (NOR)		$\overline{A + B} = Y$	0	0	1
			0	1	0
			1	0	0
			1	1	0

Self-contained logical elements are indicated on circuit diagrams as special
signs (see Table 1.1). A signal strong enough for transition of a logical element
from one state to another is usually marked as 1. No signal (or a very weak
signal incapable of affecting the system state) is usually marked as 0.

The same signs are used for indication of the state of the circuit elements:
1 means switched-on, and 0 means switched-off. Such bistable (that is, having two stable states) devices are called *triggers*. When supply voltage is
applied to such a device (see Figure 1.32), one of the transistors will be immediately enabled and the other one will remain in a blocked state. The process
is avalanche-like and is called *regenerative*. It is impossible to predict which
transistor will be enabled because the circuit is absolutely symmetrical and
the likelihood of unblocking both transistors is the same.

This state of the device remains stable just the same. Repeated switching
ON and OFF of the voltage will cause the circuit to pass into this or that

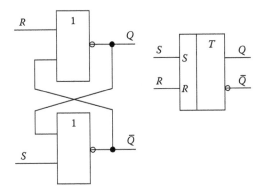

FIGURE 1.33
Asynchronous reset-set (RS) trigger formed by two logical elements NOR.

stable state. The essential disadvantage of such a trigger is no control circuit, which would enable us to control its state at a permanent supply of voltage.

In practice, the so-called Schmitt triggers are often used as electronic circuits with relay characteristics. There are a lot of variants of such triggers, each possessing special qualities. In the simplest variant, such a trigger is a symmetrical structure formed by two logical elements connected in a cycle of the type AND-NOT or OR-NOT (Figure 1.33); it is called an *asynchronous reset-set (RS) trigger*.

One of the trigger outlets is named *direct* (any outlet can be named this because the circuit is symmetrical) and is marked by the letter Q, and the other one is called *inverse* and is marked by the letter $\bar{Q}$ (Q under the dash), to signify that in a logical sense, the signal at this output is opposite of the signal at the direct output. The trigger state is usually identified with the state of the direct output, which means that the trigger is in the single (that is, switched-on) state when Q = 1, $\bar{Q}$ = 0, and vice versa.

Trigger state transition has a lot of synonyms, such as *switching, changeover, overthrow,* and *recording,* and is carried out with the help of control signals applied at the inputs R and S. The input by which the trigger is set up in the single state is called the *S input* (from *set*), and the output by which the trigger turns back to the zero position is called the *R input* (from *reset*). Four combinations of signals are possible at the inputs, each of them corresponding to a certain trigger position (Table 1.2).

As can be seen from the table, when there are no signals on both of the trigger inputs on the elements AND-NOT, or when there are signals on both of the trigger inputs on the elements OR-NOT (NOR), the trigger state will be indefinite, which is why such combinations of signals are prohibited for RS triggers.

From the time diagram of the asynchronous RS trigger (Figure 1.34), it can be seen that after transfer of the trigger to the single state, no repeated

TABLE 1.2

Combinations of Signals at the Inputs and the RS Trigger Position

Input			Output for Logical Element Type			
			AND-NOT		OR-NOT	
S (set)	R (reset)	Notes	Q	$\bar{Q}$	Q	$\bar{Q}$
0	0	Forbidden mode for *AND-NOT*	Uncertainty		Without changes	
1	0		1	0	1	0
0	1		0	1	0	1
1	1	Forbidden mode for *NOR*	Without changes		Uncertainty	

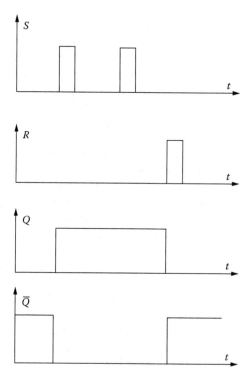

FIGURE 1.34

Time diagram of an asynchronous RS trigger.

signals on the triggering input S are capable of changing its state. The return of the trigger to the initial position is possible only after a signal is applied to its "erasing" R input.

The disadvantage of the asynchronous trigger is its incapacity to distinguish the useful signal of starting from noise occurring in the starting input

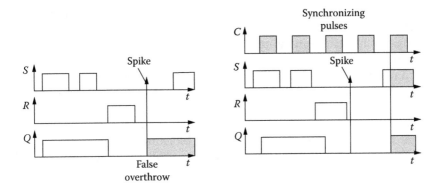

FIGURE 1.35
Time diagrams of the operation of an asynchronous trigger (on the left) and synchronous trigger (right) when there is noise.

by chance. Therefore, in practice so-called *synchronous* or D triggers, distinguished by an additional so-called *synchronizing input*, are frequently used.

Switching of the synchronous trigger to the single state (ON) is carried only with both signals: starting signal at the S input and also with a simultaneous signal on the synchronizing input. Synchronizing (timing) signals can be applied to the trigger (C input; see Figure 1.35) with certain frequencies from an external generator.

A simple amplifier with two transistors, with positive feedback, also has the properties of a trigger (Figure 1.36). In the initial position, when there is no voltage (or when voltage is very low) at the input of the circuit, transistor VT1 is closed (locked up). There is voltage on VT1 collector, which opens transistor VT2.

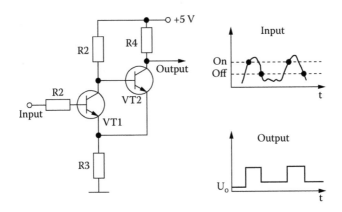

FIGURE 1.36
A simple two-transistor trigger.

The emitter current of transistor VT2 causes a voltage drop on the resistor R3, which blocks transistor VT1 and holds it in closed position. If input voltage exceeds the voltage in the emitter on transistor VT1, it will be opened and will become saturated with very small collector-emitter junction resistance.

As a result, the potentials of the base and the emitter of transistor VT2 will be equal. Transistor VT2 will be blocked. At the output, there will be voltage equal to the supply voltage. When input voltage decreases, transistor VT1 leaves the saturation mode, and an avalanche-like process occurs. The emitter current of transistor VT2, causing blocking voltage on resistor R3, accelerates closing of the transistor VT1. As a result, the trigger returns to its initial position.

1.6 Thyristors

The history of development of another remarkable semiconductor device with relay characteristics begins with the conception of a "collector with a trap," formulated at the beginning of the 1950s by William Shockley, and familiar to us from his research on p-n junctions. Following Shockley, J. Ebers invented the two-transistor analogy (interbounded n-p-n and p-n-p transistors) of a p-n-p-n switch, which became the model of such a device (Figure 1.37).

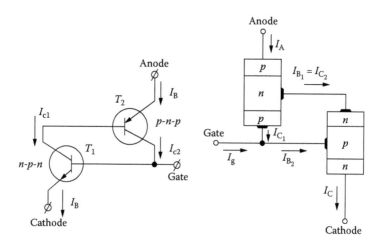

FIGURE 1.37
Two-transistor model of a thyristor.

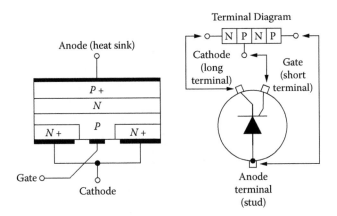

FIGURE 1.38
Structure and symbolic notation of a solid state thyratron, or *thyristor*.

The working element of this new semiconductor device with relay characteristics was a four-layer silicon crystal with alternating p- and n-layers (Figure 1.38). Such a structure is made by diffusion into the original monocrystal of n_1-silicon (which is a disk 20–45 mm in diameter and 0.4–0.8 mm thick, used more for high-voltage devices) admixture atoms of aluminum and boron from the direction of its two bases to a depth of about 50–80 microns. Injected admixtures form p_1 and p_2 layers in the structure.

The fourth (thinner) layer n_2 (its thickness is about 10–15 microns) is formed by further diffusion of atoms of phosphorus to the layer p_2. The upper layer p_1 is used as an anode in the thyristor, and the lower layer p_2 is used as a cathode.

The power circuit is connected to the main electrodes of the thyristor: the anode and the cathode. The positive terminal of the control circuit is connected through the external electrode to the layer p_2, and the negative one to the cathode terminal.

The volt-ampere characteristic (VAC) of a device with such a structure (Figure 1.39) much resembles the VAC of a diode by form. As in a diode, the VAC of a thyristor has forward and reverse areas. Like a diode, the thyristor is blocked when reverse voltage is applied to it (minus on the anode, plus on the cathode), and when the maximum permissible level of voltage (U_{Rmax}) is exceeded there is a breakdown, causing strong current and irreversible destruction of the structure of the device.

The forward area of the VAC of the thyristor does not remain permanent, as does that of a diode, and can change, being affected by the current of the control electrode, called the *gate*. When there is no current in the circuit of this electrode, the thyristor remains blocked not only in reverse but also in the forward direction; that is, it does not conduct current at all (except small

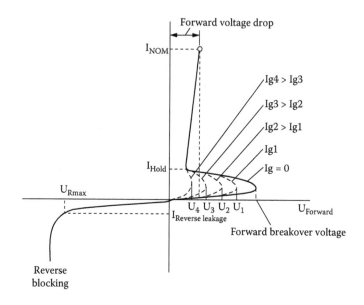

FIGURE 1.39
Volt-ampere characteristic (VAC) of a thyristor.

leakage current, of course). When the voltage applied in the forward direction between the anode and the cathode is increased to a certain value, the thyristor is quickly (stepwise) enabled and only a small voltage drop (frictions of a volt) caused by irregularity of the crystal structure remains on it. If low current is applied to the circuit of the gate, the thyristor will be switched ON to much lower voltage between the anode and the cathode. The higher the current, the lower the voltage that is required for unblocking of the thyristor. At a certain current value (from a few milliamperes for low power thyristors, up to hundreds of milliamperes for power ones), the forward branch of the VAC is almost fully rectified and becomes similar to the VAC of a diode. In this mode (i.e., when control current constantly flows in the gate circuit), the behavior of the thyristor is similar to that of a diode that is fully enabled in the forward direction and fully blocked in the reverse direction. However, it is senseless to use thyristors in this mode: there are simpler and cheaper diodes for this purpose.

In fact, thyristors are used in modes when the working voltage applied between the anode and the cathode doesn't exceed 50–70% of the voltage, causing spontaneous switching ON of the thyristor (when there is no control signal, the thyristor always remains blocked), and control current is applied to the gate circuit only when the thyristor should be unblocked and of such a value that would enable reliable unblocking. In this mode, the thyristor functions as a very high-speed relay (unblocking time is a few or tens of microseconds).

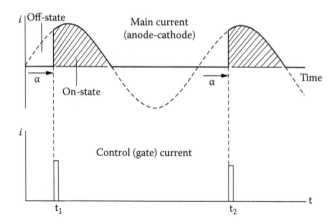

FIGURE 1.40
Principle of operation of a thyristor regulator.

Perhaps many have heard that thyristors are used as basic elements for smooth current and voltage adjusting, but if a thyristor is only an electric relay having two stable states like any other relay (a switched ON state and a switched OFF one), how can a thyristor smoothly adjust voltage? The point is that if nonconstant alternating sinusoidal voltage is applied, it is possible to adjust the unblocking moment of the thyristor by changing the moment of applying a pulse of control current on the gate with regard to the phase of the applied forward sinusoidal voltage. That is, it is as though a part of the sinusoidal current flowing to the load were cut off (Figure 1.40). The moment of applying a pulse of unblocking control current (such pulses are also called *igniting* by analogy with the control pulses of the thyratron) is usually characterized by the angle α.

Taking into account that average current value in the load is defined as an integral (that is, the area of the rest part of the sinusoid), the principle of operation of a thyristor regulator becomes clear. After unblocking, the thyristor remains in the opened state, even after completion of the control current pulse. It can be switched OFF only by reducing forward current in the anode-cathode circuit to a value less than *hold current* value.

In AC circuits, the condition for thyristor blocking is created automatically when the sinusoid crosses the zero value. To unblock the thyristor in the second half-wave of the voltage, it is necessary to apply a short control pulse through the gate of the thyristor. To control both half-waves of alternating current, two thyristors connected antiparallel are used. Then one of them works on the positive half-wave, and the other one works on the negative one. At present, such devices are produced for currents of a few milliamperes to a few thousand amperes, and for blocking voltages up to a few thousand volts (see Figure 1.41).

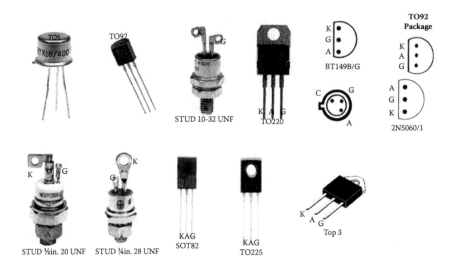

FIGURE 1.41
Modern low-power and power thyristors.

1.7 Optocouplers

Optocouplers (optical couplers) are designed to isolate electrical output from input for complete elimination of noise. All digital inputs of microprocessor-based protective devices (MPDs) are connected to the internal circuit of the MPD and to the central processing unit (CPU) through optocouplers.

The blocked *n-p* junction in semiconductor devices (diodes, transistors, and thyristors) may begin to allow electric current to pass under the effect of energy of photons (light). When the *n-p* junction is illuminated, additional vapors of charge carriers—electrons and holes causing electric current in the junction—are generated within it. The higher the intensity of the luminous flux on the *n-p* junction is, the stronger the current is. Optoelectronic relays (Figure 1.42) comprise a light-emitting element which is usually made on the basis of a special diode (a light-emitting diode, or LED), an *n-p* junction emitted by photons when current passes through it, and a receiver of the luminous flux (a photodiode, a phototransistor, or a photothyristor).

Optocouplers typically come in a small six-pin or eight-pin IC package, but are essentially a combination of two distinct devices: an optical transmitter, typically a gallium arsenide LED, and an optical receiver such as a phototransistor or light-triggered diac (see Figure 1.42). The two are separated by a transparent barrier which blocks any electrical current flow between the two, but does allow the passage of light. The basic idea is shown in Figure 1.43, along with the usual circuit symbol for an optocoupler.

Package Dimensions

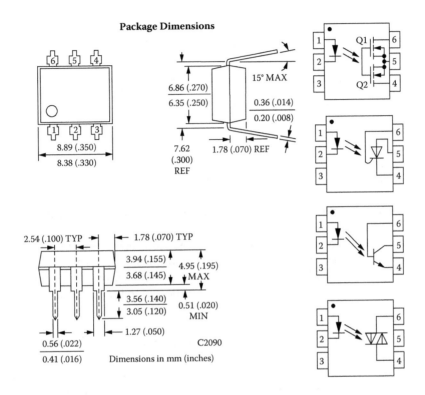

FIGURE 1.42
Octocouplers may be contained in a standard DIP case.

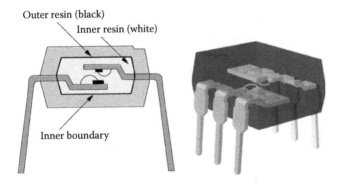

FIGURE 1.43
Internal design of the optocoupler.

Usually the electrical connections to the LED section are brought out to the pins on one side of the package, and those for the phototransistor or diac to the other side, to physically separate them as much as possible. This usually allows optocouplers to withstand voltages of anywhere between 500 V and 7500 V between input and output. Optocouplers are essentially digital

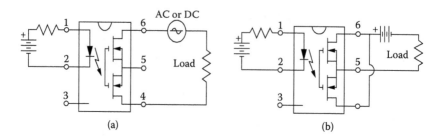

FIGURE 1.44
Connection of bidirectional switched photo-metal-oxide semiconductor optocoupler.

or switching devices, so they're best for transferring either ON-OFF control signals or digital data.

Usage of the two series-connected photo-MOS transistors (connection (a) in Figure 1.44) as output elements allows the optoelectronic relay to switch either AC and DC loads with nominal output current rating. Connection (b) with the polarity and pin configuration, as indicated in the schematic, allows the relay to switch DC load only, but with current capability increases by a factor of 2.

There is a great diversity of circuits and constructions of power opto-couplers (optoelectronic relays), including those containing built-in invertors or amplifiers (see Figure 1.45). A similar principle serves as a basis not only for miniature devices in chip cases, but also for practically all power semi-conductor relays and contactors.

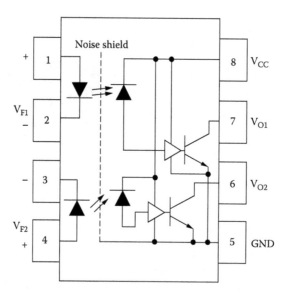

FIGURE 1.45
Twin optoelectronic relay with built-in amplifiers of power.

FIGURE 1.46
Modern semiconductor optoelectronic relays for currents of 3–5 A, produced by different companies.

It should be noted that the external designs of not only miniature opto-couplers in chip cases, but also more powerful devices of various firms, are very much alike (Figure 1.46).

Such relays are usually constructed according to a similar scheme (Figure 1.47), with only some slight variations. As a rule they comprise an RC circuit (the so-called snubber) and a varistor protection outlet, protecting the thyristors from overvoltages. They often contain a special unit (a zero-voltage detector) controlling the moment when the voltage sinusoid passes through the zero value and allowing it to enable (and sometimes to disable) the thyristor at the zero value of voltage (so-called synchronous switching).

An optocoupler characterizes by the current transfer ratio (CTR), which is defined as the ratio of the output current (I_C) to the input current (I_F). The brightness of the LED slowly decreases in an exponential fashion as a function of forward current (I_F) and time. A 50% degradation is considered to be the failure point; therefore, an optocoupler's lifetime is defined as the time at which the CTR has fallen to 50% of its original value. At LED current equal to 0.5 mA and ambient temperature 25°C, the optocoupler lifetime will be equal to 2.10^5 hours (~22 years)[1]. In some critical applications, the malfunction

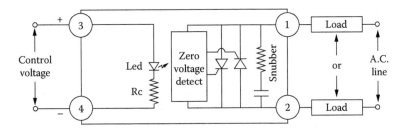

FIGURE 1.47
Standard scheme of a power single-phase optoelectronic AC relay.

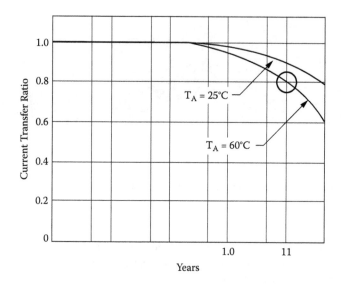

FIGURE 1.48
Current transfer ratio (CTR) degradation over time at 25°C and 60°C ambient temperature.

appears already when CTR decreases to 70–80% (see Figure 1.48), that is, upon 10–11 years, which is during the normal electronic equipment's lifetime.

1.8 Electromagnetic Relays

An electromagnetic neutral relay is the simplest, oldest, and most widespread type of relay. What are its basic elements? As a rule, when asked this question most people would probably name the following: a winding, a magnetic core, an armature, a spring, and contacts.

This all is true, of course, but if you begin to analyze how a relay works, it might occur to you that there is something missing. What is the purpose of a magnetic system? Apparently it is used to transform input electric current to the mechanical power needed for contact closure. And what does a contact system do? It transforms the imparted mechanical power back to an electric signal!

Don't you think that there is something wrong here?

Everything will become more obvious if the list of basic components of a relay includes one more element, which is not so obvious from the point of view of the construction of a relay, for example a coil or contacts. Very often, it is not just one element but several small parts that escape our attention. Such parts are often omitted on diagrams illustrating the principle of relay operation (Figure 1.49).

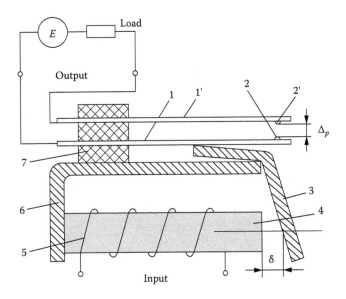

FIGURE 1.49
Construction of a simple electromagnetic relay: 1: springs; 2: contacts; 3: armature; 4: core; 5: winding; 6: magnetic core; and 7: insulator.

I am referring to an insulation system providing galvanic isolation of the input circuit (winding) from the output circuit (contacts). If we take such an insulation system into account, it becomes clear that an input signal at the relay input and an output signal at the relay output are not the same. They are two different signals that are completely insulated from each other electrically.

Note Figure 2.1, which is often used to illustrate principles of relay operation. If you use this figure as the only guide while constructing a relay, the relay will not operate properly since its input circuit (the winding) is not electrically insulated from the output circuit (the contacts).

In simple constructions used for work at low voltage, insulating bobbins with winding (shown on the Figure 1.50) provide basic insulation (apart from an insulator).

The magnetic system of a typical low-voltage electromagnetic relay comprises first of all a control winding (6) made in the form of a coil with insulated wire, a magnetic core (1), and a movable armature (3) (see Figure 1.50).

Elements of the magnetic circuit of the relay are usually made of soft magnetic steel, which is a type of steel that has a small hysteresis loop (see Figure 1.51). *Hysteresis* can be translated as a *lag*. A hysteresis loop is formed from a magnetization curve and a demagnetization curve. These curves are not superimposed on each other because the same magnetic field strength is not enough to demagnetize the magnetized material. An additional magnetic field is required for demagnetization of the material to the initial (non-magnetized) state. This happens because the so-called residual induction

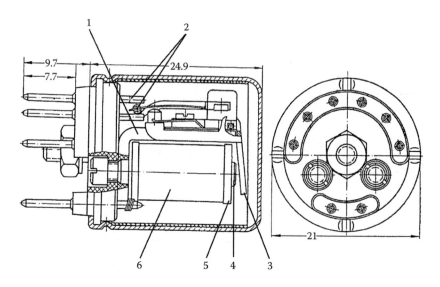

FIGURE 1.50
Real construction of a low-voltage miniature relay. 1: magnetic core; 2: contacts; 3: armature; 4: pole of core; 5: insulating bobbin; and 6: coil.

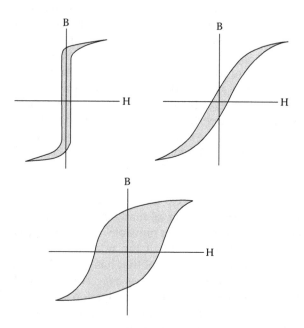

FIGURE 1.51
Hysteresis loops for soft and hard magnetic materials. B: flux density; and H: magnetic field strength.

still remains within the previously magnetized model, even after removal of the external magnetic field. Such a phenomenon takes place because magnetic domains (crystal structures of ferromagnetic material) expanding along the external magnetic field lose their alignment when the external magnetic field is removed. The demagnetizing field necessary for a complete demagnetization of the previously magnetized model is called the *coercitive force*. The coercitive force of soft magnetic materials is small, which is why their hysteresis loop is also small. That means that when the external magnetic field is removed (deenergizing the winding of the relay), the magnetic core and armature do not remain magnetized, but return to their initial stage. Apparently this lack of retentivity is a very important requirement for materials used in magnetic circuits of typical neutral relays. Otherwise, relay characteristics would not be stable and the armature might seal.

Electric steel used for the production of relays is composed of this very soft magnetic material. This is steel with a lean temper (and other admixtures like sulfur, phosphor, oxygen, and nitrogen) and rich silicon (0.5–5.0). Apart from enhancing the qualities of the steel, the silicon makes it more stable and increases electrical resistance, which considerably weakens eddy currents (see below). The hardness and the fragility of the steel are mostly dependent on the silicon content. With a content of 4–5% of silicon, steel usually can withstand no more than 1–2 folds of 90°.

For a long time, only the hot rolling method was used for the production of such steel, until in 1935 Goss discovered the superior magnetic properties of cold-rolled electric steel (but only along the direction of rolling). That gave such steel a magnetic texture and made it anisotropic. The utilization of anisotropic cold-rolled steel requires such construction of the magnetic core to enable magnetic flux to pass only in the direction of rolling. Machining of parts of a magnetic core entails high internal stress, and consequently higher coercitive force. That is why after processes of forming, whetting, and milling, parts must be annealed at 800–900°C, with further gradual reduction of temperature to 200–300°C.

Sometimes Permalloy is used for magnetic cores of highly sensitive relays. It is a ferro-alloy with nickel (45–78%) alloyed with molybdenum, chrome, copper, and other elements. Permalloy has better magnetic properties in weak magnetic fields than electric steel; however, it cannot be used for work at great magnetic fluxes since its saturation induction is only half as much as that of electric steel.

Magnetic cores for big AC relays are made of sheet electric steel, which is 0.35–0.50 mm thick. Several types of the magnetic system are used in modern constructions of relays. The clapper-type (attracted-armature) magnetic system is the oldest type of magnetic systems. Its construction was already described in Edison's patents. It was used first in telephone relays, and later in industrial and compact covered relays. Today this type of magnetic system is widely used in constructions of middle- and small-sized relays, with a plastic rectangular cover that is often transparent. They are mostly designed for

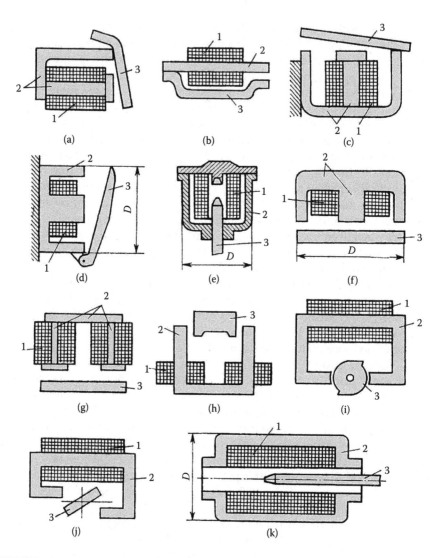

FIGURE 1.52
Types of the magnetic system of modern electromagnetic relays. 1: control coil; 2: magnetic core; 3: armature; a–d: clapper-type (attracted-armature) systems; f and g: direct motion magnetic systems; e, h, and k: systems with a retractable armature (solenoid-type); and i and j: systems with a balanced turning armature.

work in systems of industrial automation and the power generation industry (Figure 1.52). The disadvantage of relays with this type of magnetic system is sensitivity to external mechanical effects.

Further research in this direction led to the invention of a new J-type magnetic system. In this system, a balanced turning armature was widely used in miniature hermetic relays. In this magnetic system, the rotation axis of

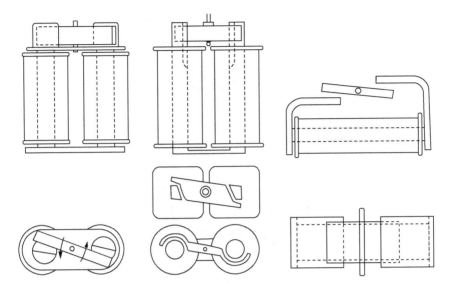

FIGURE 1.53
Some variants of construction of the J-type magnetic system.

the armature goes through its center of mass. As a result, the relay becomes resistant to external mechanical impact and can sustain linear accelerations of 100–500 g, reiterated shocks at an acceleration of 75–150 g, and separate shocks at an acceleration rate of up to 1000 g. There are a lot of variants of this J-type magnetic system (Figure 1.53).

Not only most modern miniature relays in plastic shells but also hermetic relays in metal shells (Figure 1.54), which have been produced for many years, have similar magnetic systems. Miniature and microminiature relays of this type are really very small in size: for example, the RES49 type relay is 10.45 × 5.3 × 23.2 mm with a weight of 3.5 g, and the RES80 type relay is 10.4 × 10.8 × 5.3 mm with a weight of 2 g.

Another kind of electromagnetic relay widely used in MPDs is a latching relay. This is one that picks up under the effect of a single current pulse in the winding, and remains in this state when the pulse stops affecting it (i.e., when it is locked). Therefore, this relay plays the role of a memory circuit. Moreover, a latching relay helps reduce power dissipation in the application circuit because the coil does not need to be energized all the time (Figure 1.55).

As illustrated in Figure 1.55b, the contacts of a latching relay remain in the operating state even after an input to the coil (set coil) has been removed. Therefore, this relay plays the role of a memory circuit. The double-coil latch-type relay has two separate coils, each of which operates (sets) and releases (resets) the contacts.

In latch relays, two types of latching elements are usually used: magnetic and mechanical. Like standard electromagnetic relays, latching relays are

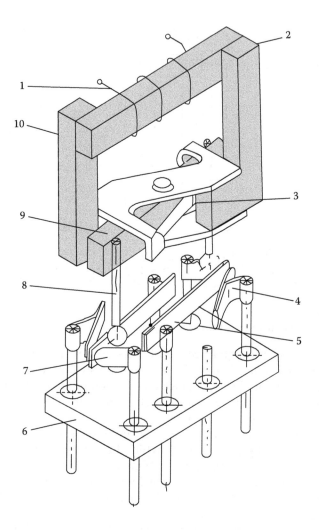

FIGURE 1.54
One of the most popular variants of construction of the turning-type magnetic system, widely
used in miniature relays. 1: winding; 2: pole of the magnetic core; 3: restoring spring; 4: fixed
normally open contact; 5: moving contact; 6: heel piece of relay; 7: fixed normally closed contact;
8: pusher with a small insulating ball, weighted at the end; 9: turning armature; and 10: core.

produced for all voltage and switched power classes: from miniature relays
for electronics, with contact systems and cases typical of standard relays of
the same class according to switched power, up to high-voltage relays and
high-current contactors.

Magnetic systems of most miniature latching relays are constructed on
the base of a permanent magnet (Figures 1.56 and 1.57). The smallest latching relays in the world in standard metal cases of low-power transistors are
produced by the American company Teledyne Relays (Figure 1.58).

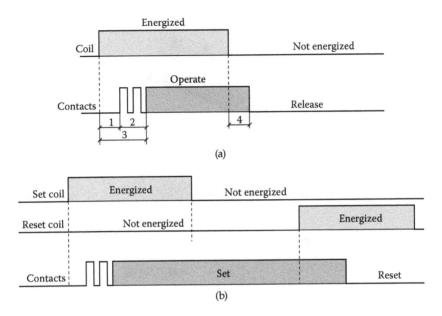

(a)

(b)

FIGURE 1.55
(a) Time chart of non–latch relay. 1: current-rise time; 2: bounce time; 3: full operate time; and 4: release time. (b) Time chart of double-coil latch relay.

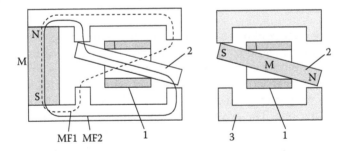

FIGURE 1.56
Popular types of magnetic systems of cheap miniature latching relays, in plastic cases produced by many companies. M: permanent magnet; MF1: magnetic flux in first position; MF2: magnetic flux in second position; 1: coil; 2: rotating armature; and 3: yoke.

Another popular type of latching relay with magnetic latching is the so-called *remanence* type. This relay consists of a coil and an armature made of a special ferromagnetic material on a nickel base, with admixtures of aluminum, titanium, and niobium, capable of becoming magnetized quickly under the effect of a single current pulse in the coil, and of remaining in its magnetized state when the pulse stops affecting it.

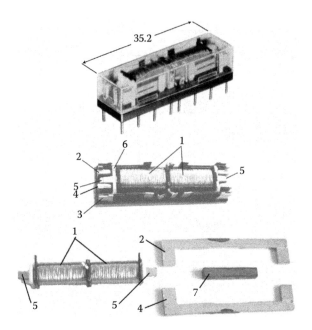

FIGURE 1.57

Construction of a miniature latch relay with a magnetic latching of the DS4 type, produced by Euro-Matsushita. 1: set and reset coils; 2 and 4: plates of the magnetic core; 3: contacts; 5: ferromagnetic pole lugs; 6: plastic pushers put on the pole lugs; and 7: a permanent magnet placed in the center of the coil.

This type of relay contains a coil with one or two windings wound around it. In the first case, magnetization and demagnetization of the material of the core are carried out by current pulses of opposite polarities; and in the second case, they are carried out by two different windings on the same bobbin, one of which magnetizes the core while the other one, which demagnetizes it, is a disabling one. The advantage of the type of relay is that it does not require any special construction. One has only to make a core in the already existing construction of a standard relay of any type from remanent material, and the latching relay is complete!

Miniature single-coil latching relays for operating voltage of 3.0–5.0 V are found in many applications, including MPDs. In these relays, coil current must flow in both directions—through a single coil (Figure 1.59). Current flowing from pin 8 to pin 1 causes the relay to latch in its reset position, and current flowing from pin 1 to pin 8 latches the relay in its set position.

A simple integral circuit of the MAX4820/4821 type (+3.3 V/+5 V, eight-channel, cascadable relay drivers with serial-parallel interfaces), produced by Maxim Integrated Products Ltd., driving up to four such single-coil (and also four ordinary dual-coil) latching relays, includes a parallel-interface relay driver (UI) with open-drain outputs (Figure 1.60), and inductive-kickback

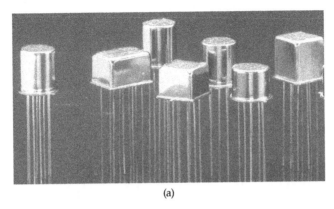

(a)

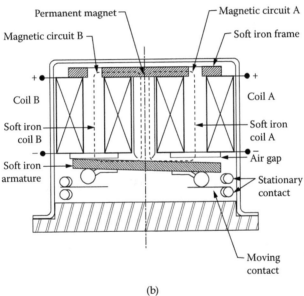

(b)

FIGURE 1.58
(a) The smallest latching relays in the world, produced by the American company Teledyne Relays. External design of relays in Centigrid® and TO-5 cases. (b) Construction of miniature latching relays produced by the American company Teledyne Relays.

protection. Latch any of the four relays to their set or reset positions by turning on the corresponding output (OUTX). That output is selected by asserting its digital address on pins A2 to A0 while the CS is high. Activate the output by toggling CS. Both devices feature separate set and reset functions that allow the user to turn ON or turn OFF all outputs simultaneously with a single control line. Built-in hysteresis (a Schmitt trigger) on all digital inputs allows this device to be used with slow rising and falling signals such as those from optocouplers or RC power-up initialization circuits.

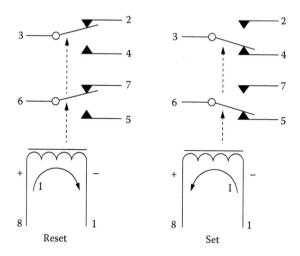

FIGURE 1.59
Principle of operation of a miniature single-coil latching relay. The AROMAT AGN210A4HZ relay, for example, has such a principle.

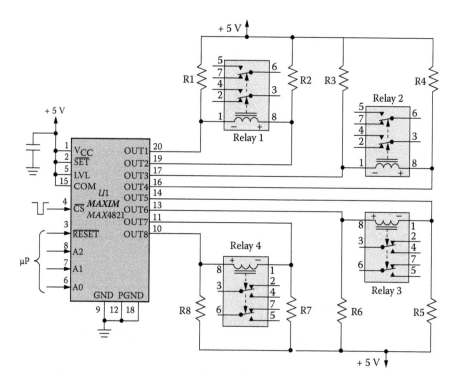

FIGURE 1.60
The MAX4821 integral circuit easily drives four, single-coil latching relays.

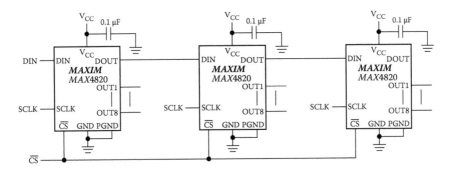

FIGURE 1.61
Daisy-chain configuration for MAX4820.

The MAX4820 features a digital output (DOUT) that provides a simple way to daisy-chain multiple devices. This feature allows the user to drive large banks of relays using only a single serial interface (Figure 1.61). Electromechanical relays used in MPDs are amplifier-driven relays, that is, relays controlled by electronic switches. Very often, single bipolar (Figure 1.62), or field-controlled (Figure 1.63) transistors are used as amplifiers. Diodes switched parallel to the winding of the relay are necessary for preventing damage to transistors by overvoltage pulses occurring on the winding of the relay at the moment of blocking of the transistors (transient suppression).

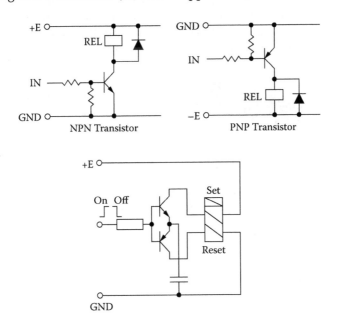

FIGURE 1.62
Amplifier-driven relays on bipolar transistors.

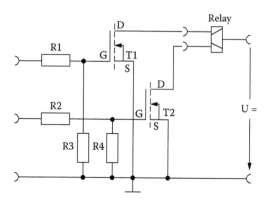

FIGURE 1.63
Amplifier-driven latching relay on FETs.

Especially for electromagnetic relay control, a set of amplifiers on the basis of Darlington's transistor is produced in a standard case of an integrated circuit (Figure 1.64). The eight *n-p-n* Darlington connected transistors in this family of arrays are ideally suited for interfacing between low-logic-level digital circuitry (such as transistor-transistor logic [TTL], complementary metal-oxide semiconductor [CMOS], or *p*- or *n*-channel MOSFET [PMOS or NMOS, respectively]) and the higher current or voltage relays for a broad range of computer, industrial, and consumer applications. All these devices feature open-collector outputs and freewheeling clamp diodes for transient suppression. The ULN2803 is designed to be compatible with standard TTL families, while the ULN2804 is optimized for 6–15 V high-level CMOS or PMOS.

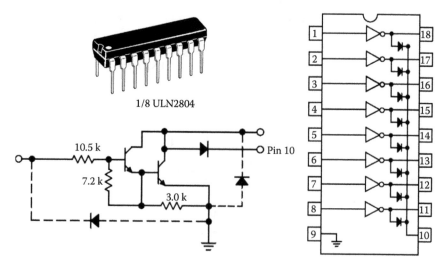

FIGURE 1.64
UNL2804-type chip (Motorola).

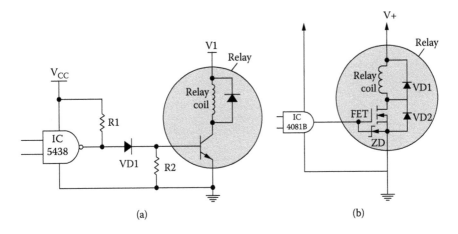

FIGURE 1.65
Amplifier-driven super miniature relays on (a) bipolar transistors and (b) FETs, produced by Teledyne Relays. (a) circuit diagram for relay series 411, 412, 431, and 432; and (b) circuit diagram for relay series 116C and 136C.

Amplifier-driven relays are produced by some firms in the form of independent, fully discrete devices with the electronic amplifier built-in directly to the case of the electromagnetic relay. Even the smallest electromagnetic relays in the world, such as those for military and industrial applications produced by Teledyne, are placed in transistor cases (TO-5 type packages; see above) and are produced with built-in amplifiers (Figure 1.65).

Reference

1. Ben Hadj Slama, J., H. Helali, A. Lahyani, K. Louati, P. Venet, and G. Rojat. Study and Modelling of Optocouplers Ageing. *Journal of Automation and Systems Engineering*, vol. 2, no. 4, 2008, http://hal.archives-ouvertes.fr/hal-00368838/en/

2

Design and Functional Modules of Digital Protective Relays

2.1 Overall Structure and Design of Digital Protective Relays (DPRs)

A digital protective relay (DPR) consists of the following basic materials: an analog input block (internal current and voltage transformers), input filters (anti-aliasing filters and sample-and-hold circuits), a multiplexor, an analog-to-digital converter, a microprocessor, different memory devices, a logic (digital) inputs block, and a relay-controlled output block (see Figure 2.1).

The structure of the DPR is made up of a set of printed circuit boards (PCBs) corresponding to different functional parts, which are installed in enclosures of different types and sizes (see Figure 2.2). There are several ways to lay out the PCBs in the DPR enclosures. One of them is the so-called stacked module arrangement, where PCBs are located one above the other. The PCBs should be attached to each other with thread bushes, forming a single structural module similar to a bookcase (see Figure 2.3). Then, the module is to be installed into the DPR enclosure. The PCBs should be fastened to each other by way of connectors and a flat flexible lead. An obvious drawback of this design is the impossibility of replacing a single module without disassembling and dismantling the whole DPR.

Another kind of DPR enclosure is an "open-cube" body (see Figure 2.4). In this case, three PCBs are fastened to each other with special angular connectors, so that they form the side and rear walls; and they are fixed to the metal front panel, which forms the fourth wall.

After assembling, the whole structure should be installed into the main enclosure. Multiple slide-in boards are the most widely used devices (see Figure 2.5). This type of structure has an aluminum body with guide rails for installing individual (module) PCBs of DPRs. The boards may be installed vertically or horizontally. One more additional base PCB with a set of connectors and the copper paths connecting these connectors is at the bottom of this case.

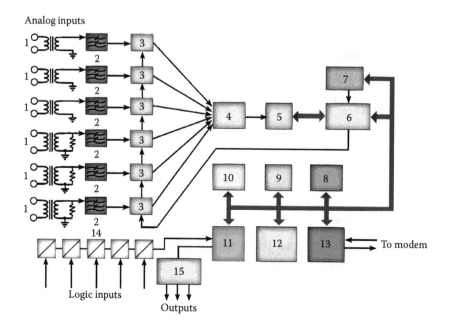

FIGURE 2.1

Structural diagram of the DPR. 1: internal current and voltage transformers; 2: anti-aliasing filters; 3: sample-and-hold circuits; 4: multiplexor; 5: analog-to-digital converter (ADC); 6: microprocessor; 7: timer; 8: electrically erasable programmable read-only memory (EEPROM); 9: random access memory (RAM); 10: read-only memory (ROM); 11: logic inputs-outputs; 12: keyboard and monitor; 13: serial communication port; 14: photo couples; and 15: output relays.

FIGURE 2.2

Appearance of the modern DPR in different enclosures.

When sliding the boards into the DPR enclosure on the guide rails, the board connectors plug into the sockets on base board, and thus the boards are connected with each other. The base board can be made as a so-called cross board with sockets and cooper paths only (Figure 2.6a), and as a so-called motherboard with a microprocessor, memory, and auxiliary devices serving

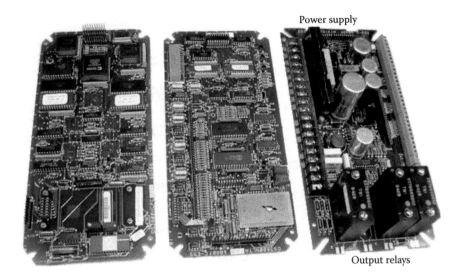

FIGURE 2.3
Individual printed circuit boards constituting a three-level stacked relay module LodTrack
(General Electric).

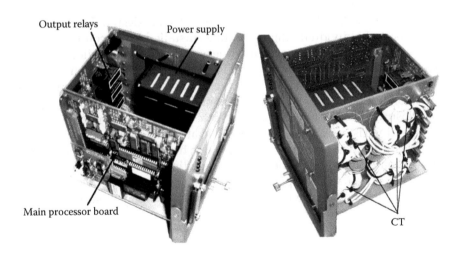

FIGURE 2.4
Construction style "open cube" of the DPR type MPRS (ABB).

the microprocessor (Figure 2.6d). Sometimes as cross boards, the flexible flat multicore cable with sockets (Figure 2.6d) is used.

In some poorly designed units (see Figure 2.7), you have to take out several modules to get to the power supply module. You also have to unsolder outputs of all current transformers from the rear connector block, which then

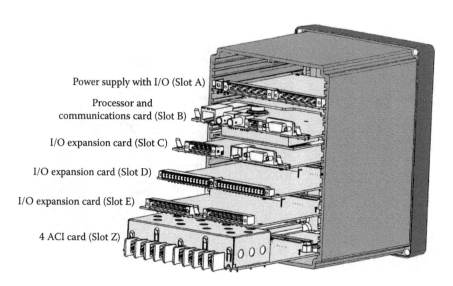

Power supply with I/O (Slot A)

Processor and
communications card (Slot B)

I/O expansion card (Slot C)

I/O expansion card (Slot D)

I/O expansion card (Slot E)

4 ACI card (Slot Z)

FIGURE 2.5
Horizontal slide-in boards.

must be resoldered again, in order to slide out the module to replace the
power supply.

Relay type T60 (see Figure 2.8) has a very strange design. Relays of this
type consist of individual plug-in modules installed in the common enclosure. Unlike all other DPRs, each module of T60 model is installed in a steel
casing that makes the unit heavy (not less than 15 kg). When you open the
body, you can see the PCB with a high-capacity connector at the end. This
connector also has a very strange design and is equipped with a large plastic enclosure, divided into large cells with electronic components, output
relays, and varistors inside. This casing is attached to the connector with
eight plastic latches, four on each side, which have to be opened simultaneously. Attempting to open the casing immediately led to the breakage of
one of the latches, so I stopped. Since this plastic casing has no functional
load, in my opinion it is required only to make the relay unserviceable.

2.2 Analog Input Modules

Analog input modules are the simplest elements of the DPR. They consist of
a set of current and voltage transformers (see Figure 2.9).

The design of voltage transformers is similar to that of conventional low-
power transformers. Current transformers contain an isolated multiturn
secondary coil wound on a frame and covered with insulating film. The

FIGURE 2.6
Various designs of DPRs with slide-in boards.

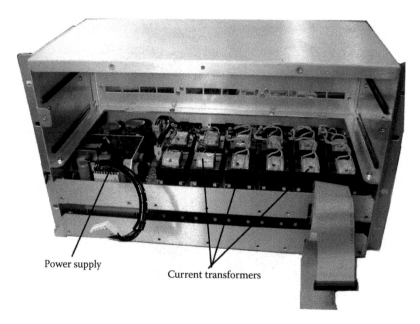

FIGURE 2.7
SEL-487 relay enclosure with partially extracted modules (Schweitzer Engineering Laboratories).

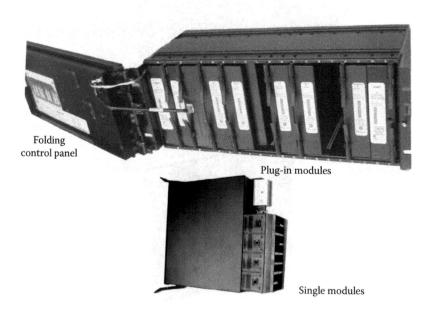

FIGURE 2.8
Modular DPR type T60 (General Electric).

FIGURE 2.9
Analog input modules consisting of a set of current and voltage transformers.

primary winding has a few turns (typically five turns for a nominal primary current of 1 A, and one turn for a rated current of 5 A) of conventional stranded insulated bonding conductor wound on the secondary winding (see Figure 2.9). Such transformers, in fact, are the current-to-voltage converters. If during DPR operation the rated input current of the analog inputs needs to be changed from 1 A to 5 A (or vice versa), you need only add (or, conversely, wind off) several turns of the conductor. Usually this DPR assembly causes no problems during the operation, constituting as it does the most reliable part of the unit.

FIGURE 2.10
Analog input modules of DPR type SIPROTEC 7U6125 (Siemens) with signal preprocessing electronic components.

In most types of DPRs, this set of transformers is represented by a separate module, although there are some designs where this module contains input filters, analog-digital converters, and other elements for analog signal preprocessing (see Figure 2.10).

There are several types of DPRs equipped with miniature toroidal current and voltage transformers encapsulated with an epoxy compound (see Figure 2.11). This provides better moisture protection while obstructing heat dissipation.

However, this design is unserviceable and does not allow changing the transformation ratio.

Also, we should keep in mind that even if such a design seems to be more reliable, its actual operational reliability might be even lower than that of conventional nonencapsulated transformers. This results not only from obstructing the dissipation of the heat but also from the internal mechanical stresses at the windings arising during solidification and shrinkage of the epoxy compound. Such problems are typical for thin-conductor multiple-turn windings (as in voltage transformers).

FIGURE 2.11
Analog input module with encapsulated current transformers on a toroidal core.

2.3 Output Relay Modules

The design of output relay assemblies is a little more complicated. Seldom are they performed as separate modules, such as the relays manufactured by the Chinese firm Nari-Relays (see Figure 2.12).

An output relay module contains a number of relatively high-power electromechanical relays designed for directly enabling a high-voltage coil deactivating switch or high-power intermediate relay with a mechanical lock and several low-power relays whose contacts are designed for activating the external alarm devices and circuits.

During the study of the type RCS-9681 (Nari-Relays) relay, I found a rather strange engineering solution. This solution, aimed at enhancement of the running speed, was realized with two electromechanical relays: one a rather powerful unit (ST type) with an operating time of 10–12 ms, and the other

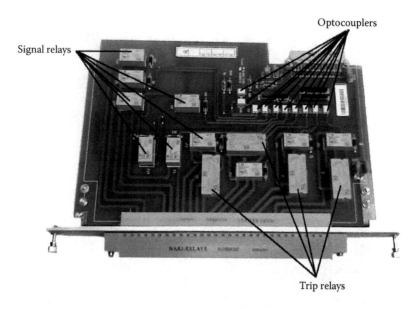

FIGURE 2.12
Individual output relay module manufactured by Nari-Relays.

a small high-speed relay (DS–P type) with a response time of 3–4 ms; see Figure 2.13.

The contacts of both relays were connected in parallel in order to combine, according to the developer, the high speed of the low-power relay with a sufficiently large switching capacity of the conventional unit. In practice, the small high-speed relay (which has small parts and small gaps between the contacts) is not suitable at all for switching loads under 220 V DC (its maximum switching voltage is 125 V).

In an interview with the developers, it was revealed that they also did not take into account the fact that the process of closure is always accompanied by a contact bouncing (i.e., multiple interruptions of switched circuit after the first closing of contacts). Thus, the contacts of the miniature relay will be overloaded at the moment of closure and can simply burn out before shunting with more high-capacity contacts.

It is important to note that the double-break contact systems are widely used in engineering. However, in such systems a contact with a larger gap is closed first as it has higher resistance to electric arcing, and then it is shunted with the standard silver contact (see Figure 2.14).

Since the use of miniature electromechanical relays with settings inappropriate for DPR operating conditions has become rather common, some companies have been trying to find a way out by connecting varistors in parallel to contacts in order to facilitate the switching of the inductive load (see Figure 2.15). In this case, readers should refer to DPR types SEL-787, SEL-751, and some others, including the JS series, as miniature relays with a

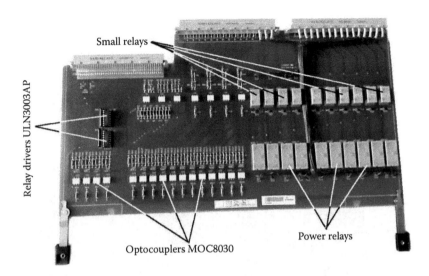

FIGURE 2.13
Module of DPR output relays type RCS-9681 (Nari-Relays) with two relays connected in parallel (high power and low power) in each channel. 1: Low-power output relay type DS-P; 2: high-power output relay type ST; 3: photocoupler type MOC8030; and 4: control drivers of the output relays type ULN3003AP.

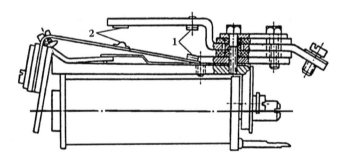

FIGURE 2.14
Two-stage electromagnetic relay contact system. 1: auxiliary switching tungsten contact; and 2: the main silver contact.

maximum switching DC voltage of 150 V; and varistors of the type 14D431K with a clamping voltage of 710 V. It should be noted that this solution is not very effective as the overvoltage above 700 V resulting from switching the DC load contains a significant inductive component.

At low induction coefficients, when overvoltage at the contacts does not exceed 700 V, the varistor just does not work and voltage is sufficient to maintain DC electric arcing on the contacts of the relay. In addition, it is impossible to perform standard conformity tests of the relay contacts shunted with the varistors (e.g., insulation resistance, and withstanding voltage).

Varistors

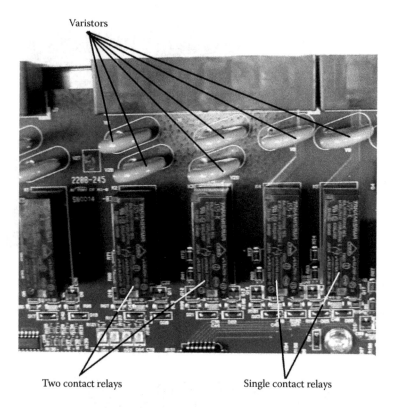

Two contact relays   Single contact relays

FIGURE 2.15
Detail of the output relay module showing contacts shunted by varistors; DPR types SEL-787 and SEL-751.

Type T60 DPRs (General Electric; see Figure 2.16) are equipped with both conventional electromechanical and solid-state output relays. According to the product description (T60 Revision, 5.6x), the semiconductor output relays are equipped with special circuits "monitoring the DC voltage at open contacts and DC current at closed contacts." This is stated as if everything is clear and obvious ... but further text brought me to a standstill: "The voltage is registered as a logical unit when the contact circuit current exceeds 1–2.5 mA and the current is considered as a logical unit when it exceeds 80–100 mA." This is an extremely strange explanation, to put it mildly. It is not only about the text, but also about the engineering solution. First, it provides monitoring of direct current only, which limits the scope of use. Second, the load current may be very small (1–3 mA) such as, for example, the current at the logical input of another DPR or sensitive electromechanical slave relay. How will the current monitoring system work in this situation? It turned out that the developers of this system took this possibility into account and offered customers the option to connect an additional external resistor in parallel to the contacts. If the voltage is 48 V, they recommend a 500 Ohm 10 W resistor.

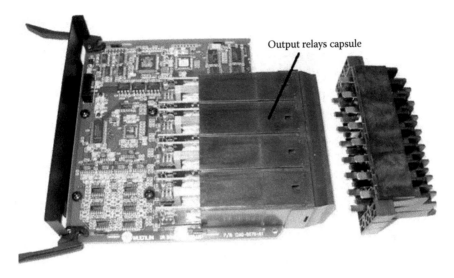

Output relays capsule

FIGURE 2.16
DPR type T60 output relay module with cover removed; one may ask, "Why are the output relays put in the casing?"

This is a very large resistor! Can you imagine the size of the resistor you would have to use for 220–250 V? And where it would have to be installed? The developers of the T60 have kept silent about this.

Another "invention" was the automatic cleaning of the contacts (auto-burnishing) of external relays that transmit signals to the logic inputs of T60. The designers took care of the fact that if input current at the logic inputs is very low (less than 3 mA) and the external relay contacts are oxidized, the signal "cannot pass" through them.

For such self-cleaning purposes, the inputs of T60 are equipped with special nonlinear elements (obviously, they are similar to posistors) with low resistance in the switched-off (cold) mode and rapidly increasing resistance under voltage (the temperature is also increasing).

As a result, at the first moment after closure of external relay contacts, the current of 50–70 mA passes through them, and then (within 25–50 ms) it rapidly drops to 3 mA. It sounds like a nice idea if you have little knowledge of the processes at the contacts. "Obstruction" of contacts due to the oxidation occurs in low amperage circuits with switching voltages lower than 20–30 V. At higher voltages, black and unsightly contacts, resulting from the breakdown of very thin oxide layers, still conduct very well even the smallest currents (this is known as the *freaking effect*). Therefore, in terms of the actual DPR operating voltages, the problem is completely contrived, and its technical implementation is quite senseless.

Different types of modern DPRs manufactured by AREVA are equipped with electromechanical type G6RN-1 relays (see Figure 2.17) as a standard and/or a solid-state relay. Either of the relay types can be ordered. Areva

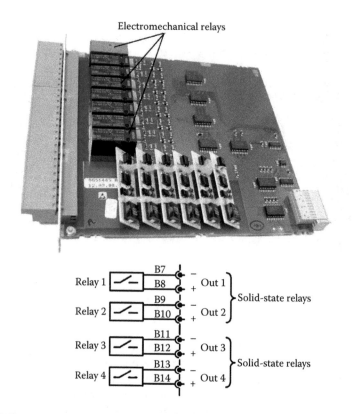

FIGURE 2.17
Output relay module with conventional electromechanical relays, and indication of the high-speed solid-state relays in Areva technical documentation.

stated that the standard relay is able to switch a 250 A load within 30 ms or a 30 A load within 3 seconds at 300 V. We suggest that the reader assess the stated switching capacity of Areva DPR using the diagram taken from the relay G6RN-1 technical documentation (see Figure 2.18).

According to Areva, the solid-state relay withstands steady loads of up to 10 A. It is not really clear how small semiconductor elements without heatsinks are capable of conducting a long-term 10 A current if it is known that small semiconductor devices without heatsinks are usually heated to a very high temperature with subsequent breakdowns under long-term currents exceeding 2–3 A.

Another solution is to use hybrid output relays, arranged in parallel connections to the electromechanical relay contact and semiconductor switching element (see Figure 2.19). IGBT transistors with appropriate opto-isolated drivers protected from overvoltage by varistors are used as switching elements. Usually, transistors have a large current margin (40–90 A) for providing the necessary resistance against pulsed currents and for improving

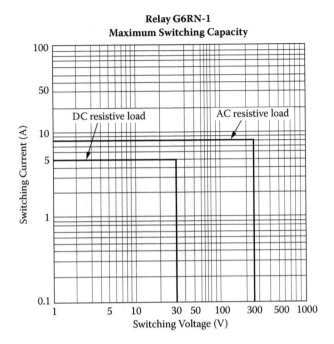

FIGURE 2.18
Maximum switching capacity of miniature relay type G6RN-1 installed in AREVA DPR.

reliability. In the hybrid scheme, transistors stay under current only until the contacts of electromechanical relays are closed (10–15 ms), and therefore they do not warm up even if heatsinks are missing. This idea is used in a DPR manufactured by SEL. Protection of the IGBT transistor is provided by shunting the reverse-switched diode. In the event of wrong polarity of the external load (e.g., the switch trip coil), it is activated immediately upon supply of an external voltage, which can lead to serious problems.

The manufacturer of the same DPR type, SEL-487, uses a sophisticated hybrid relay supplemented by a diode bridge (type KBU4M, 1000 V, 4 A; see Figure 2.20). In this case, the IGBT transistor (type IRG4PF50, 900 V, 50 A) connected to the bridge diagonal provides switching both AC and DC load current. Diode bridges of this type have more than twentyfold short-time overload capacity (before relay contacts are closed), which makes them suitable for switching the significant current loads.

In addition to the output relay, the module also contains a set of optocouplers acting as a buffer between the control microchips and output relays, as well as resistors that preset the photocouplers' operating mode.

SEL has declared that this design of the output relay makes it extremely fast acting (with a response time of 10 microseconds). The question is, who needs it if the time spent by the DPR for processing an input signal and

Varistors

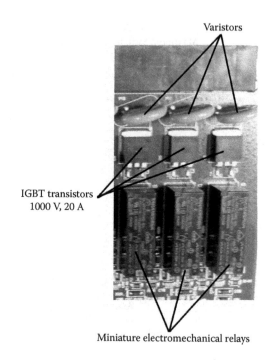

IGBT transistors
1000 V, 20 A

Miniature electromechanical relays

FIGURE 2.19
Detail showing the hybrid output relay module.

issuing an internal instruction to the output relay comes up to 20–40 ms? In addition, such a high speed may result in major troubles caused by malfunctions due to the short impulse and high-frequency noise.

2.4 Digital (Logic) Input Modules

More often, you can meet with the structures where output relays are installed on the same common board with other electronic parts of the DPR, such as, for example, logic (digital) input assembly; therefore, such combined assemblies are often called *input-output modules* (I/O modules); see Figure 2.21.

Usually, the DPR logic (digital) input assembly is a combination of a certain number (5–40) of identical input channels. In the design, each input channel is generated by a high-power resistor (R1) quenching the principal component of the 100–250 V input voltage, an optocoupler (Opt2) providing galvanic isolation of the DPR's internal circuits from the 100–250 V input voltage, an overvoltage protection element (R_U) at the input, and several auxiliary elements (see Figure 2.22).

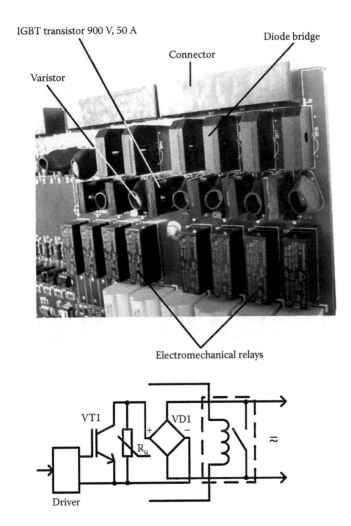

FIGURE 2.20
Detail showing the diagram of the DPR hybrid output relay module type SEL-487 designed for AC and DC switching.

In this diagram, the logic element *disable* is realized on the basis of the VT transistor. The optocoupler-based (Opt1) inhibit input of the element allows blocking the logic input upon the microprocessor command activating the Opt1 optocoupler.

The digital input units in various DPRs, manufactured by Areva (see Figure 2.23), also have a similar set of elements. However, unlike those described above, in these units there is a possibility of a change in voltage threshold level to activate a logical input. It can be realized by changing the biasing potential on the base of the switching transistor.

FIGURE 2.21
Combined input-output module of DPR type REC-316. 1: input elements assembly; and 2: output relay assembly.

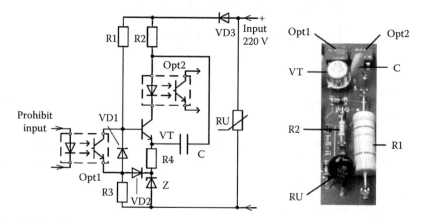

FIGURE 2.22
Schematic circuit diagram of a logic input single channel of relay series 316 (REL, REC, RET, REG, etc.), manufactured by ABB.

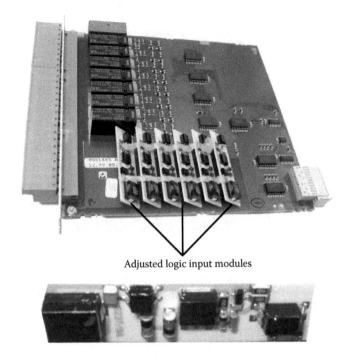

Adjusted logic input modules

FIGURE 2.23
Adjusted logic input modules in AREVA-manufactured DPRs, and enlarged detail showing the single module.

Sometimes logic input channels have less sophisticated designs. They contain one quenching resistor, an optocoupler, and a small ceramic capacitor with the input of some hundreds of picofarads. It should be noted that such a capacitor cannot absorb the energy of the impulses coming to the input of the logical channel under the transition processes followed by surge overvoltages within a powerful and very long DC network at the substations.

Also, an important element of such a channel is a diode connected in parallel to the input in the opposite direction (or in series in the forward direction) and preventing voltage reverse polarity at the DPR input (see Figure 2.24). This prevents false activation of the logic inputs due to the voltage impulse reverse polarity occurring at the DPR inputs upon the switching of inductive loads (switched trip coils, and intermediate relay windings). Unfortunately, not all DPR types are equipped with such diodes and bidirectional input photocouplers (see Figure 2.25), leading to the activation of DPR logic inputs under the voltage of any polarity.

In one of the new DPR models (SIPROTEC 7U6125; see Figure 2.26), all electronic parts except the input voltage and current transformers block are

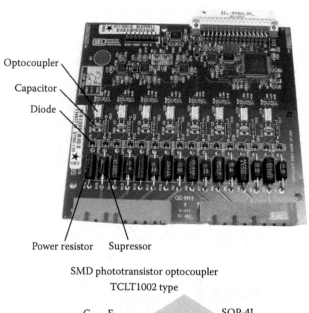

Optocoupler

Capacitor

Diode

Power resistor   Supressor

SMD phototransistor optocoupler
TCLT1002 type

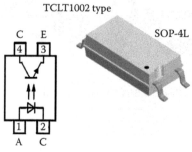

FIGURE 2.24
Logic input module of DPR type SEL-787.

installed on a common PCB in such a way that the switching power supply is located, literally, end to end with the central processing unit (CPU) arranged under the communication module (see Figure 2.26). This is the only arrangement of this sort that I have ever seen. The effectiveness of this arrangement in relation to the electromagnetic compatibility raises doubts, considering the high-frequency electromagnetic emissions at the switching power supply. The method employed for improving the switching capacity of the miniature relay contacts also looks strange. The relays are shunted with 4.7 nF, 250 V capacitors. This raises two issues: first of all, one capacitor is not able to improve the switching capacity of the contacts, because you also need a resistor connected in series; and, second, a capacitor with a rated voltage of 250 V is not sufficient for 220–250 V mains with significant switching overvoltage.

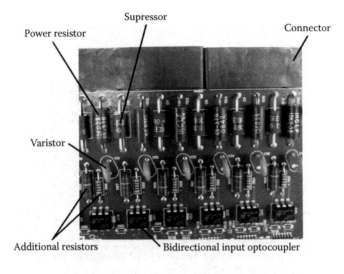

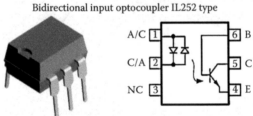

Bidirectional input optocoupler IL252 type

FIGURE 2.25
Detail showing the logic input assembly of a SEL-487 relay.

Resistors quenching most of the input voltage at the logic input assembly are located on the back side of the PCB. Voltage-regulator diodes and transistors constitute the input voltage stabilizer (its operating principle is similar to the one discussed above and implemented in DPR series REC, REL, and RET from ABB), providing an operation over a wide range of input voltages (24–250 V), while diode bridges of the logical inputs make them insensitive to the polarity of the input voltage. Both of these features of the arrangement are more likely disadvantages than advantages (see above). Besides, the complete absence of high-voltage surge protection (such as varistors or suppressors) at the inputs of this assembly is a significant design flaw, in our opinion.

The current protection relay, type 7SJ8032, has a similar design (see Figure 2.27). Unlike the previous design, the analog input assembly is installed on the common PCB, while the central processor is installed on a separate board, placed in line to the main board. The logic inputs' design concept, output relays, and protective element at the relay contacts are the same as those detailed above.

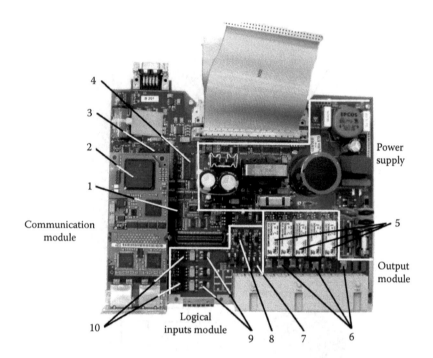

FIGURE 2.26

Combined DPR module type SIPROTEC 7U6125 (Siemens). 1: 32 Mb complementary metal-oxide semiconductor (CMOS) flash memory type S29AL032090; 2: communication controller type MPC860EN; 3: CPU type MCF5280CVM66 (Freescale); 4: synchronized dynamic random access memory (DRAM) type 48LC2M3282; 5: output relays type V23061-A1001-A302; 6: capacitors (4.7 nF, 250 V) connected in parallel to the relay contacts; 7: diode bridge type DF10S (1000 V, 1 A); 8: a voltage-regulator diode; 9: transistors, type BSP135 (600V, 0.1A); and 10: optocouplers, type SFH601.

2.5 Central Processing Units (CPUs)

The main DPR module, often called the *CPU module*, is the most sophisticated and expensive PCB, and it is crammed with elements (see Figure 2.28). The DPRs are based on surface-mount technology (SMD technology). The board contains the microprocessor, memory elements, an analog-to-digital converter (ADC), a multiplexor, auxiliary (peripheral) microprocessors, a communication controller, a communication port, and other elements.

In the past 15 years since the release of relay series 316, the design of the main CPU has not undergone any significant transformation (see Figure 2.29). We only noticed the smaller size of the main microprocessor and associated operating elements as well as a smaller number of elements. This has resulted from recent advancements in the field of nanotechnology, which has led to a significant reduction in the size of the semiconductor elements

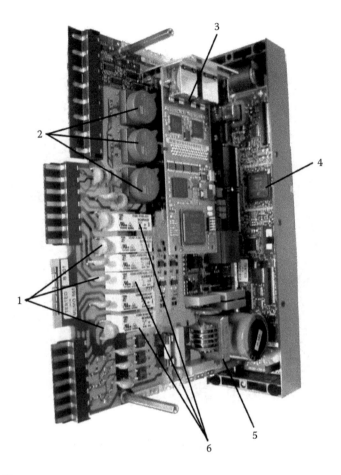

FIGURE 2.27
SIPROTEC 7SJ8032 (Siemens). 1: protective capacitors (2.2 nF, 250 V) connected in parallel to relay contacts; 2: encapsulated current transformers; 3: communication module; 4: the main microprocessor; 5: elements of the power source; and 6: electromechanical output relays.

(reaching units and even fractions of a micron), a reduction of thickness of semiconducting and insulating layers, a reduction of operating voltage, and increases in operating speed and the number of elementary logic cells in one device. All this has resulted in a sharp rise in the sensitivity of semiconductor components and, particularly, memory cells to ionizing radiation.

The problem is compounded by the ever-growing use of memory elements in modern microprocessor structures. A great many modern highly integrated circuits constituting microprocessor devices contain large embedded memory elements with completely uncontrolled operability. In recent years, the problem of the sharp rise of sensitivity to ionizing radiation has become critical not only for the memory elements but also for high-speed logic gates, comparators, and so on (i.e., practically for the whole of modern

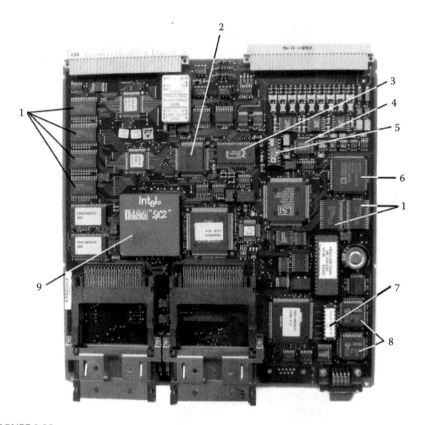

FIGURE 2.28

Main DPR module type REL316 manufactured in the 1990s. 1: Static random access memory (SRAM) type TC551001BFL-70L (Toshiba); 2: peripheral microprocessor type NG8237016 (Intel); 3: flash memory (Intel); 4: 16-channel multiplexor type ADG506AKP (Analog Devices); 5: analog-to-digital converter AD677 (Analog Devices); 6: peripheral microprocessor type ADSP-2105 (Analog Devices); 7: permanent storage with the algorithm of protection ("software key"); 8: communication controllers type Z85C3016VSC (Zilog); and 9: main processor series 486 (Intel).

microelectronics). Besides the ionizing radiation, modern microelectronic devices may be affected by intentional high-frequency electromagnetic emissions implemented in many kinds of modern electromagnetic weapons.

2.5.1 Analog-to-Digital Converters (ADCs)

The process of analog-to-digital conversion (i.e., conversion of the input analog signal to digital code) involves two operations: sampling by time and by level. Therefore, the ADC is a measurement of the instantaneous values of the analog input signal at given intervals (sampling) and the coding of the measured values. To encode the complete range of possible changes of the continuous (analog) signal (U), U is evenly divided into a finite number

FIGURE 2.29
Main DPR module type SEL-787, manufactured in 2009.

of discrete levels ΔU, called *sampling levels*. Each value (in our example, from 1 to 9) of the signal level can be coded (in a binary number). In our case, these are the numbers from 0001 to 10001. After the coding, the analog signal operations can be replaced by the corresponding binary number operations performed by the microprocessor. Recorded binary numbers are transmitted to the microprocessor at regular intervals t. During level sampling, the measured signal may vary from the sampling level ΔU (see Figure 2.30). In this case, it should be rounded to the next integral quantum number, thus causing rounding error.

An important feature of ADCs is the capacity (length) of the generated binary number. The capacity of the ADC is the amount of discrete signal levels defined and encoded by the ADC. For example, *8-bit capacity* means that the ADC is able to sample and encode signals as 256 discrete values ($2^8 = 256$). In order to convert analog values (e.g., changing the voltage at the current input of DPR within 0–10 V) with an error less than 2–5%, DPR inputs are equipped with 12-bit ADCs generating $2^{12} = 4096$ sampling levels.

In the above example of the DPR input analog signal, the resolution capability of the ADC should be 10/4096 = 2.44 mV. In practice, it is not always possible to reach such a high capacity due to noise at the input (in the example,

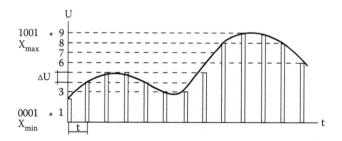

FIGURE 2.30
Sampling of signal by ADC.

it is referred to as units of millivolts). In this case, the ADC is not able to distinguish between adjacent levels of the input signal, so the least significant bits of the output signal are almost useless. The actual achievable capacity (effective capacity or effective number of bits [ENOB]) is always less than what is rated.

It is obvious that the time sampling generates an error and thus causes the actual loss of information. In fact, the sampled signal does not show the behavior of the original instantaneous continuous signal, for example between t_1 and t_2, t_3 and t_4, and so on (see Figure 2.30). In other words, this process results in generating an error that depends on sample interval T: for small values of a sample interval (i.e., high-frequency sampling, or sampling frequency), the number of measurement points is high and the loss of information is small, while the situation is reversed under large sampling intervals. Usually in DPRs, uniform sampling is applied with an interval of 1/12, 1/20, or 1/24 of the basic operating frequency period (i.e., sampling at a frequency of 600, 1000, or 1200 Hz for the operating frequency of 50 Hz, or, as appropriate, 720, 1200, and 1440 Hz for the base frequency of 60 Hz). The higher frequency is used in the microprocessor-based fault recorders For oscillographic testing of the emergency mode in a network with a bandwidth of 0–1000 Hz, the frequency of sampling should be at least 2000 Hz.

As ADC selects the input values at fixed intervals, it is clear that there is no way to determine the value of the input signal in the intervals between these samples (intervals 1–6 of Figure 2.31). If the input signal alternates more rapidly (under higher frequency) than measurements are made (i.e., than the sampling frequency), then the exact ADC signal recovery is impossible while false low-frequency signals appear at its output. These false signals depend on the difference of the frequencies mentioned above, also known as *aliasing*. For example, a sine wave signal of 3000 Hz measured at a sampling frequency of 2000 Hz would be reproduced as a 1000 Hz sine wave. The optimum ratio between signal frequency and sampling frequency is described by the Nyquist-Shannon theorem (or the Kotelnikov-Shannon theorem in Russian technical literature).

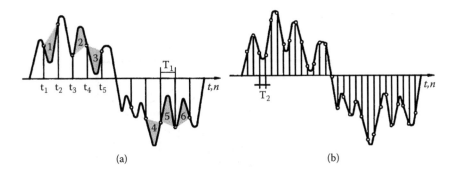

FIGURE 2.31
Generating sampling error within the sampling intervals.

Generally, to improve conversion precision, the sampling frequency must be at least twice the highest frequency component in the signal, while all frequencies exceeding the sampling rate should be removed (filtered) from the input signal. This removal is provided by an anti-aliasing filter.

The so-called aperture distortion introduces an additional error due to the clock jitter of the signal that determines the moment of input signal measurement. As real AD converters (see Figure 2.32) do not provide immediate conversion, the input signal value should be kept constant during the conversion. For this purpose a special unit, called a *sample-and-hold circuit* or *S/H*, should be connected to the ADC input. The S/H circuit consists of a capacitor that periodically switches between the analog signal source and the ADC input. The switching frequency is controlled by the semiconductor switch S1. A simplified S/H circuit in theory can consist of four basic components (see Figure 2.33); however, in reality the interconnection of the components obviously differs from the one shown in the figure.

The amplifier A1 (input buffer) has high input impedance while providing sufficient charging current to charge-storage capacitor C_H. The control circuit opens and closes the S1 switch contact between the buffer amplifier output and storage component (capacitor) to store the last received signal as a capacitor charge. The amplifier A2 converts the charge-storage capacitor with high impedance into a voltage source with low internal impedance required for outer load. When S1 is closed, the output signal should be equal to input signal, but is exposed to noise and errors due to the large amplifier bandwidth. When S1 is open, the voltage stored on the capacitor (in theory, it should stay the same) goes to the device output until the S1 switch is closed again.

Virtually, actual oscilloscope records as shown in Figure 2.34 demonstrate that S/H circuits function as an additional filter cleaning the signal sent to the ADC input from the emission, transient processes, and high-frequency noise.

The internal structure of ADC devices is rather sophisticated. It includes a variety of supplementary service units improving the quality of conversion performed as single multilead microchips (see Figure 2.32). Usually, the AD con-

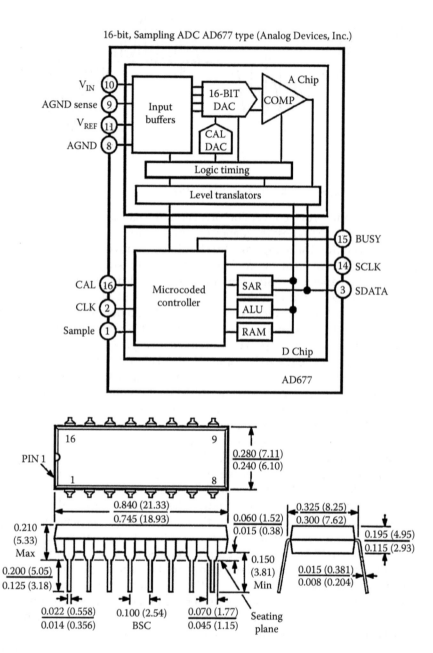

FIGURE 2.32
16-bit ADC AD677 (Analog Devices, Inc.).

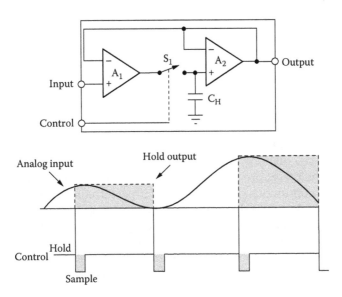

FIGURE 2.33
Sample-and-hold circuit.

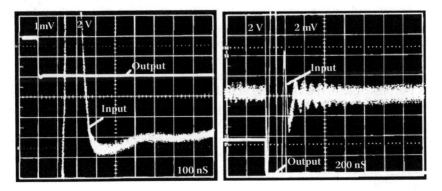

FIGURE 2.34
Actual oscilloscope records of signals at input and output of sample-and-hold circuit.

version path also includes the multiplexor. A *multiplexor* is an electronic switch that sends signals from numerous DPR inputs to ADCs in turns (see Figure 2.36). Thanks to the multiplexor, we can use one ADC, which is rather sophisticated and expensive, to process several signals (inputs) at the same time.

2.5.2 Memory Devices

In microprocessor systems, the data are stored in special devices called *memory*. There are two main types of memory devices: *read-only memory* (ROM), which permanently stores in the microprocessor operating program, and

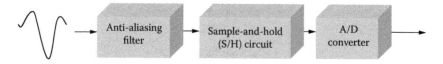

FIGURE 2.35
DPR analogue signal input circuit.

random access memory (RAM), which is designed for temporary storage of intermediate calculations. As the permanent storage device is designed only for reading the preset program, it is ROM. Opposed to this is RAM, which is designed for continuous data exchange with external devices (i.e., for high-speed reading and recording of new data at any random moment upon an external device query). There are many subtypes of these two major memory devices. For example, some types of ROM can be reprogrammed as required during operation. For example, you can erase the old relay protection settings and input the new ones. These types of ROM are called *erasable programmable read-only memory* (EPROM). The S/H circuit should be connected to the ADC input in series with an anti-aliasing filter (see Figure 2.35).

EPROM was invented (U.S. patent 3,660,189, 1972) by Dov Frohman, the founder and former general manager of Intel Israel who is widely respected as a leader and innovator in the worldwide technology and semiconductor industry (see Figure 2.37). EPROM remained Intel's most profitable product well into the 1980s. At first, you could erase the contents of the EPROM only by using ultraviolet radiation sent through a special window in the chip body within 10–15 minutes. There were special devices designed for this purpose. Later, the electrical erasable programmable read-only memory (EEPROM or E²PROM) allowed erasing and rewriting the contents of the memory with electric signals sent to special inputs. This has turned out to be very convenient for reprogrammable microcontrollers and microprocessor-based protective devices (MPDs). However the disadvantage was that conventional EPROM had a far higher capacity and longer life than even very large EEPROM.

In recent years, a new type of permanent memory, so-called flash memory, has become available. Unlike EEPROM, which provides bit erasing and bit overwriting of data, flash memory erases and writes data by large blocks. Since the erasing of memory takes a good deal of time, such large-block operations significantly accelerated the process compared to that of old EEPROM devices.

Flash memory was first introduced by Fujio Masuoka, DSc, of the Toshiba Corporation in 1984 (see Figure 2.38). The flash memory is based on metal-oxide semiconductor field effect transistors (MOSFETs). These transistors are equipped with insulated control electrodes, called *gates*, and are controlled through supplied electric potential.

Thanks to very effective sealing, the electrical charge (defining the transistor state) is stored in the MOSFETs for a very long time. In order to change

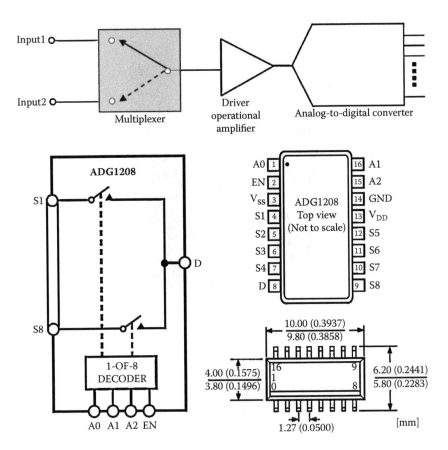

FIGURE 2.36
8-input multiplexor type ADG1208 (Analog Devices, Inc.).

the state of the transistor (i.e., to erase the memory of the cell formed by such transistor), you only have to discharge its control junction, that is, remove the electric charge. Memories based on this principle can store data for years and allow up to 50,000–100,000 rewrites. It is quite clear that the integrity of memory based on the electric charge depends on the memory cell's self-discharge rate, which is highly affected by various adverse factors such as ionizing and radiation, among others.

All types of permanent memory are nonvolatile, that is, those in which the information is not initialized upon power-off. In fact, the part of the relay protection program defining the MPD algorithm (i.e., determining the type of defense: differential, distance, current, etc.) should not be modified under any circumstances while the other part of the program can be changed by the customer (e.g., the part responsible for operation settings and protection mode). In order to separate these two parts, sometimes they are designed as two separate chips, one of which stores rewritable data (EEPROM), and

FIGURE 2.37
Dov Frohman, inventor of the EPROM.

FIGURE 2.38
Fujio Masuoka, DSc, inventor of flash memory (Toshiba Corp.).

the other stores nonrewritable data (the so-called software key [SWK] based on ROM). In order to modify DPR operation, it is required to replace the SWK chip with the new one. For automatic control of ROM and EEPROM serviceability, the recorded memory array is summed and encoded as a certain number, which is referred to as the checksum, recorded in a specially reserved cell. In test mode (usually at microprocessor loading), it scans the contents of the memory and compares it with the checksum. If a discrepancy is detected, further work is blocked by the microprocessor. In some advanced systems, there are two EEPROM devices running in parallel. If one of them finds such a discrepancy (i.e., memory damage), the memory is automatically

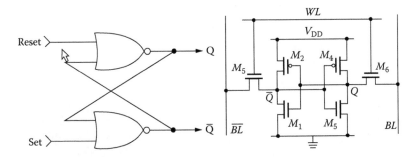

FIGURE 2.39
Standard static RAM (SRAM) cell based on six CMOS transistors.

overwritten from the second intact EEPROM. Also, there are several types of RAM, such as static RAM (SRAM) and dynamic RAM (DRAM).

A standard cell of a static binary memory is a binary transistor-based trigger consisting of two cross-connected (ringed) inverters (logical NOT elements) and switching transistors providing access to the cell (see Figure 2.39).

Such a design is bistable, that is, when switching from one state to another, it retains its position until receipt of a modifying electric impulse. Thus, the memory based on this principle (see Figure 2.40) does not require periodic reloading for saving data, but it remains volatile and entire data are lost upon power-off.

Sometimes, an external lithium battery is installed on the motherboard and used to keep the data safe upon power failure. Typically, such cells last many years, but when the voltage is reduced they must be replaced. However, if you simply remove this cell, all the data recorded in such memory will be

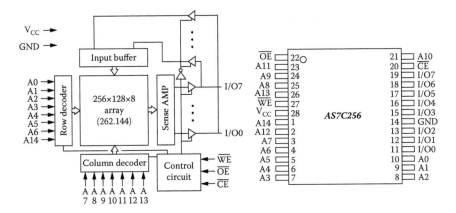

FIGURE 2.40
Volatile static RAM (SRAM) type AS7C256 (32k × 8 bit) based on CMOS transistors (Alliance Semiconductor).

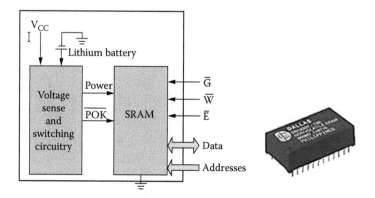

FIGURE 2.41
The design and appearance of nonvolatile SRAM.

lost. Therefore, such a battery should be replaced when the external power supply is connected in parallel to the battery holder.

There is also a special kind of memory called nonvolatile SRAM (nvSRAM). This memory can store data after disconnection of external power thanks to a built-in tiny lithium battery (see Figure 2.41). This memory is faster than EPROM and EEPROM, so sometimes nvSRAM is used as a permanent rewritable memory (i.e., another form of EEPROM). Standard SRAM can be used if we need a relatively simple low-capacity memory with low power consumption. For example, it is used for registers and cache memory.

Large operating memory in devices is realized as DRAM. Each cell of the memory contains a low-capacity capacitor C and a solid-state switch VT, installed in a single chip (see Figure 2.40). The capacitors are charged upon inputs of single bits into the cell or discharged upon input of zero bits into the cell. The solid-state switch "locks" the cell and retains the charge inside the capacitor. To access a particular cell, you select its address in rows and columns (see Figure 2.42).

In practice, the capacitor and solid-state switch functions in DRAM chips are performed by CMOS microtransistors that, like capacitors, are capable of accumulating and storing the charge within a certain period of time thanks to the efficient internal sealing. Advantages of this design include relatively low cost and large capacity of memory. However, because the capacitance generated by microtransistors is very small, the stored charge diffuses rapidly and the data must be updated within regular intervals to avoid data loss (actually, this is why this type of memory is called *dynamic*). This process is called *memory regeneration* (memory refresh) and is implemented by a special controller.

The entire string of "cells" in DRAM is rewritten within a certain period of time called the *regeneration step*, and in 8–64 milliseconds all rows in the memory are updated. Such memory regeneration process significantly "slows

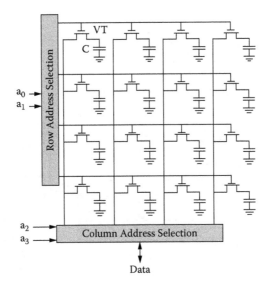

FIGURE 2.42
DRAM schematic circuit diagram.

down" the system, because during this process the exchange of data with the memory is impossible.

Therefore, the regeneration based on the conventional row search is not applied in modern DRAM units. There are several more cost-effective versions of the process, including supplementary operating assemblies installed inside the DRAM chip. It is clear that the actual design of the DRAM unit (see Figure 2.43) is much more sophisticated than its simplified schematic circuit diagram.

In recent years, a variety of advanced DRAM types (EDRAM, FPM DRAM, EDO DRAM, SDRAM, DRRAM, etc.) have been developed. Mere enumeration of them can take a lot of space, and consideration of them is far beyond the scope of this chapter.

To sum up, I'd like to emphasize only one very important feature of the DRAM: its high sensitivity to electrical noise and to radiation. Electrical interference caused by internal circuits, or penetrating from the outside, can lead to spontaneous switching of dynamic memory cells, containing a single bit, into the opposite value. Initially, it was assumed that the effects of radiation were caused by alpha particles emitted from plastics insulating a memory chip, pollutants contained in the chip itself, and packaging materials under the influence.

As mentioned above, recent progress in the field of nanotechnology has led to a significant reduction in the size of semiconductor elements (reaching microns and even fractions of a micron), a reduction of thickness of semi-conducting and insulating layers, a reduction of operating voltage, and an increase in operating speed and the number of elementary logic cells in one

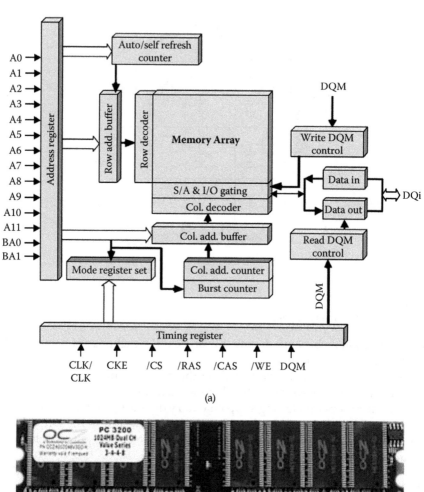

(a)

(b)

FIGURE 2.43
(a) Realistic design of 64 Mb synchronous DRAM: electrical flowchart. (b) Realistic design of DRAM: the appearance of certain DRAM module types, which are widely used in PCs.

device. All this has resulted in a sharp rise of the memory cells' sensitivity to ionizing radiation. This sensitivity is so high that ordinary (that is, completely normal) background radiation at sea level is dangerous for the memory cells. High-energy particle currents coming from the space are especially dangerous. Even one such particle penetrating the memory cell generates secondary streams of electrons and ions, causing spontaneous switching of a simple transistor or capacity discharge at charge memory elements. The problem is compounded by the ever-growing use of memory elements in modern microprocessor structures. A great many modern highly integrated circuits constituting microprocessor devices contain large embedded memory elements with completely uncontrolled operability.

The problem of the sharp increase in sensitivity to ionizing radiation is relevant not only to the memory elements, but also to high-speed logic gates, comparators, and so on, that is, practically for the whole of modern microelectronics. The worst thing is that the random microprocessor failures caused by electromagnetic noise and radiation may be temporary, such as spontaneous changes in the contents of memory (RAM) and registers, while internal damages may not be overt. Both of these types of failures cannot be detected by any tests, and can show up at the most unexpected moments.

2.5.3 Microprocessors

The microprocessor is the central DPR assembly controlling the operation of all other assemblies and performing arithmetic and logical data operations. The modern microprocessor is practically a complete control system. It has a sophisticated internal architecture and represents a very large integrated circuit built on silicon base layers. It is manufactured by special technology involving chemicals, gases, and radiation.

The processor contains a number of microtransistors interconnected with ultrathin aluminum connecting channels enabling interaction for recording and processing of data, and providing the performance of numerous functions. The first model of the 4004 microprocessor, announced by Intel in 1971 (see Figure 2.44), contained "only" 2300 transistors and performed about 60,000 computations per second. The 486 series processor, widely used today in DPRs, contains 1.6 million transistors, while the Pentium IV processor has 42 million transistors and performs hundreds of millions of operations per second.

Today, microprocessors are the most complex element among electronic devices. Hundreds of manufacturing steps with surpassingly strict requirements to the purity and accuracy of each are required to produce modern microprocessors. Initially, the first very thin layer of silicon dioxide is grown on the base under a high temperature. Then the base is covered with photographic emulsion that is disintegrated by ultraviolet radiation, which is then covered with the so-called template (a template with the circuit pattern).

During the photolithography process, ultraviolet radiation, passing through the template, forms the circuit pattern on the base. Light-exposed

Ted Hoff   Federico Faggin

Masatoshi Shima

FIGURE 2.44
Inventors of the world's first microprocessor.

areas of photolayers become soluble and are washed out with a special solvent during the further processing, thus exposing the corresponding part of the silicon dioxide layer not protected by the template. Chemicals etch out the unprotected areas of silicon dioxide, revealing the silicon dioxide pattern on the base. In order to separate the finished layer from the new one, an additional thin silicon dioxide layer has to be grown on the circuit pattern. Then it has to be covered with a microcrystalline silicon layer and another photolayer. Subsequently, the second layer has to be grown in the same way. A special template is used for the exposure of each microprocessor layer.

The formation of semiconductors with desired conductivity and p-n transitions of the transistors from the pure silicon is performed during the all-ion-implant process, where the areas of the silicon base exposed to ultraviolet light are bombarded with ions of different impurities. The ions penetrate into the base, providing the necessary electrical conductivity of these areas.

Coating and etching have to be repeated several times. After this, very small gaps in the layers are left for the interlayer compounds. These gaps are filled with metal atoms joining the layers of the future microprocessor (in today's microprocessors, the number of layers may reach 20 or more.

The ultrafine layers of metal are left at the edges of the chip for attaching the external microprocessor pins. The total manufacturing cycle consists of more than 250 stages, after which the microprocessor has to pass thorough testing and then be integrated into the protective enclosure.

The microprocessor performs the following functions:

- Reading and decoding commands contained in the main memory
- Reading data from the main memory and registers of external device adapters
- Data processing and entry into the main memory and registers
- Generating control signals for DPR output devices

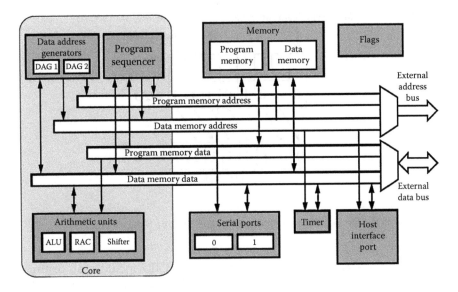

FIGURE 2.45
An example of the internal structure of a commercial microprocessor with integrated memory and some supplementary elements.

Microprocessor software determines the specific tasks performed by the microprocessor. Microprocessor types differ by memory type and size, set of instructions, processing speed, the number of input and output lines, and data capacity. In general, a microprocessor flow diagram can be as that depicted in Figure 2.45.

The *CPU* is the binding assembly of any microprocessor device, its core. Some of the superior modern microprocessors have several cores working in parallel under the control of a master core (see Figure 2.46). They are called

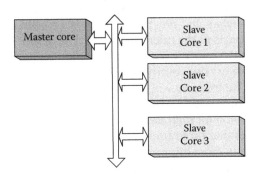

FIGURE 2.46
Structure of multicore processor. Master core: control (lead) core; and slave core: (executive) leaded core.

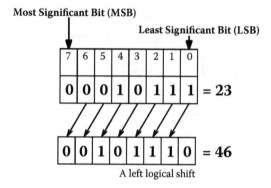

FIGURE 2.47
Doubling with shifter. 10111: a binary image of 23; and 101110: a binary image of 46.

multicore processors. The core of a conventional CPU includes an arithmetic logic unit (ALU), a register-accumulator (RAC), a multibit shifter, data address generators and program sequencers, and an internal bus.

The *arithmetic logic unit* performs arithmetic or logical operations on the data presented in binary or binary-coded decimal code. The result is stored in the so-called register-accumulator.

RAM cells represent the RAC, but, unlike the main memory, they use shorter commands for data exchange (i.e., the RAC is the fastest memory device of the microprocessor).

The *multibit shifter* with a set of multiplexors provides processing of shift logical instructions and multiplication and division operations. In a binary computing unit, a one-position shift to the left of the binary number has the same effect as multiplying by 2 (see Figure 2.47), and a right shift has the same effect as dividing by 2 (zero is shifted to a new position). Since the shifting speed significantly exceeds the speed of multiplication and division operations, it is widely used as a tool for program optimization.

Data address generators and program sequencers coordinate the interaction between different parts of the microprocessor. The device includes the clock speed generator, clock-pulse driver, and program sequencer. The framing bit is generated by a crystal-controlled oscillator equipped with an external crystal oscillator. The frequency of the clock speed generator determines the speed of the microprocessor.

The *program sequencer* (or *controller-sequencer*) provides for temporarily halting the execution of one program in order to promptly fulfill the other program that is currently more important. Program sequencer services interrupt procedures, receive interrupt requests from external devices, determine the priority of the request, and send the interrupt signal to the microprocessor.

Microprocessor memory is designed for short-term storing, recording, and transmitting data used in the calculations directly to the next operating ticks. Microprocessor memory is based on registers and provides high-speed DPR operation, as the main memory is not always capable of providing the required speed of recording, searching, and data reading.

The microprocessor *interface system* is designed for communication with other DPR devices. It includes the following:

- Internal interface of the microprocessor
- Buffer storage registers
- I/O control circuit and system bus (the I/O port is interface equipment providing connection of the other device to the microprocessor)

In order to extend and improve microprocessor functionality, both standard peripherals and additional circuit boards with chips can be connected to the microprocessor and system bus. These include the math coprocessor, I/O coprocessor, and interrupt controller, among others.

The *math coprocessor* accelerates operations with binary floating-point numbers, binary-coded decimal numbers, and trigonometric functions. The math coprocessor is controlled by the main microprocessor. It has its own instruction system and operates in parallel with the latter. The result includes the acceleration of operations in dozens of times over. Typical modern microprocessors are, as a matter of course, equipped with math coprocessors.

The *I/O coprocessor* works parallel with the microprocessor. It significantly accelerates the execution of I/O procedures when servicing multiple external devices, and deallocates the microprocessor from I/O procedures, including direct memory access mode.

The *system bus* provides communication between various elements of the microprocessor. The bus is a group of conductors used as communication lines for transmitting digital data. There are three main types of buses in the microprocessor: a data bus, address bus, and control bus. The data bus provides data transfer between processor assemblies. The address bus is used to transfer the memory cell address to obtain data from the permanent storage or random access memory. The control bus transmits control signals from the microprocessor to other system elements.

The most important features of the microprocessor are as follows:

Clock speed defines the performance of the processor. The processor operation mode is defined by a special chip called the *master clock generator*. Each operation of the processor is performed in a certain number of cycles. The clock speed dictates the number of elementary operations performed by the microprocessor per 1 second. The frequency of the first microprocessor type 4004 was 108 kHz, the 486 series microprocessors had 33 MHz, and the frequency of the Pentium-IV is 1.5 GHz. Microprocessors with codes DX2 or DX4 provide internal doubling (×2) or quadrupling (×4) of the clock frequency.

Processor word is the maximum number of binary digits (bits) that can be processed by the processor at the same time. That is, if the processor is capable of processing 8 bits at a time, it is called an 8-bit processor; if it can process 32 bits, it is called a 32-bit processor; and so on. The higher the processor word, the more information it can process per unit of time and the higher the device performance under otherwise equal conditions. For example, the 486 series microprocessor that is often used in DPRs is a 32-bit processor. Often, the less sophisticated process controllers are equipped with cheaper 8- and 16-bit processors. As already noted, the central processor is connected to other devices through the system bus. Since each bus has a specific bit width, which may differ from the CPU word, sometimes the CPU word is represent by two digits. For example, the digital symbol *32/64* means that the processor has a 32-bit data bus and a 64-bit address bus. Processors with capacities exceeding 32 bits are not used in DPRs. Today, 64-bit processors are used primarily in corporate network servers for operating high-resource consuming applications and providing exceptional reliability of systems for banking, manufacturing, engineering, and research.

Port is a special device providing communication between the microprocessor, external devices, and peripherals. Communication controllers, located on the main board, control the data transmit-receive process. This requires protection against unauthorized or unqualified access to the internal logic and DPR settings.

In the era of electromechanical relays, all connections were made with bundled rigid conductive wire. After installation, the relay is covered with protective caps and sealed. This guarantees protection against unauthorized or unqualified access to the relay.

The internal operating logic, functions, and settings of DPRs can be easily changed from an external computer or remote access local area networks (LANs, on Ethernet). The consequences of such intervention are unpredictable and dangerous, so some manufacturers of DPRs take measures to prevent such intervention. One such measure is so-called *hard logic*—the irreversible operating algorithm previously approved by the customer. This principle is basic to DPR types SPAC-800 and SPAC-810, produced in Russia under the license of ABB. *Semihard logic* (i.e., an algorithm allowing the input-output of particular functions and protection settings without access to basic logic change) is the most feasible principle used in many types of DPRs.

However, DPR devices with the so-called freely programmable logic have become widely used lately. According to the manufacturers, this logic provides the best flexibility and versatility for relay protection, and it grants considerable room for adaptation of DPRs to specific customer needs and characteristics. These are the latest series of SIPROTEC from Siemens, SEPAM-80 from Schneider Electric, and many others. Programming of these devices is formalized and includes operations with special tables, matrices, logic elements, logic equations, and logic modules (often, they are rather imperfect and require a thorough analysis for the correct choice). Naturally, multiple access-level passwords should be entered in the DPRs of this type. For example, SIPROTEC has more than ten access levels, and none of them allows completely separate access to the logic from access to the relay settings input. Therefore, during on-site DPR debugging, you have to obtain full programming access, which makes password-level separation senseless. Moreover, during the operation the entire embedded protection logic and settings can be easily destroyed and new ones can be uploaded. Reloading the RAM with a firmware update results in the "erasing" of passwords and all other data; thus, it is possible to reinstall any DPR logic.

It stands to reason that the greatly increased risk of unauthorized or unqualified access to the relay is the reverse of flexibility and versatility. In this regard, some attempts have been made to return to restricting access to the internal DPR logic with more reliable methods previously applied in electromechanical relays. For example, in Nari-Relays DPR type RCS-9671 (transformer differential protection), we have to connect two points at DPR output terminals with conductive wire (i.e., an appropriate jumper) in order to activate any function. This jumper supplies positive potential (from a separate, specially designed 24 V low-power internal power supply) to the input activating the required function. It's a very good solution, in our opinion.

2.6 Internal Power Supplies

The power supply is the principal component of DPRs upon which the reliability of the device's working capacity depends. In the modern DPR, it uses exclusively switch-mode power supplies (SMPSs).

In the 1960s, the first SMPS was introduced. And since then, it has been intensively developed until today the SMPS has almost completely eclipsed the older linear power supply (LPS) from all areas of technology. What is the difference between these two types of secondary power supplies, and is SMPS superior to the LPS?

Widely applied everywhere in technical equipment during many decades, the LPS is a rather simple and even primitive device (see Figure 2.48) consisting of only a few elements: a voltage transformer, the rectifier, a filter based on a capacitor, and a semiconductor stabilizer (the Zener diode with the powerful transistor, or a single-power solid-state element with an analogous function). Unlike the LPSs, the SMPSs are much more complex devices

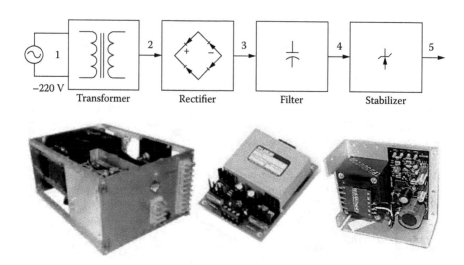

FIGURE 2.48
Structure and appearance of linear power supplies.

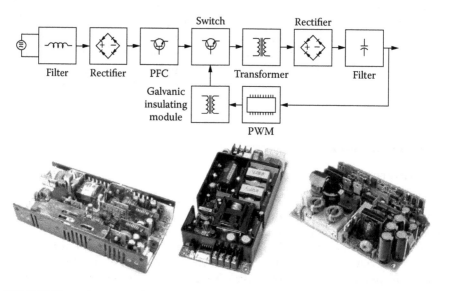

FIGURE 2.49
Structure and appearance of switching-mode power supplies.

working at high frequency and consisting of hundreds of active and passive elements (see Figure 2.49).

What are the basic differences between these two types of power supplies? In the LPS, the input alternating voltage is transformed to a necessary level (or levels, in the case of multiple secondary windings in the transformer) by means of the transformer, then rectified by a diode bridge, filtered by

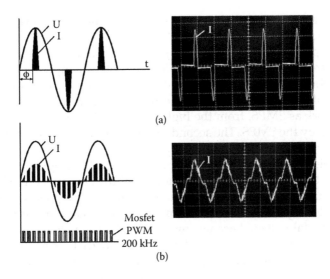

(a)

(b)

FIGURE 2.50
The form of a current and shift of phases between a voltage and a current consumed SMPS,
(a) without PFC and (b) with PFC.

means of the electrolytic capacitor, and stabilized by a nonlinear electronic element. The voltage applied to the stabilizing element must be greater than the nominal output voltage of the power supply, and its excess is dissipated in the form of heat on this stabilizing element (this sometimes demands the use of a heatsink).

The presence of some excess of a voltage on the stabilizing element enables carrying out stabilization of an output voltage with decreased or increased input voltage due to a change of the share of the energy dissipation on the stabilizing element. For this reason, the coefficient of efficiency of such power supplies is always greatly below volt.

In the SMPS, the input alternating voltage is first of all rectified by a diode bridge (or simply passed through the diodes of this bridge without change in the case of feeding a secondary power supply from a DC network). Then it is smoothed out and acts on the switching element (usually based on a MOSFET) by means of which the constant voltage is "cut" into narrow strips (switching frequency is from 70 up to 700 kHz for high-power supplies and from 1 to 3 MHz for low-power supplies). The rectangular high-frequency pulses that are generated are applied to the transformer, which outputs voltage matching the demanded level of a voltage which then is rectified and smoothed. The stabilization of the level of the output voltage at changes of the level of the input voltage is carried out by means of a feedback circuit consisting of a special driver which provides a pulse-width modulation (PWM) control signal of the switching element through a galvanic decoupling unit (it is usual to include an additional isolation transformer). This driver is small, but contains a complex integral circuit to change the width of control pulses according to the power supply output voltage level in order to compensate for deviations.

Low-cost power supplies have such structures. Better and more expensive SMPS devices contain at least two additional units: the input high-frequency filter and the power factor corrector (PFC). The first unit is necessary for protection of the network, that is, all other consumers are connected to the same network as SMPS, from the high-frequency harmonics oscillated into the network by the SMPS. The second unit is used to increase a power factor of the power supply.

The problem of the correction of a power factor (PF) originates due to the presence of the rectifier bridge with the smoothing capacitor on SMPS input. In such a circuit, the capacitor consumes a current, by pulses, from a network only during those moments of time when an instantaneous value of the input sinusoidal voltage becomes more than the DC voltage on the capacitor (which depends on that discarded from the load). During the rest of the time when the voltage on the capacitor is more than the instantaneous input value, diodes of a rectifier bridge appear locked by a reverse voltage applied from the capacitor, and the consumption of a current is absent. As a result, the current consumed by the SMPS appears essentially out of voltage phase (see Figure 2.50).

When a great number of SMPS devices are connected to an AC network, the combined decrease of the PF in the network becomes appreciable (a typical PF value for a single SMPS without correction is 0.65). In this connection, active PF correction is employed by means of the so-called PFC in the SMPS. The PFC is an independent voltage converter, the so-called boost converter (BC), supplied with the special control circuit (see Figure 2.51). Basic elements of the BC are choke L, diode VD2, capacitor C2, and fast-switching element VT (based on a MOSFET). The functioning of this device is based on the production of a high-voltage pulse with a reversed polarity on the inductance (L) at the breaking of a current in its circuit.

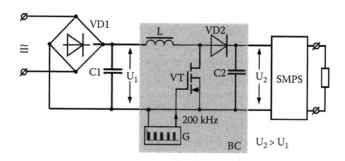

FIGURE 2.51
Connections diagram the buster converter (BC) with the SMPS.

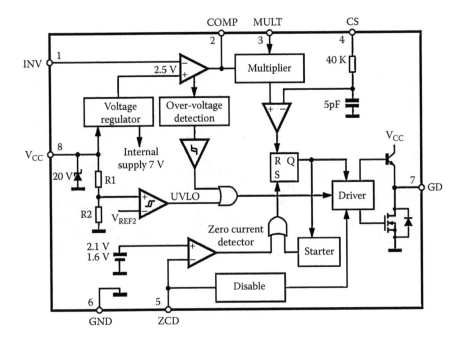

FIGURE 2.52
A microcircuit L6561 type (STMicroelectronics) for booster converter controls and power factor corrector.

Transistor VT switches the current in the inductance L on and off at a high frequency (it is usually 200 kHz), and the high-voltage pulses formed during the switching process charge the capacitor C2 through the diode VD2 from which the load (in our case, actually, the SMPS) feeds. Thus, a voltage on the capacitor C2 always exceeds the input BC voltage. Owing to this property, the BC is widely used in electronic devices as the voltage level converter: from the standard voltaic cell voltage level of 1.2 to 1.5 V to the other standard voltage level of 5 V, which is necessary for integral microcircuit control. In our case, the capacitor C2 is charged up to a voltage of 385–400 V.

Owing to that, capacitor C1 has a very small capacity (it is, per se, only a high-frequency filter) and the control circuit of the switching element with a PWM constantly traces a phase of the input alternating voltage and provides a matching binding of the control pulses (that is, pulses of a current carried though the switching element) to a phase of an applied voltage. It is possible to eliminate completely the phase shift between the current and the voltage consumed by capacitor C2 (see Figure 2.51). In addition, the same control circuit provides rigid stabilization of a level of voltage charge on the capacitor C2. Despite the small dimensions of the PFC control microchip, it has a complex internal structure (see Figure 2.52), and the entire PFC unit is

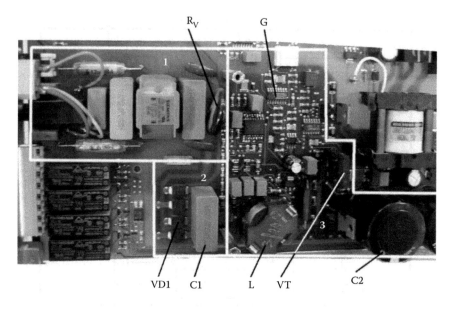

FIGURE 2.53
A fragment of printed-circuit-board with PFC and BC. 1: input filter; 2: input rectifier bridge with the filtering capacitor; 3: PFC; RV: varistor; and G: microcircuit for booster converter controls.

FIGURE 2.54
The linear (at the left) and switch mode (at the right) power supplies with identical characteristics. T: the transformer.

considerably complex and occupies an appreciable area on the SMPS PCB, because of the number of added passive elements (see Figure 2.53).

The question arises: Why have complex devices such as SMPSs expelled such simple and well-proven LPS devices from the market?

The basic advantages of the SMPS over the LPS that are usually specified in the technical literature are as follows:

1. A significant decrease in size and weight due to the smaller main transformer (the high-frequency transformer has considerably smaller dimensions and weight in comparison with the transformer of a commercial frequency of the same power)
2. The very wide range of a working input voltage
3. Considerably higher coefficient of efficiency (up to 90–95%, against 40–70% for LPS)

In addition to the above, we would also add one more important advantage: the possibility of working in a network of both AC and DC voltages.

At first glance, the differences between two devices equal in power and properties are readily apparent: the LPS is much simpler, but contains a much larger and heavier transformer (T).

The flat module SMPS (Figure 2.54) is the universal power supply of microprocessor-based protective relays of such series as SPAC, SPAD, SPAU, and so on, which is moved in the relay case. Naturally, to use the relay design of the LPS with the large transformer is inconvenient. But, what prevents using three separate small transformers instead of one large transformer with three secondary windings? There certainly is enough space on the PCB of the LPS. In this case, the overall dimensions of the LPS will not differ much from those of the SMPS. Even in the case of a powerful source with one level of an output voltage, it is possible to use some of the flat transformers connected between themselves in parallel. So, the presence of the small transformer is not an absolute advantage of SMPSs.

In connection with the SMPS's operational ability over a very wide range of input voltages due to the use of a PWM in a control system of the switching element, this advantage is touted to us as the essential reason for selecting SMPS over LPS. Well, really, in practice is it important that the SMPS can work at the input voltages changing within the limits of from 48 V up to 312 V? In fact, this range comprises at once some rows of rated voltages, such as 48, 60, 110, 127, and 220 V. It is abundantly clear that in concrete equipment,

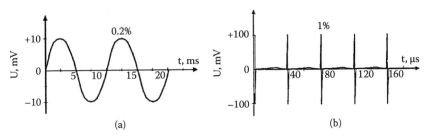

FIGURE 2.55
Typical levels of output voltage pulsations for LPS (0.2%) and SMPS (1%) with an output voltage of 12 V.

the SMPS will work at any one rated voltage (changing within the limits of no more than ±20%), instead of on all of them simultaneously. And if it is necessary to use the equipment with both 110 V and 220 V voltages, for example, there are well-known solutions in the form of the small switch and tap winding of the transformer to handle this.

The coefficient of efficiency is an important parameter in the case of a powerful power supply, but not in the case of a power supply as small as 25–100 watts, which we are considering. A high coefficient of efficiency and absence of a heat dissipation (which is characteristic of SMPS) can be of some importance in miniature portable power supplies of completely closed systems, for example as the power supply of laptops. But in many other cases, for example in controllers and electronic relays for industrial purposes, the coefficient of efficiency of power supply is not that important.

The possibility of functioning when fed from a DC network is the only major and absolute advantage of the SMPS. The LPS cannot operate in a DC mains' power. Here, in brief, we have given an analysis of the advantages of the SMPS in contrast to the LPS. We shall now turn our attention to the disadvantages of the SMPS.

One of the disadvantages of the SMPS is a high level of impulse noise on the power supply output (see Figure 2.55). Unlike the LPS, with its weak 50 Hz pulsation (Figure 2.55a), the pulsations of an output voltage in the SMPS (Figure 2.55b), as a rule, have much greater amplitude and cover a wide frequency range from several kilohertz up to several megahertz. This creates problems with high-frequency radiations that influence circuits within the electronic equipment in which SMPSs are mounted.

These radiations also affect the wires that in turn affect electronic equipment that is external to the SMPS. Besides, in SMPSs it is necessary to take steps to prevent the penetration of high-frequency radiations in the feeding power network (from which they are passed and can disturb the operation of other electronic devices) by using special filters (see Figure 2.56).

The presence of a high-frequency component in the output voltage and in the internal SMPS circuits has led to increased requirements for the numerous electrolytic capacitors that are available in the internal SMPS circuits. Unfortunately, these requirements are seldom considered by engineers in the development process of the SMPS. As a rule, the types of these capacitors

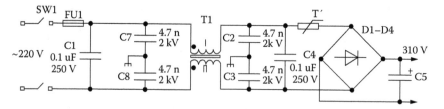

FIGURE 2.56
The circuit diagram of the typical input filter of an SMPS.

FIGURE 2.57
Destruction of copper streaks on the printed circuit board, which is taking place under capacitors because of electrolyte leakage.

are selected only on the basis of their capacitance, operating voltage, and dimensions, without taking into account their high-frequency characteristics. However, not all types of capacitors have long life under the effect of a high-frequency voltage; rather, only special types having low impedance at high frequencies. As a result, such nonsuitable electrolytic capacitors often heat up noticeably because of high dielectric losses at high frequency.

The rise in temperature of an electrolyte intensifies chemical reactions inside the capacitor that lead to a speed-up in the dissolution of the capacitor shell and even to an outflow of the electrolyte directly onto the PCB, which, in very dense installations, leads to shorting outlets of other elements or the breakdown of circuits owing to the dissolution of copper paths of the PCB (even despite the presence of a strong covering of paths by a special mask) (see Figure 2.57).

Other well-known kinds of SMPS faults that are caused at high temperatures of an electrolyte are the slow desiccation (over several years) of an electrolyte and the significant (on 30–70%) decrease of the capacitance that leads to a sharp decline in the characteristics of the power supply, and sometimes full loss of its working capability.[1]

For maintenance of effective work of the PFC, the power-switching element (it is usually a MOSFET) should possess lower impedance in the conductive state. The value of this impedance largely depends on the maximum operating voltage of the transistor. For transistors with a maximum operating voltage of 500–600 V, this impedance achieves 0.05–0.30 Ohms, whereas for transistors operating at higher voltages (1000–1500 V), this impedance is one to two exponents higher (for example, 12 Ohms for the transistor 2SK1794 for voltage 900 V, 17 Ohms for transistor IXTP05N100 for voltage 1000 V, and 7 Ohms for transistor STP4N150 for voltage 1500 V). It is explained by choosing the low-voltage (maximum operating voltage 500–600 V) transistors for

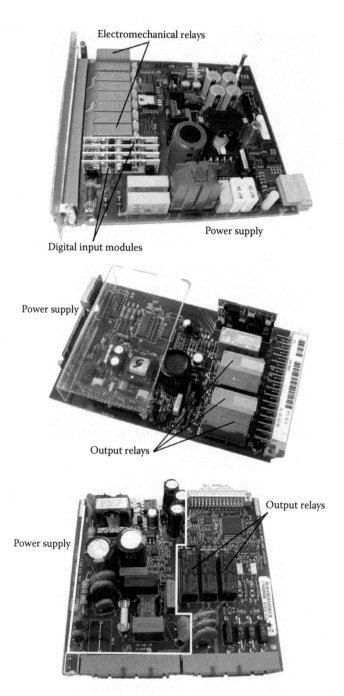

FIGURE 2.58
Power supplies installed on a PCB common with other functional modules.

the PFC. For example, in an actual SMPS design of such crucial devices as microprocessor protective relays and microprocessor-based emergency modes recorders, the following transistor types are widely used: IRF440, APT5025, and the like with the maximum operating voltage of 500 V. This is definitely not enough for functioning in an industrial electric network with a rated voltage of 220 V because of the presence of significant switching and atmospheric voltage spikes. As is known, for protection against such spikes, electronic equipment is supplied, usually with the varistors. However, because of insufficient nonlinearity of the characteristic near an operating point, the varistors are selected so that between their normally applied operating voltage and a clamping voltage there will be essentially enough difference. For example, for varistors of any type intended for operating at a rated voltage of 220 VAC, the clamping voltage is 650 to 700 V. In the power supplies of microprocessor devices mentioned above, varistors of type 20K431 are used with the clamping voltage of 710 V. This means that at spikes with amplitudes below 700 V, the varistor will not provide protection for electronic components of the power supply, especially power transistors (500 V) that have been connected directly to a main network.

Both the transformer and the coil in the PFC have high impedance at high frequencies that limits a current carried through them and through switching elements. However, a malfunction in the integral microcircuits providing control of the power-switching element in the PFC or the basic power-switching element of the SMPS (e.g., as a result of an impulse spike) leads to a transition from high-frequency AC operating mode to DC mode (that is, with very low impedance), a sharp current overloading of the power solid-state elements, and their instant failures. Considering the high density of the SMPS PCB, this often leads to a fault of the nearby elements and to burnout of whole sections of printed circuit strips. Generally, regarding reliability, it should be absolutely clear that the reliability of such complex device as SMPSs—containing, as they do, an assemblage of complex microchips and power solid-state elements working at high voltage in a pulsing mode with a high rate of current and voltage rise—will always be appreciable below the reliability of such simple devices as LPSs containing only some of the electronic components which work in a linear mode.

The density of elements on the PCB and the specific power of SMPSs constantly increase; for example, an EMA212 type power supply (Figure 2.49, on the right), with dimensions as small as 12.7 × 7.62 × 3 cm, has an output power of 200 watts. These are found in use in miniature electronic components having surface mount technology (SMT), with a very dense installation of high-power elements, and a constant increase of switching frequency. In the past, this frequency did not exceed 50 to 100 kHz. Today, many powerful SMPSs with output currents of up to 20 A operate at frequencies of 300 to 600 kHz; and less powerful ones, for example those controlled by an ADP1621-type integral circuit, operate at frequencies of 1 MHz and more. This promotes the further decrease of SMPS mass and dimensions. For the

FIGURE 2.59
Cabinets with DPR installed.

same purpose, the SMPSs in DPRs are sometimes placed on a common board with other DPRs devices, for example with output relays (see Figures 2.58 and 2.26). This tendency in SMPS evolution has been advertised as the great advantage of SMPSs. The downside of this is the practically full loss of maintainability of the SMPS.

However, are such power supplies necessary in the devices, installed in the control cabinet? The question can be put even more broadly: are the built-in power supplies in the electronic devices for industrial purposes intended for installation in control cabinets together with tens of other analogous devices necessary in general? Why not release in completely automatic systems (in the control cabinet) such devices as control units, electronic relays, electronic transducers, and so on without power supplies, and only with a connector intended for connection of an external power supply?

This external power supply installed in the control cabinet (see Figure 2.59) should be, in our opinion, linear, to have a good power reserve, and should be supplied by necessary elements for overvoltage and short circuit protection. Moreover, in the control cabinets in reference to automatic systems with increased reliability, such LPSs should be doubled for increasing reliability and may connect among themselves through a diode (a so-called hot reserve). It may seem strange, but in the epoch of SMPSs there are many

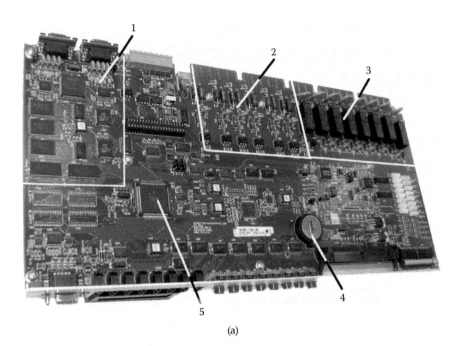

(a)

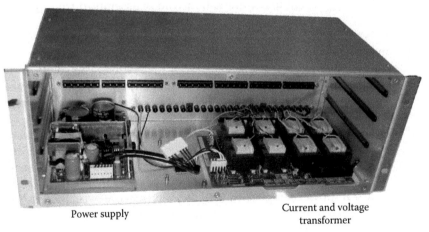

Power supply   Current and voltage
                                                      transformer

(b)

FIGURE 2.60

(a) Universal PCB with all DPR modules (SEL-311L) except power supply and input transform-ers. 1: connection section; 2: digital input section; 3: output relay section; 4: battery for memory backing; and 5: main processor (highly integrated 32-bit microcontroller MC68332ACEH25 type). (b) Power supply and current and voltage transformers made as separate modules (SEL-311L, Schweitzer Engineering Laboratories).

companies (VXI, Lascar, Calex Electronics, Power One, HiTek Power, and R3 Power, among many others) that are continuing to manufacture LPSs. This testifies to their popularity in certain areas of techniques and to their accessibility to practical applications. In my opinion, the approach specified above would allow considerably increasing the reliability of the automatic systems such as remote control, relay protection, and so on (with a mains supply of an alternating current) without increasing its cost (owing to the smaller cost of electronic devices without the built-in power supplies).

An analogous approach can be used, and in the case of feeding the electronic equipment (e.g., the same microprocessor protective relays) installed in the control cabinets from a network of a direct current, with only the difference that two power supplies should be SMPSs instead of LPSs. Thus, these SMPSs should be subjected to serious redesign. First, the power factor correctors (as absolutely senseless units for mains supply from direct current) should be dropped; this in itself will increase the reliability of SMPSs. Second, those SMPSs intended for installation in a control cabinet should be large enough (in such SMPSs, it is senseless to pursue compactness) and convenient for repairing; they should not contain miniature SMT elements. Third, the numerous electrolytic capacitors that are available in SMPSs should be concentrated on a separate PCB intended for its simple replacement after each five years of maintenance (i.e., before the capacitors start to fail).

The main filter should be used as a complete device (hundreds of models of such filters are present in the market) instead of assembled from separate elements, so it is possible to replace simply and quickly if needed. In some novel DPRs of the last generation with integrated PCB modules, the power supply made again as a separate module as in early DPR constructions (see Figure 2.60(b)). Such construction eases idea realization. The solutions we offer, in our opinion, will allow lowering the dependence of stationary electronic industrial equipment (such as DPRs) on secondary power supplies, and will considerably increase its reliability.

In summary, some words about the newest trends which have appeared in the design of secondary sources of power supplies: the attempt to use microprocessors both in LPSs[2] and in SMPSs.[3] It may be that our view seems excessively conservative to the reader, but, in our opinion, microprocessors are necessary in power supplies as well as toilet seats, where microprocessors are employed to measure the exact temperature of a matching part of a body and heat up the seat to that temperature for comfort.

It has become abundantly clear, as we have pointed out, that the presence of functionally unnecessary complex units in the equipment is the one way to decrease its reliability. If it comes to future DPR designs, we should mention the previous idea that in the future, DPRs should be built on the principle of the PC, that is, they should be produced and distributed freely in the market as a set of separate modules (PCBs), standardized by function, size, and type of connectors. These modules should be designed to slide into special metal cabinets separated into sections with partitions. Each section should be assigned to one DPR and should be equipped with connector blocks for connecting external cables. The numbering of the connectors should also be standardized. Grounded metal partitions should be installed between the separate PCBs of each DPR.

Cabinets and installed modules should be produced by special technology preventing the penetration of external electromagnetic interferences (such cabinets are widely available on the market) and must be fitted with special filters installed on each entering cable (such filters are also widely available on the market). This trend should lead, in our opinion, to the sharp reduction of DPR cost due to the launching of new players offering separate universal DPR modules, and to a significant improvement of DPR quality, reliability, and resistance to external electromagnetic effects. When purchasing a DPR, the customer will no longer be tied to the whims and the exorbitant parts prices of a single company (which usually stops production of obsolete models, also canceling the production of parts for the models which are still in operation for 15–20 years), and will be able to buy the necessary modules on the market, as well as update and change the configuration of separate DPR modules. In the future, the universal modules will support the development of a common software platform (kind of a simplified Windows platform) with a set of applications and libraries for specific types of protection. In its turn, the common software platform will also help to automate the dynamic diagnostics of the DPR by loading completely relevant settings and test algorithms into the DPR and mode simulator (Retom, DOBLE, Omicron, etc.) that will significantly facilitate DPR checking and drastically reduce the number of errors.

This is how we see the future of microprocessor relay protection. And the first step in this direction will probably be the development of international standards formulating the requirements to the DPR design and software. But this is another story.

References

1. Gurevich, V. Reliability of Microprocessor-Based Relay Protection Devices: Myths and Reality. *Engineer IT*, part 1: no. 5, 2008, pp. 55–59; and part 2: no. 7, 2008, pp. 56–60.
2. Sadikov, Y. Power Supply Adapter with Output Voltage Adjustment in the Interval 1.5 to 15V and Output Current up to 1A. *Electronics Info*, no. 12, 2008, pp. 42-43.
3. TDK-Lambda. EFE-300/EFE-400: 300/400 Watts: Digital Power Solution. Datasheet. Tokyo: TDK-Lambda, 2009.

3

Problems with the Reliability
of Digital Protective Relays

3.1 Introduction

Malfunction of relay protection is one of the main causes of the heavy failures that periodically occur in power systems all over the world. According to the North American Electric Reliability Council,[1] in 74% of cases the reason for heavy failures in power systems was the incorrect actions of relay protection in trying to avoid the failure. Thus the reliability of a power system depends on the reliability of relay protection in many respects.

It is a fact that intensive research and development in the field of electromechanical protective relays (EMRs) have been completely frozen for the past 30–35 years and all efforts of developers have been redirected to the development of electronic protective relays, and then digital protective relays (DPRs). Meanwhile, EMRs completely provided and continue to provide reliable protection of all objects in the electric power industry. The reason for the full disappearance of the EMR and transition to the DPR is not the inability of EMRs to carry out the functions; rather, it is something completely else. Due to the large expenditures by leading DPR manufacturers in promoting the microprocessor-based protective device (MPD), the development of new materials and technologies has not affected the EMR in any way. After decades of years in operation, today's EMRs have finally worn out and become outdated and, consequently, are the cause of a fair amount of discontent among protective relay experts. On the other hand, demounting EMRs and transitioning to DPRs in the electric power industry are connected with the necessity for investing significant amounts of money, not only for purchasing DPRs, but also for computers and special expensive test equipment, as well as for the replacement of expensive DPR failed units, which cannot be repaired. Significant capital investments are required as well for reconstructing grounding systems of substations, training relay personnel, and so on. All this essentially disrupts the process of the transition to DPRs. According to Reference 2, in 2002 in Russian power systems there were in operation

98.5% EMRs and only 3% of various electronic devices of relay protection. According to Reference 3, DPRs constitute about 3% of the total quantity of relay protection devices in Russia. In the West, rates of replacement of relay protection in working power objects also are not so high (excluding new erected power objects, of course). According to Reference 4, at existing rates about 70 years are required for the replacement of all old protective relays that are microprocessor based. Such low rates for updating protective relays in working power objects all over the world cause intensive advertising activity of MPD manufacturers and their distributors.

One of the main reasons usually presented in the vindication of DPR advantages is the considerably higher reliability of DPRs, ostensibly, in comparison with electromechanical and electronic relays. This thesis is represented as being so obvious, that, usually, it does not cause objections and frequently is repeated by managers and even by technical specialists of the power engineering companies. However, in a deeper analysis of a situation, it appears that the basis of this thesis is made with a whole set of widespread myths about microprocessor protection.[5]

3.2 Reliability Myth

Myth 1: DPR reliability is higher than EMR reliability because DPRs do not contain moving internal elements.[6]

EMR malfunction is usually associated with ageing and damage of wiring insulation (wear, drying), corrosion of the screws and terminal clips, and deterioration of the mechanical parts of the relay. However, the number of operation cycles (i.e., movement of mobile parts) over the entire EMR service life under real operating conditions in power systems does not exceed several hundreds. So to speak about mechanical deterioration of EMR mobile parts, for all practicality, it is only in the case of evident defects of the factory or manufacturer or in the case of use of improper materials for these purposes. As for corrosion of the metal elements or drying out of the insulation, this is a consequence of using poor-quality materials by the relay manufacturer. Such defects are typical for EMRs of Russian manufacturers that do not come close to meeting the products of the leading Western companies which have been in operation for 30–40 years even in a tropical climate.[7] Thus, to speak of EMRs as an insufficient mechanical resource, *as a kind of the relay*, is absolutely unreasonable. On the other hand, if the moving elements of EMRs are in movement *only at the moment of operation (pickups) of the relay*, the thousands of electronic components in the DPR *constantly are at work*: signal generators, numerous transistor switches, amplifiers, comparators, timers, counters, logic elements, and voltage stabilizers constantly work; the microprocessor constantly exchanges signals with elements of memory; the analog-to-digital converter (ADC) constantly conducts processing input signals; and so on. Many elements constantly are under the influence of high working voltages (120–250 V) and voltage spikes which periodically arise in the input circuits and external power supply circuits, as well as constant power dissipation (i.e., they are heated) and the like. In especially heavy work in MPD conditions, switching power supplies very frequently is the reason for MPD malfunction.

> **Myth 2: Reliability of semiconductor (solid-state) relays on discrete components is higher than reliability of electromechanical relays.[8] Reliability of electronic protective devices based on integral microcircuits (ICs) with a high degree of integration is higher than reliability of devices on discrete electronic components.[8] The reliability of microprocessor-based relays is higher than the reliability of electronic nonmicroprocessor devices.**

The unconditional statement about the greater reliability of semiconductor relays over electromechanical relays is a popular mistake.[9] Semiconductor relays possess increased reliability only at very large numbers of switching cycles (hundreds of thousand, or millions) or at high switching frequencies. In many other cases, the reliability of semiconductor relays is essentially lower than that of electromechanical relays.[10]

Discrete electronic elements have a much higher capability for withstanding voltage spikes and other adverse influences than integrated microcircuits.[11] According to Reference 12, 75% of all damages to microprocessor devices are the result of voltage spike impact. Voltage spikes with amplitudes from tens of volts up to several kilovolts, arising from switching transients in circuits[13] or the impact of electrostatic discharges, are "fatal" for internal microcircuits and processor microcells. According to Reference 12, normal transistors (discrete elements) can withstand a voltage of electrostatic discharges almost 70 times higher than, for example, a microchip of memory (such as erasable programmable read-only memory [EPROM]) in a microprocessor system.

The most calamitous of temporary failures caused by electromagnetic noise that occur in the microprocessor functioning can be time,[14] such as spontaneous changes of the operative memory (random access memory [RAM]) and register contents, and internal damages can have the latent character.[15] These kinds of damage do not come to light in any tests and can appear at the most unexpected moments.

In Reference 16 it is mentioned that in connection with low stability to transients and voltage spikes, DPRs demand especially rigid requirements for their protection level against electromagnetic influences. Attempts to use a DPR without strengthened electromagnetic protection frequently leads to its malfunctioning.[16,17] Electronic devices with discrete components contain

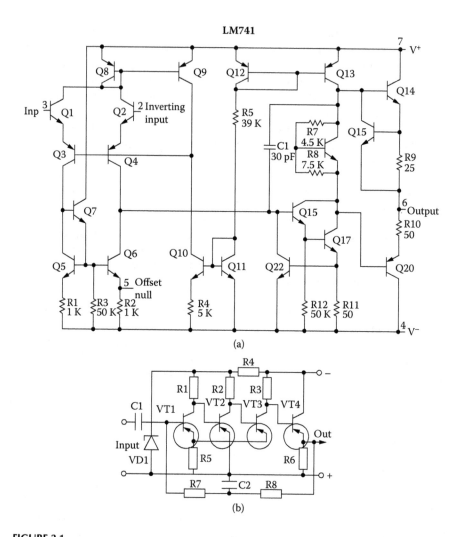

(a)

(b)

FIGURE 3.1

Circuit diagrams of two amplifiers. (a) A widely used IC LM741 type containing 20 transistors; and (b) the amplifier on the discrete elements with same parameters, containing only four transistors.

fewer components than similar devices on ICs (see Figure 3.1). This does not seem to promote higher reliability of ICs.

The statistics on damages of MPD elements, collected by engineers of some MPD manufacturers (see Figure 3.2),[18] very persuasively deny the myth about the higher reliability of ICs. According to the statistics submitted in Reference 8, it is extremely visible that the protective relays with electronic elements have three times the damageability of electromechanical relays, and microprocessor-based relays have 50 times the damageability!

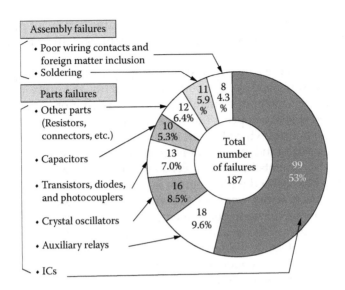

FIGURE 3.2
The statistical data on damages to MPDs based on investigations by leading Japanese companies.

Reliability of the microprocessors of such manufacturers as Intel and AMD can actually be very high, and in fact the microprocessor though not big is a very important part of the DPR, which contains very many ICs. In Reference 19, it is affirmed that the main processor unit (i.e., the printed circuit board [PCB] with the microprocessor, memory, the ADC, the library of programs, and all auxiliary elements) part of the DPR is the most subject to malfunctioning.

Another strike against microcircuits lies in not only physical damage of the microprocessor but also software failures—damages not known earlier for electromechanical and electronic relays. As is pointed out in Reference 19, program bugs are not always detected during DPR testing. An additional source of problems is the necessity for periodically upgrading the DPR program versions. Frequently during this process, software-to-hardware incompatibilities[19] appear. Such problems can show up during the most unexpected moments and can lead to very heavy consequences for a power network. As is known, one of the reasons for a large failure in a power supply system of the United States and Canada that occurred in August 2003 was a computer-related problem, a "lag" of a computer control system in the power company First Energy.[20]

> **Myth 3: Reliability of DPRs is much higher than reliability of all other types of protective relays due to the presence of the built-in self-diagnostics. With the self-diagnostics in DPRs, 70–80% of all internal DPR elements are covered.[21,22]**

This thesis is very widespread and is found in practically all publications devoted to DPR advantages. We shall consider features of this self-diagnostics more in detail.

We already observed the self-testing design principles in Chapter 2. We shall add now some additional notes. In particularly, the problem with reliability of the DPR memory elements is actually much more complex than it was described above. It appears that memory elements are subject to random unpredictable failures which have not been connected to physical damage of memory cells. Such random, temporary malfunctions caused by spontaneous changes of the contents in memory cells are referred to as *soft failures* or *soft errors* (not to be confused with program failures, which are *software programming errors*). Such errors were not known earlier for the electronic devices based on discrete semiconductor elements or on usual microcircuits. Progress in the last few years in the area of nanotechnology has led to a dramatic reduction in the sizes of semiconductor elements (along the order of microns and even parts of a micron), reduction of thickness of layers in semiconductor and insulation materials, reduction of working voltage, increase in the working speed (working frequency), reduction of electric capacity of separate memory cells, and increase in the density of placement of elementary logic cells in a single chip. All this taken together has led to a sharp increase in the sensitivity of memory to ionizing radiation.[23,24] This sensitivity became so high that usually (i.e., in completely normal situations) the radiating background at sea level became dangerous to memory cells. Streams of the high-energy elementary particles coming from space are especially dangerous. Even one such particle that hits a memory cell gives rise to secondary streams of electrons and ions, causing spontaneous switching of the elementary transistor or discharging the capacity in charge storage memory elements.

The problem is aggravated in modern MPDs with the tendency toward the ever-expanding use of memory elements.[24] Many modern integrated microcircuits with high integration levels, included in the MPD, contain complex structures with embedded memory of such volume that serviceability, in general, is not supervised in any way. As shown in References 25 and 26, the problem of the sharp increase in sensitivity to ionizing radiations is actual not only for memory but also for high-speed logic elements, comparators, and so on—that is, for practically all modern microelectronics.

Power supply. DPRs of all types are supplied with switching power supplies. As was mentioned above, the ICs of family ADM 691-695 can be used for the continuous monitoring of the power supply voltage level. As well as in a case with the watchdog timer, the ADM 691-695 chip generates a signal for locking the CPU at inadmissible voltage supply reductions. The locking signal remains until the voltage level is restored. Is it really possible to count such voltage-level monitoring as self-diagnostics of the power supply, raising its functioning reliability? Hardly; therefore, it is only the internal technological locking which prevents failures in the CPU. Such monitoring has no relation to the reliability of the power supply.

And meanwhile, power supplies are the most unreliable unit of the DPR. First, elements of the power supply work in a high load mode; they are constantly subjected to the influence of high values of a voltage and a current, voltage spikes, and dissipated rather high power on the elements. Second, they contain a lot of aluminum electrolytic capacitors that rather badly carry the influence of high-frequency currents on which switching power supplies work, and frequently are the reason of full breakdown of the power supply (and, consequently, of the whole DPR). Can monitoring of an output voltage level in this case help? Can it signal beforehand about the deterioration of a capacitor and thereby prevent sudden DPR failure?

Output electromechanical relays. As shown in previous research[27,28] that I conducted, contacts of the miniature electromechanical relays (usually used in all types of modern DPRs as output elements) directly control the trip coils of high-voltage circuit breakers or the coils of auxiliary relays with a significant overload. Therefore, reliability of these relays is essentially reduced to the comparison with the value which was normalized by the relays manufacturer. On the other hand, in promotional brochures of various manufacturers about DPR advantages, it is usually emphasized that serviceability of such important elements as output relays is continuously supervised by means of a self-diagnostics system of the DPR, but due to complexity of relay contact supervision, the coil continuity is only supervised generally. This is done by passing a constant weak current through the coil. But what if the most intense and unreliable part of the electromechanical relay is not the coil at all, but the contacts?! We shall observe the problem with contact of output electromechanical relay in more detail in Chapter 5.

In conclusion of this section, it is necessary to note that, contrary to popular belief, internal self-diagnostics actually is not the means intended for decreasing MPD failure rate (i.e., for increasing its reliability). The purpose of self-diagnostics is the locking of the DPR and delivering an alarm signal before the occurrence of emergency mode in a power network, instead of during a DPR failure.

> **Myth 4: DPRs are essentially more reliable in comparison with relay protection devices of the previous generation as they contain considerably fewer components and these components are much less subject to physical ageing. DPRs also contain a smaller number of internal connections.[29]**

As for the claim that DPRs contain a fewer number of components than relay protection devices of the previous generation, it turns out that actually the number of components making up some DPRs is greater. As for the claim of more intensive physical ageing of elements of the protective relays of the previous generation, this thesis also does not bear up. The author of this thesis compares the modern elements and technologies used in DPRs with materials (impregnation and cover varnishes, plastics, insulation materials, and contact and anticorrosion materials) that were in use in the USSR 50 years ago and employed in the protective relays for several decades. As we already remarked above, old electromechanical relays of Western manufacturers (BBC, Westinghouse, General Electric, etc.) in which high-quality materials and coverings were applied still work successfully and show no signs of ageing.

Besides, the progress in the field of materials over the last decade has not been slower than the progress in the field of microelectronics. On the other hand, not all is so bright with the ageing of the electronic components widely used in DPRs. Even high-quality electrolytic capacitors of Japanese manufacturing start to change the parameters after 7–10 years of high-frequency operation in DPR switching power supplies. As a result, a change of parameter of one such capacitor only (see Figure 3.3) completely stops power supplies from functioning. For example, power supplies such as SPGU240A1, used in DPR types SPAC, SPAD, SPAU, and SPAJ, have been shown to cause this. In other cases, not only destruction of the electronic components takes place, but also even dissolution copper streaks on PCBs under action of the electrolyte which has leaked out from capacitors (see Figure 3.4).

One more problem is the aspiration of manufacturers for DPR miniaturization at any cost. This has led to using electronic elements in DPRs working with an overload and dissipated increased value of heat that does not promote an increase in DPR reliability and a reduction of element ageing. This problem for circuits of digital inputs on which voltage up to 250 V is applied[30] is especially prevalent.

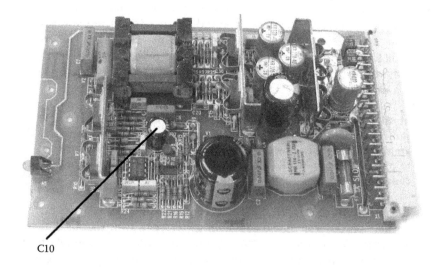

C10

FIGURE 3.3
Switching the power supply SPGU240A1 type used in various types of DPRs. C10: the capacitor, where a change of parameters results in full loss of the working ability of the power supply.

FIGURE 3.4
Fragment of DPR PCBs with damages caused by electrolyte leakage from capacitors.

Multilayered PCB DPRs involve a huge number of contact transitions (crosspieces) between layers. From the author's personal experience, there have been cases of faulty DPR actions due to the increase of transitive resistance of these transitions.

The design of many types of DPRs comes with a motherboard with multicontact sockets and functional boards with the reciprocal sockets connected with a motherboard. The flexible multicore trunks with the numerous contact sockets connecting among themselves and separate PCBs are sometimes used instead of a motherboard. These contact connections do not always provide a reliable transfer of low-voltage and low-current signals between boards. In any case, contrary to the widespread myth, DPRs contain many more interconnections than the relays of all previous generations.

3.2.1 One More Class of Problems Which DPR Manufacturers Prefer Not to Mention

In view of the increased sensitivity of modern microelectronics to electromagnetic radiations, there is a problem for DPRs in connection with electromagnetic compatibilities (EMCs). Many experts have noted the frequent incompatibility between real parameters of grounding systems in substations and the requirements showed by DPRs[31,32] and, as a result, DPR failures. But little is known by the experts in the field of relay protection about the problem of "electromagnetic terrorism," the powerful electromagnetic radiations[33] that intentionally impact electronic devices, and also about the problem of hacker attacks (i.e., cybersecurity).[34] These problems were unknown earlier in relay protection and became a reality only in connection with DPR applications, as their sensitivity to electromagnetic noises is 10,000 times higher than in electromechanical relays.[31] The built-in DPR software is also subject to external influences. And if, in addition to all the aforesaid, one takes into account that one DPR carries out the functions of 3–5 EMRs, the situation with DPR reliability is aggravated even more, as damage of one of the common DPR elements is equivalent, in consequence, to the simultaneous damaging of several kinds of protection at once. In this connection, in Reference 35 transition of microprocessor protection is offered to provide additional independent, simple, inexpensive, but *not microprocessor* reserve protection for extreme situations.

3.3 The Real State of Affairs with DPR Reliability

In previous publications I have already analyzed the transition from electromechanical to microprocessor-based protective relays, and considered the prospects and problems of microprocessor relay applications.[36,37] The rather sharp reaction of the readers often arising after these publications, on the one hand, and my detailed answers to the criticism of opponents, on the other, show that among the specialists in this area, there is no common opinion about the prospects of microprocessor protection: there is no unequivocal understanding that, as with any other complex device, the microprocessor protection not only possesses obvious advantages, but also has serious weaknesses.

3.3.1 Myth about the Extreme Importance of Microprocessor-Based Protective Devices

One of the widespread fables[36–39] justifying the inevitability of transition to microprocessor relay protection is the myth that electromechanical protective relays do not provide for the performance of the technical requirements for relay protection, and thus the continuing existence of today's electric power industry is not possible without DPRs.

Actually, no new functions in relaying DPRs have been introduced. The parameters and facilities of the high-quality electromechanical relays and semiconducting relays (that is, the static analog devices constructed on the basis of discrete solid-state elements and integrated microcircuits) completely provide all relay protection requirements. In relaying, there are no actual problems that could not be solved by means of electromechanical or static relays (note: recording emergency modes is not a relay protection function). Confirmation of this is the fact that branched and complex electrical networks and systems exist and successfully function all over the world and have done so for more than 100 years, whereas microprocessor-based relay protection has appeared in use in not very appreciable numbers just 10–15 years ago. Thus, with the beginning of the use of DPRs, the functioning logic of an electric power system has not changed, the number of operations that are carried out by an electric power system has not increased, the quantity of the produced electric power has not changed, and the principles of transmission and distribution of the electric power have not changed.

3.3.2 Why Have Digital Protective Relays Become So Popular?

In spite of the absence of any principal problems in electromechanical relays in providing reliable protection of power devices, progress in the development of electromechanical relays completely stopped 30–35 years ago since the efforts of developers were directed first to the creation of electronic, and then to microprocessor-based, protection. The matter is that the production expense of a completely robotized (down to automatic testing) DPR manufacturing process using cheap high-integrated electronic components is far less expensive than the manufacturing and manual assembly of precision mechanical elements of electromechanical relays; therefore, it is in manufacturers' interest to push MPDs. For example, the ordinary electronic component-mounting machine, CM402-M/L, can install 60,000 components an hour (see Figure 3.5). Yes, 60,000 components an hour! It is abundantly clear that

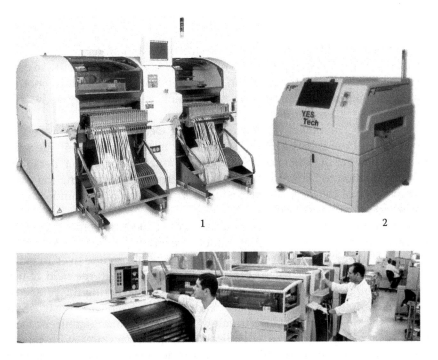

FIGURE 3.5
PCB manufacturing. 1: component mounting machine CM402-M/L type (Panasonic Factory Solutions Co., Ltd.) with the capability to install of 60,000 components an hour; and 2: automatic optical inspection machine (Yes-Tech Co.) with the capability to test 250,000 components an hour.

with such high-efficiency fully automatic manufacturing of PCBs, of which one is the DPR, manufacturers can reap fabulous profits in comparison with the manufacturing of mechanical relays. In the manufacturing sphere, we see that the most important advantage that DPRs have are enormous profits for the manufacturers. Apologists for the widespread use of DPRs often bring up such reasons in favor of DPRs as the ability to record emergency modes which is absent in electromechanical relays, the ability to interchange information between the relay units, and so on. But all these are advertising gimmicks which have no connection with reality. Today in the market, there are hundreds of versions of microprocessor recorders of emergency modes capable of transmitting data over Ethernet networks, which record emergency modes much better and more fully than DPRs. There are information transfer systems, such as supervisory control and data acquisition (SCADA), that have worked well for many years with electromechanical relays. Unlike the relay of protection, microprocessor-based recorders are not capable of affecting the reliability of power supply and initiating collapses in a network of failures.

In many electric power systems, electromechanical relays until now reliably protected many crucial power installations of all voltage classes and other utilities equipment. Sometimes, electromechanical protective relays include working in parallel with microprocessor-based relays for maintaining greater reliability of the important electric installations and especially crucial equipment (see Figure 3.6). Thus, often it appears (especially in cases of complex damages with transition of one kind of short circuit to another) that electromechanical protection works noticeably more quickly than microprocessor-based protection.

In many electric power systems employing electromechanical relays for a long time, these relays already are coming to the end of their life span, many of them are in rather pitiable condition, and the operational personnel see the transition to DPRs as the only alternative for maintaining the working capability of relaying because of the dictatorship of the manufacturers (as discussed in this section).

Today in the world market, there simply are no electromechanical protection relays being developed using modern materials and technologies, and all leading world protection relay manufacturers have gone over to exclusively manufacturing DPRs. At the same time, progress in the field of new materials, components, and technologies allows constructing the protective relays on completely new principles in which it is possible to construct, for example, hybrid relays.[40,41] Unfortunately, today's DPR manufacturers, faced by the increasing functional complication of their products with no significant means to decrease DPR manufacturing costs, are not interested in investing in any alternative kinds of relays to compete with the profitability of the DPRs. And DPR profitability stems not only from the wide difference between the production price and sale price, but also from the use of new production technology (surface mounting of superminiaturized elements

FIGURE 3.6
Fragment of the panel of protection at a distance of crucial lines of 161 kV containing electro-mechanical relays of LZ31 type (above) included for working in parallel with an MPD type MiCOM P437.

and highly integrated microcircuits on the multilayered PCB) that presupposes no repairing of DPR modules. It is now common to throw out failed DPR modules made using this technology and replace them with a new one. Such an approach is advertised by DPR manufacturers as the high maintainability of their products. But when considering that the whole DPR costs $10,000–15,000 and consists of 4–5 such modules (separate PCBs), it becomes clear what such "maintainability" means to the consumer (that is, to electric power systems).

The ageing and service life of protection devices are directly connected with DPR reliability and their costs. For DPRs (as well as for electromechanical relays) in many countries, the normal life expectancy is 20–25 years. Actually, many electromechanical relays are in service about 30 or even 40 years, while the computer-based devices age much more quickly. Keep in mind the physical ageing of electronic components, such as electrolytic capacitors (the service life of which does not exceed 7–10 years) and others, and especially the software. So, according to Reference 8, the life expectancy of designed obsolescence (Figure 3.7) has sharply decreased from 30 years, for

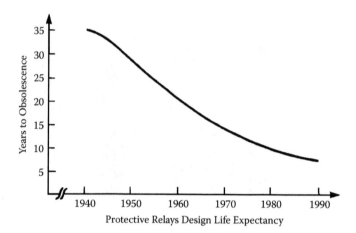

FIGURE 3.7
Protective relays design life expectancy.[8]

the traditional electromechanical relays to approximately 5 years for modern DPRs. This means that DPR users have to spend much greater sums in the future for updating relaying (both hardware and software) and much more often than they had to earlier when using electromechanical protection.

Despite the problems noted above, the tendencies in relay protection development are such that widespread and increasing use of DPRs is made inevitable. The DPR expansion is connected not only with the necessity of replacing the old electromechanical relays with finished normative terms, but also with installing in-service new power elements; the last 10–15 years all over the world have seen the gradual transition to relaying of the new generation based on microprocessors. To "push" DPRs on the market, the manufacturers of these devices, and their numerous sales representatives, have engaged in strong advertising campaigns in eulogizing DPRs in every possible way while belittling the advantages of other types of relays. The basic thesis of these advertising campaigns is the statement that DPRs provide very high reliability relaying unlike the old and worn-out electromechanical relays which are approaching their age limit. At the same time, it is abundantly clear that DPRs are a complex technical system consisting of many thousands of components. Like any other complex electronic systems, they will have failures and cannot possess absolute reliability, especially if one is to consider the "hothouse" operating conditions in power electrical networks. This being so, one would expect there should be many publications in the technical literature considering the technical problems of microprocessor relays. But how many such articles considering DPR problems have you read? It is a significant fact that the overwhelming majority of publications in the technical journals devoted to DPRs are written by engineers of DPR-manufacturing companies. Naturally enough, these publications represent direct or veiled advertising, and not serious analysis of problems with reliability or other quite real DPR problems. Since the DPR manufacturers are the advertisers generously paying for significant areas of journal pages, the journals are extremely reluctant to accept articles devoted to the criticism of DPRs, and sometimes do not hesitate in declaring this. One gets the feeling that there is a certain taboo imposed on discussion on this theme. If an author happens to break by chance through this "iron curtain," there is a squall of criticism including personal attacks and even charges of attempts to halt the technical progress.

3.3.3 The Actual Problems with the Reliability of Digital Protective Relays

In References 36 to 39 we already considered, in detail, problems with the reliability of each of the basic functional units of DPRs and have shown, through concrete examples, that the so-called self-diagnosis by which 80% of DPR units are ostensibly captured is, by and large, an advertising gimmick and a widespread myth. While it is true that self-diagnosis in DPRs can reveal some internal damages, for example a failure of the internal power supply or the CPU, how is it possible to speak seriously about this as a great "advantage" of DPRs against electromechanical relays if, in the electromechanical relays, there are no internal power supplies and CPUs—that is, there is simply nothing to "self-diagnose"?!

As brought out in Reference 38, the analog input modules (current and voltage transformers), digital inputs, and output relays are not captured by a self-diagnosis in DPRs. In addition, as shown in Reference 38, the system of a self-diagnosis is constructed on microprocessors and memory elements, so it is an additional source for malfunctions of DPRs.

Actually, the self-diagnostics is not an advantage of DPRs against electromechanical relays, and is only a partial compensation for very serious DPR disadvantages: concentration of many protective functions in the single module. For example, only a single DPR type M-3430 (see Figure 3.8) provides a full protection of the generator on power stations from all possible emergency modes and combined functions of 14 separate protective relays. It is only possible to speculate what will occur if this DPR malfunctions at emergency mode due to the fault of any cheap internal component in the power supply or CPU. The high-power and very, very expensive generator *will stay without any protection!*

It is absolutely clear that without self-diagnostics, it would be impossible to admit such a complex protection device on a gun shot to the protection of electrical power installations. So, the self-diagnostics in DPRs is a forced measure and a not-so-beautiful application; therefore, to advertise it as a great achievement in relaying is absolutely not justified.

Strangely enough, opponents of my position have not denied this position on the problems of the DPR units, but rather have concentrated only on criticism of some general opinions and reasons about DPR reliability, borrowed

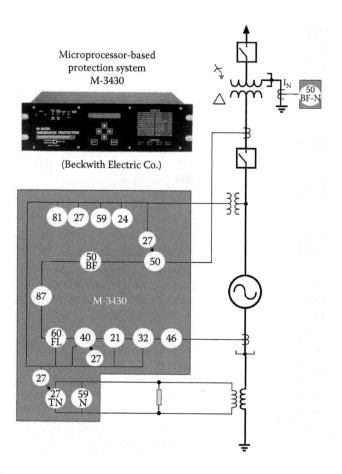

FIGURE 3.8
Structure of the microprocessor-based system M-3430 type (Beckwith Electric Co.) for complete protection of the power generator.

by me (with corresponding numerous references) from others who have investigated the problem. We decided to carry out our own research by putting to use statistical data on protective relay malfunctions for 2007–2008 of one of the electrical power companies (for ethical reasons, I do not publish the name of this company). Initial statistical data on relay protection failures and calculations are given in Tables 3.1 and 3.2.

It is possible to come to two important conclusions (which can seem paradoxical to some) resulting from our calculations:

1. The yearly intensity of failures for microprocessor-based protective relays is much more than that of electromechanical relays.
2. The yearly intensity of failures of protective relays significant increased over the past few years in connection with the usage of new

kinds of protective relays. That is, for the past few years, the tendency of a decrease in DPR reliability (see Figure 3.9) has taken place.

Actually, there is nothing unusual in these conclusions. According to other statistic data, presented in Reference 8, it is quite visible that electronic (static) relays have three times greater damageability than electromechanical ones, and microprocessor-based relays have 50 times greater damageability (see Table 3.3).

TABLE 3.1

Failure Rate of Protective Relays of Various Kinds

Parameter/Relay Kind	Electromechanical		Static		Microprocessor Based	
	2007	2008	2007	2008	2007	2008
Total number of relays in service	2312		2745		3787	
Number of failures	1	4	8	8	43	51
Relative failures, %[a]	0.043	0.173	0.291	0.291	1.135	1.347
Average relative yearly failures, %[b]	0.11		0.29		1.24	
Yearly intensity of failures[c]	1		2.6		11.3	

[a] *Relative failures* is the relation of failure numbers for some relay kinds to the total number of relays of the same kind.

[b] *Average relative yearly failures* is the average number of relative failures for two years (2007 and 2008).

[c] *Yearly intensity of failures* is the ratio of average numbers of relative yearly failures of different kinds of relays to the same parameters of electromechanical relays (defined as 1).

TABLE 3.2

Increasing of Relay Protection Failures at Usage of New Kinds of Relays

Starting Service Year	Relay Kind	Total Number of Relays	Total Number		Relative, %		Average Relative Yearly, %	Failures, Yearly Intensity
			2007	2008	2007	2008		
1970 1975	Electromechanical	2312	1	4	0.043	0.173	0.11	1
1975 1980	Electronic (static)	2745	8	8	0.291	0.291	0.291	2.6
1990 1995	DPR type 1	1423	19	25	1.33	1.76	1.54	14
2000 2005	DPR type 2	342	6	5	1.75	1.46	1.61	14.6
2003 2005	DPR type 3	49	3	1	6.12	2.04	4.08	37
2005 2008	DPR type 4	10	3	1	30	10	20	182

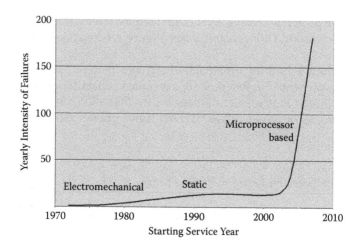

FIGURE 3.9

The tendency of increasing failures for DPRs of new types (according to Table 3.2).

TABLE 3.3

Typical Failure Rates of Protective Relays (According to Reference 8)

Relay Kind/Characteristic	Failure Rate per Year, %	Life without Obsolescence
Electromechanical	0.1	>30
Electronic (static) single function	0.3	>20
Digital	5.0	>20

However, as has been noted, insofar as one microprocessor protection incorporates the functions of several relays, this should be taken into account when making a comparative estimation of reliability. For example, if one DPR carries out protective functions of 10 single electromechanical relays, the difference between them in damageability will be only five times, not 50. At first sight, such an approach is quite logical; however, it does not consider the fact that DPRs contain such common units as power supplies, CPUs, input analogue electronic circuits, and so on, faults of which lead to failure at once of all these ten virtual relays. That is to say, the weight factor of a single fault in a multifunction DPR is more (in our instance, ten times more) than in the single-functional electromechanical relay. For this reason, it is possible for us, in order not to complicate the business, to continue to compare the failure rates of microprocessor-based and electromechanical relays without taking into account the difference in the number of functions carried out by them.

Important factors, such as mistakes of personnel (that is, the so-called human factor), were not considered in programming the DPR and in working with it. Modern multifunction DPRs contain hundreds of parameters and set points, and tens of inputs and outputs, and can generate thousands of various messages. According to Reference 8, "[T]raditional methods of assessing relays by hardware inspection and testing are no longer adequate, since up to 80% of the engineering design content of contemporary digital relays [is] in the software area." It has therefore become increasingly important for the new generation of relay engineers to have basic knowledge in computers, software, and programming. Absence of such knowledge leads to repeatedly increasing the number of mistakes related to the "human factor." According to Reference 2, in 2000 in Russia, the share of guilt of operational personnel in wrong actions of relay protection was 61.6%. Also, the explanation of the reasons for this is bright: "Insufficient qualification of the personnel of the power enterprises for service of the equipment on new element base."

An additional aggravation of the condition is the presence in a single power system of many types of DPRs of different manufacturers with very essential differences from each other of the program interface, programming principles, and testing. All this leads to further complication of the process of transition from electromechanical to microprocessor-based protection. In Reference 43, this is directly underscored by stating that "the situation becomes complicated also that the purpose of such transition—substantial increase of efficiency of relay functioning—as a rule, is not attained"; and, further, "The percent of wrong acts of modern relay panels and cabinets often appears much more than for the old electromechanical relays." This is confirmed in Reference 43, which states that "the statistics show, that use of digital protective relays (DPRs), despite its essentially best technical characteristics in comparison with previous generations of protective devices, has not increased, and in many cases even has decreased the number of correct acts of relaying of power equipment."

3.3.4 Criteria for the Estimation of Reliability (Failures) of Microprocessor-Based Protective Devices

In attempting to carry out a similar analysis on failures of relaying in Russia, we have run into an unforeseen problem: it appears that in Russia, a base parameter of a reliability assessment in relaying is the percentage of correct (or not correct, i.e., faulty) operations,[42] instead of the number of relay damages, as in the case considered above. So, for example, in Reference 44 it is noted that in the most advanced Russian power company, Mosenergo (in Moscow), at the end of 2001 there were already 2332 DPR units of four different firms in service, and during four years only eight cases of faulty operation of DPRs have been registered. On this basis, the authors conclude that "it specifies their high reliability and high service characteristics." In Reference 2 it is also noted that the percentage of their correct operations is accepted as the basic reliability index for DPRs.

More complete data are published on the Internet by Alexey Vladimirov, deputy chief of relaying of the Central Dispatcher Control Service of a power pool system of Russia. His data, covering the period from 2000 to 2009 for 110–750 kV power networks, recorded 2913 cases of DPR operations. In 89.5% of the cases they operated correctly, and in 10.6% they were faulty. For the same time period, electromechanical relays operated 17,529 times. In 93.53% of the cases they operated correctly, and in 6.48% incorrectly. Microelectronic protection devices operated 5685 times. They operated correctly in 92.91% of the cases, and incorrectly in 7.07% of the cases.

But these bare figures don't tell the whole story. We need to translate them into more informative data. For this purpose, we suggest a new parameter: "a specific coefficient of incorrect operations," or $K_{FO} = N_{i\text{-}FO}/N_i$, where $N_{i\text{-}FO}$ is the number of incorrect operations, i is the kind of relay protection, and N_i is the total number of operations of a relay of a given relay type. The number of incorrect operations of each relay type is computed based on the given percentages: for DPR, 309; for microelectronic, 402; and for electromechanical, 1136. From this, we can present specific coefficients for incorrect relay operations, that is, DPRs operated incorrectly 1.7 times more often than electromechanical relays.

We need to bear in mind that this only speaks of faulty operations; it does not give us the number of serious internal DPR damages that occur without (before) faulty operations. But why is the reliability of the devices and systems estimated by the frequency of their faulty operations instead of by the number of damages of their basic internal elements, thereby making impossible proper functioning of the device or system? If the signal about damage of its internal power supply (meaning the incapability of the DPR to perform its functions) from DPRs installed in a protection system has been received, but there were no emergency modes in a power network controllable by this DPR (that is, there were no faulty actions of the relaying), this event should not be fixed as a failure of DPRs and neglected in the analysis of DPR reliability. Only if the internal damage of the DPR coincides with the time of the emergency mode in a protected network will this damage be considered in a reliability assessment; and if does not coincide, it will not be.

Some well-known definitions for *reliability* and *failure*[45] are as follows:

Reliability: the ability of an item to perform a required function under stated conditions for a stated period of time.

Failure: the state or condition of not meeting a desirable or intended objective, and may be viewed as the opposite of success.

Failure rate: the number of failures experienced or expected for a device divided by the total equipment operating time.

However, accident in a power system is the *result* of relay protection failure, yet the reliability and failure definitions don't even take into account the *result* stemming from low reliability or a high failure rate. It is just not clear why the failure of a single protective unit is taken into consideration only in the case that it is the *result* of the accident in the power system without any consideration of the accident itself. It is difficult to see the logic in such an approach. Such an approach simply does not lead to the proper analysis of the protective relay failures, similar to the analysis that we have used above.

In our opinion, in the estimation of the relay of protection, it is necessary to consider three types of events:

1. The damages (D) of the relay which have been not connected with faulty actions of the relaying, but require repair or replacement of the failed elements, unit, and modules.
2. Faulty actions (FA) of a relay that comprise improper operations in the absence of an emergency mode or an inability to operate (or faulty operation also) in the emergency mode.
3. Personnel mistakes (PM) connected with operation, testing, or programming of the relay, keeping in mind the personal actions that have an influence on the relay functioning properly, but are detected before relay improper action occurs.

All these components should be taken into account, in our opinion, when calculating the generalized normalized criteria of failures F_Σ of relaying:

$$F_{\Sigma_i} = \left(\frac{F_{D_i} + F_{FA_i} + F_{PM_i}}{N_i} \right) \times 100\%$$

where $F_{D_i}, F_{FA_i}, F_{PM_i}$ are the number of failures of each type for the relay i kind for the considered period of time, and N_i is the number of the relay i kind, being in operation during the considered period of time.

The suggested parameter could serve as the tool for an estimation of the quality of the relay protection when analyzing a situation and making decisions.

3.3.5 Summary

In summary, it is desirable to cite the well-known expert in the field of MPD, E. M. Shneerson, DSc, a former leading expert of the All-Russian Relay Research, Design and Technology Institute (VNIIR), who worked for a long time at Siemens. Shneerson, in the monograph,[29] writes,

> In itself increasing technological levels of protective relays do not necessarily lead to increase efficiency in reaction on incipient faults. So, for example, out-of-date electromechanical and partly electronic static protective relays at a correct choice of protective functions and setting will certainly provide more effective protection of a network than those that are microprocessor-based without enough proven choice of the specified parameters. (p. 491)

Shneerson states further, on p. 508, "As shown in practice, the percent of the wrong actions in the use of digital protective relays at an initial stage essentially does not decrease, and in some cases even increases." And in summary, on p. 522, "Despite essentially higher technical perfection of the digital protective relays, their real operational efficiency, especially at initial stages, is lower than these protection devices of the previous generation."

3.3.6 Conclusions

1. DPR reliability is lower than the reliability of electromechanical relays and electronic relays on discrete elements.

2. Built-in self-diagnostics DPR is ineffective and is not a means at all for increasing MPD reliability.

3. Nanotechnologies, used in the manufacture of DPR elements, lead to the occurrence of problems not known earlier for relay protection. Ignoring these problems can lead to catastrophic consequences. The managers making decisions in the field of relay protection and the personnel of the power companies should be informed about these DPR features.

4. The recording function of emergency modes in a power network and the data transmission function on modern connection channels are not direct functions of relay protection, and for their realization there are separate microprocessor devices which carry out these functions much better than DPRs. Compared to relay protection, failure of these devices does not lead to heavy failures in power systems. Therefore, for relay protection devices the focus should be on other demands on reliability and, accordingly, to use other approaches when designing them that are directed toward increasing reliability and decreasing vulnerability.

5. The persons responsible for making decisions on the reconstruction of relay protection and ways of further developments should understand, precisely, the properties and features of DPRs to take into account not only widely promoted DPR advantages but also DPRs' serious lacks, one of which is lowered reliability.

3.4 What to Do with All These Problems?

To answer that question we should look at the construction of DPRs, as discussed in Chapter. 2. As is known, the DPR has a modular design. The basic functional units of DPRs are the motherboard with an ADC, a microprocessor, various types of memory, and other accessories; digital (logic) inputs with a photocoupler module; analog inputs based on the current and voltage transformers module; and the output relays module. Isn't this modular design of a DPR similar to that of a personal computer (PC)? However, there is one significant and even essential difference between PCs and DPRs: each PC module is available on the open market and you can assemble your PC with modules produced by different manufacturers from different countries. What does it lead to? It leads to price reduction and the ability to assemble a PC with the best-suited modules regarding their characteristics and price. The same applies to the software. There is some universal platform (Windows) and a huge market of applications for every taste.

But, are there any significant structural differences between the PC and the DPR? In reality, there are few essential differences: it uses the same power source, and the same main module (motherboard) connected to auxiliary modules, such as the analog inputs module (a set of CT and VT with the filter and ADC) instead of the video card, the logic inputs module instead of the TV tuner, and the output relays module instead of the sound card. What is the essential difference between the software designed for multifunctional DPRs and any other software? There is no difference! So what's the problem? Why do we have now a vast number of totally noninterchangeable DPR designs, instead of a set of universal modules in the form of PCBs? Why do we have a wide variety of DPR software versions that are inconsistent even with each other?

To answer this question, let's trace how this remarkable business works. For example, what happens if some specific module of a specific DPR type at a given substation fails? This is what happens: since there is no market for universal modules, *the user can replace the failed module only and exclusively with the same one, produced by the same manufacturer.* Thus, after you spent a small fortune to purchase the DPR from one of the manufacturers, you actually fall into the bondage of economic dependence on this manufacturer for the next 10–15 years, since after you have chosen one manufacturer, it no longer matters if there are other manufacturers on the market, as you cannot use their products. And the only way to get out is to pay a small fortune one more time for a DPR from another manufacturer (and, thus, you go from one bondage to another). And what does the manufacturer in such a situation of absolute monopoly do? Right: increase the price! The price of one spare DPR module can reach almost one-third or even one-half of the price of the entire DPR! As you have nowhere to go, you pay that price. And what happens after 8–10 years of the DPR operation? Here's what happens: the manufacturer had already developed several new designs during that time and it has become unprofitable for the manufacturer to maintain the facilities producing spare modules for the old relays, so production has been stopped. What does the consumer do in this situation? Right: throw the old DPR into the garbage, even if there is only one faulty module (PCBs of the modern DPR are developed by the irreparable technologies), and shell out money for purchasing a new one. Who wins? Right: the same manufacturer!

But why is the manufacturer allowed to cash in on the consumer in all conditions? What should have been done to turn the tide in favor of the consumer? That's right: the DPR should be realized as a set of modules—circuit boards with universal standard dimensions and connectors, just like those for a PC, and with an integrated basic program shell compatible with the software for the given type of protection or protection kit available on the market. Today each type of DPR has its own body which significantly differs from that of any other DPR type, and sometimes differs even from those from the same manufacturer. Today as a rule, a single DPR is installed in the relay cabinets: 3–5 units in each cabinet (see Figure 3.10). If a new DPR will be realized as a set of universal modules on the PCBs, then such sets (at least, in most cases) will not require the individual cases. Each DPR may be installed as a single horizontal section in the cabinet with the PCB guiderails, individual door, and rear wall with connectors and terminals for connecting external cables just as in PCs.

The relay cabinet should be made by special technology intended to protect its contents from electromagnetic influences. There are modern technologies (special cases, electrically conductive pads and greasing, filters, etc.), which could significantly reduce the effect of external electromagnetic emission of a wide frequency range on highly sensitive DPR equipment (see Chapter 8).

FIGURE 3.10
The modern method of mounting the DPR in relay cabinets.

The proposed tendency of development could open the DPR market to new "players," some of whom could produce analog input modules equipped with current and voltage transformers, and others could produce motherboards, or software. The consumer could "build" the DPR out of the separate modules from different manufacturers, just as it is today for PCs, based on the cost and quality of these modules; and the consumer could use the same software for all its DPRs. It would answer many of the questions set out above and significantly reduce the cost of relay protection. This could also enable installing two sets of identical DPRs instead of one in order to improve reliability, and using the second set as backup, starting automatically upon the "watchdog" signal of the damaged core DPR. In addition, it would eliminate the necessity of the individual power source for each DPR, and allow using one double high-capacity power source set of improved reliability for the entire cabinet. And finally, this would allow installing many service modules, capable of improving DPR reliability, in the same cabinet.

Thus the relay protection maintenance would be simpler as the service staff would not need to read thick folios (Figure 3.11) about different DPRs installed in the facility and study characteristics of the software of each DPR type. In addition to easier maintenance and reducing the new DPR lead time, it would significantly reduce the percentage of errors caused by the so-called human factor. Such DPR design would also solve the problems of testing complex DPD functions (see Chapter 7).

How can the program of DPR reconstruction described above be realized in practice? The easiest way is to start in one big country with a large market and several DPR manufacturers able to closely work with each other. In our opinion, the first step in this direction should be the setting of national standards containing requirements of the new type of DPRs, their software, their

FIGURE 3.11
Thick folios containing user information about DPRs.

test procedures, and so on. These standards must be developed, by a wide range of professionals, including scientists, potential DPR producers, their potential customers, and representatives of project organizations. It seems that such relay protection development, even in one country, would show an example to be emulated around the world.

3.5 "Intellectualization" of Protective Relays: Good Intentions or the Road to Hell?

There was an interesting story in an old science fiction novel. It all started from such an innocent thing as an odd night call to all the phones on planet Earth. The call announced the birth of Global Intelligence to all people of the Earth. It turned out that at some stage of development, the proliferation of computers escalated into a new quality: the millions of computers, integrated into a single network that controlled everything and everyone on planet Earth, suddenly became conscious of themselves as an entity capable of reproducing itself through the automated factories and robots connected to the same network, as well as capable of defending itself with computerized weapon systems designed to destroy the human race. From the perspective of Global Intelligence, humanity was nothing but a useless vestige, gobbling up the planet's resources. There are no prizes for guessing about the further development of the action.

Network-connected computers already control almost all types of modern industrial production systems, water supply and electricity systems, telecommunications systems, and networks. New terms such as *smart grid* and *relay protection with artificial intelligence* have emerged in technical literature rather than in science fiction. Today technical literature rather than science fiction refers to the creation of a so-called smart house, where even the refrigerator will analyze the stored products, and, based on the analysis of consumption, will make an order and send it over the network to the nearest supermarket. Today you can find microprocessors everywhere, even on the water closet lid.[46] Humanity is moving by leaps and bounds to the creation of all-powerful Global Intelligence prophesized in the old science-fiction novel.

But let's get back to reality. And the reality is that major failures in the energy systems that have occurred in America and Europe (the United States in 1965, 1977, and 2003; France in 1978; Canada in 1982 and 2003; Italy in 2003; the United Kingdom in 2003; and Sweden in 1983 and 2003) were caused by incorrect or, rather, unpredictable actions of relay protection during complex emergency modes due to disabling the wrong sectors of the network in every particular situation. Had the action of relay protection under these specific circumstances been different, the system failures might have been avoided. I am referring to the power systems (in the United States and Western Europe) that have already been equipped with computers and microprocessor-based protection. For comparison, I should mention that one of the world's largest power systems with a negligible percentage of computerized relay protection devices and the old worn-out equipment has never suffered from such failures. I am referring to the power system of Russia. The answer to the question about the reasons for this can be found in the book of E. M. Schneerson,[29] foremost authority in the field of modern relay protection:

> Improvement of the technical level of relay protection devices (RPD) alone does not necessarily lead to the equivalent improvement of efficiency as related to the response to emerging damages. For example, outdated electromechanical and to an extent electronic static RPD, if protection functions and settings are chosen correctly, certainly provides better protection for the network than the microprocessor RPD without rational definitions of the specified parameters.

In fact, the behavior of electromechanical and electrostatic RPDs under emergency situations was rigidly determined by their operating principle and settings. Current trends[47–52] in the development of DPRs are associated with the increase of their "independence" (that is, in fact, unpredictability) in making decisions. It relates to the relay protection self-learning capability peculiar to adaptive neural networks, as well as the use of technologies of artificial intelligence with fuzzy logic, and the like.

Another clear trend in the development of the modern DPR is the excessive complexity by including extraneous functions not typical for relay protection. Here, for example, is the list of functions performed by the so-called intelligent controller (Intelligent Protection and Automation Controller iPAC by Dynatrol Systems Inc.):

IEEE protective functions (ANSI number):
- Sync Check (25)
- Under Voltage (27)
- Directional Power (32)
- Phase Balance (46)
- Instantaneous Over Current (50)
- Inverse Time Over Current (51)
- Over Voltage (59)
- Voltage Balance (60)
- Directional (67)
- Reclosing (79)
- Under- and Over Frequency (81)
- Lockout (86)
- Differential for Transformer Protection (87)

Measurement functions:
- Voltage and Current RMS Values
- Neutral Current RMS Values
- Power Factor Measurement (power factor correction: capacitor bank switching)
- Total Power Measurement
- Real and Reactive Power Measurement
- Power Quality Measurement (FFT for harmonic measurement)
- Frequency Measurement
- Total Harmonic Distortion Measurement

Advanced features for protections:
- Coil Monitoring for relay failure detection
- Cold Load Pickup Logic to prevent protective devices from operating when cold load is put on the circuit
- Voltage Constraint with Current Pickup lowered to increase sensitivity when voltage is also collapsing during the fault
- Breaker Control Blocking for coordination with upstream and downstream protective devices via DI or peer-to-peer communication
- Directional on overcurrent devices

Add to this the monitoring of external current and voltage circuits, the registration of events, the functions of an emergency digital oscilloscope, and other routine functions of the DPR. Then there is the danger of excessive concentration of protective functions in a single terminal (e.g., a microprocessor relay-type SYMAP by Stucke Elektronik GmbH incorporates 39 separate protection functions); additional relay protection functions, extrinsic to the protection itself, lead not only to the physical complication of the device, consequently reducing its reliability, but also to the complication of its software and user interface. This in turn leads to a sharp increase in the number of software errors (again, the so-called human factor). Due to such a large number of functions, using the same internal resources of DPRs and possible conflicts of the embedded logic functions during complex emergency mode accompanied by a transition of one type of damage to another, it is not always possible to predict the behavior of the protection. Damage to one function that is common to all the internal-element DPR functions (power supply, watchdog, memory, microprocessor, its servicing subassemblies, etc.) will result in the instant failure of all protective functions at once.

Despite the obvious problems existing today due to the excessive concentration of protective functions in a single terminal, some leading experts in their philosophizing about the future of relay protection not only advocate additional "adding on" of extraneous functions to relay protection, but also even go further. They put forward the fantastic idea of "multi-dimensional relay protection"[50] and "relay protection with proactive functions,"[52] acting on the basis of its own experience, its own analysis of the current status of the protected object and prediction of its future state. In essence, this is about the relay protection capability taking completely unpredictable actions, as an independent intelligence making its own (previously not determined) solutions and changing power systems operating modes (through switches) at its own discretion before the emergency mode occurs.[53]

It must be emphasized that there is nothing wrong with the development of computer-based diagnostic and prediction methods for electrical equipment, and it could only be welcomed but for the attempts to "intercross" it with the protection relay.

Apart from the risk of losing control over the relay protection actions, current trends of its development dramatically increase the risk of hacker attacks on the grid because a computerized relay protection system is a good target for changing the state and affecting the modes of power systems. Despite the serious concern of specialists about this problem,[54,55] the trend toward greater susceptibility of power systems to hacker attacks only grows.

Another serious threat to the stability of power systems, which is based on current trends in its development, is the development of technologies of artificially produced destructive impacts on electronic and computer equipment.[56-61] The development of these technologies throughout the world contributes to increasing the spread of microprocessor technology and the memory elements with high sensitivity to external electromagnetic emission, on the one hand, and the tendency of constantly increasing the density of microelectronic components as a result of the reduction of the thickness of the operating and insulating layers in the crystals, on the other hand. These two tendencies, directed toward each other, form a very dangerous vector of development of modern technologies.

Moreover, today you do not need any special knowledge or equipment in order to create a device capable of destroying all the electronic devices of your neighbor. You can find numerous descriptions on the Internet of such devices based on common microwave ovens (see Figure 3.12).[61]

As for the special "combat" units, there have been very impressive achievements in this field (see Chapter 8). Compact mobile ultra-broadband pulse generators with a capacity of 1 billion watts, portable emitters in the form of pistols and rifles, explosive devices in the form of attaché cases destroying all the electronics within a radius of many hundred meters, special ammunition, and other weapons are designed specifically for the remote destruction of electronic equipment.

So, where are we going, and where will we arrive? Why do the current trends go unnoticed by specialists? Obviously, there are a lot of parties very much wanting this trend to continue. However, in our opinion, it is not a question of whom to blame, but of what to do.

FIGURE 3.12
Electromagnetic weapons based on a household microwave oven.

In contrast to isolated measuring and monitoring of computer systems, the protective relay is associated directly with the possibility of destructive impact on power system modes. This is the most important and fundamental difference of relay protection from all the other computerized devices and systems used in electric power engineering, preconditioning a need for a different approach to relay protection.

The above thesis, in our opinion, is the answer to this question. What's needed is a different approach in maximizing the reliability of relay protection and avoiding features unrelated to relay protection, limiting the number of functions in a single microprocessor terminal, avoiding algorithms with nondeterministic logic allowing the unpredictable action of relay protection, taking advantage of maximum simplification of the user interface, and conducting special research and development providing for the operation of relay protection against intentional destructive electromagnetic influences, for example by introducing stand-by emergency relay protection sets. Only electromechanical relays resistant to intentional electromagnetic influences, requiring no live power, and thus being always ready to work can be used as such stand-by relay protection sets. Therefore, in our opinion, it is too early to dismiss electromechanical relays. Rather, they should be improved through new technologies and materials, and their range should be updated.

Since the relay protection algorithms are not so complicated (all of them were effectively realized with electromechanical devices, which now account for over 90% of all protective relays in Russia), the current protection devices can be as simple as possible. There have been no new functions introduced in relay protection by DPRs, but only some relay protection characteristics were improved. In particular, distance protection received polygonal characteristics instead of the circular ones of old electromechanical relays. Therefore, in reality there are no objective reasons for today's substantial complication of the relay protection functions.

On the other hand, recently more and more complicated and sophisticated systems for monitoring electrical equipment modes based on the continuous monitoring of the electrical characteristics (tangent of dielectric loss angle, partial discharges in insulation, arrester leakage current, the number and composition of gases dissolved in transformer oil, etc.) have emerged with the ability of predicting the processes' progress in time. Automatic process control systems have become more complicated as well as real-time control systems for measuring vector values of current, voltage, and power; the systems for registration and oscillographic testing of emergency operation; and so on. In contrast to relay protection, all these systems don't have direct influence on the operating mode of power systems, and therefore there are no restrictions on the trends of their development.

In our opinion, the only purpose of relay protection should be the relay protection itself (i.e., identifying emergency modes and issuing instructions to the electrical devices changing the operating mode of the protected object in order to exit the emergency mode) and no more. All other problems should be solved by other systems independent of the relay protection. Therefore, further development of microprocessor relay protection and other microprocessor and computer systems in power engineering should take place in independent, unrelated, parallel courses.

In order to prevent the arbitrariness of manufacturers imposing more and more "advanced" and less reliable DPRs[62,63] to the energy sector, it is necessary, in our opinion, to formulate basic requirements for MPD design principles (not for the technical parameters, but for the principles of design) in the relevant standard. The same principles could also include earlier proposals[64] on DPR design such as a set of individual replaceable modules that are universal in functions, sizes, and contact connections (PCBs) by analogy with personal computers.

References

1. Hunt, R. K. Hidden Failure in Protective Relays: Supervision and Control. Master of science thesis, Virginia Polytechnic Institute, 1998.
2. Konovalova, E. V. Main Results of Relay Protection Devices Maintains in Power Systems in the Russian Federation. In Collection of Reports of XV Scientific and Technical Conference "Relay Protection and Automatics of Power Systems" [in Russian]. Moscow, 2002.
3. Belotelov, A. K. Scientific and Technical Policy of the Russian Open Society "EU of Russia" in Development of Systems of Relay Protection and Automatics. In Collection of Reports of XV Scientific and Technical Conference "Relay Protection and Automatics of Power Systems" [in Russian]. Moscow, 2002.

4. Johnson, G., and M. Thomson. *Reliability Consideration of Multifunction Protection.* Highland, IL: Basler Electric, 2001.
5. Gurevich, V. How to Equip the Relay Protection: Opinions of the Russian Experts and a View from the Side [in Russian]. *News in Electric Power Industry*, no. 2, 2007.
6. Projjalkumar, R. Is the Era of Electromechanical Relays Over? *Frost & Sullivan Market Insight*, March 5, 2004.
7. Gurevich, V. Microprocessor-Based Relay of Protection: An Alternative View. *Electro-Info*, no. 4, 2006.
8. Heising, C. R., and R. C. Patterson. Reliability Expectations for Protective Relays: Developments in Power Protection. Paper presented at the Fourth International Conference in Power Protection, 11–13 April 1989, Edinburgh.
9. Mahaffey, T. R. Electromechanical Relays versus Solid-State: Each Has Its Place. *Electronic Design*, September 16, 2002.
10. Tyco Electronics. Electromechanical vs. Solid State Relay Characteristics Comparison. Application Note 13c3235, Berwyn, PA: Tyco Electronics.
11. Gurevich, V. *Electronic Devices on Discrete Components for Industrial and Power Engineering.* Boca Raton, FL: CRC Press, 2008.
12. Clark, O. M., and R. E. Gavender. Lighting Protection for Microprocessor-Based Electronic Systems. *IEEE Transactions on Industry Applications*, vol. 26, no. 5, 1990.
13. Uchimura, K., J. Michida, S. Nozu, and T. Aida. Multifunction of Digital Circuits by Noise Induced in Breaking Electric Contacts. *Electronics and Communications in Japan*, vol. 72, no. 6, 2007.
14. Henderson, I. A., J. McGhee, W. Szaniawski, and P. Domaradzki. Incorporating High Reliability into the Design of Microprocessor-Based Instrumentation, *IEEE Proceedings*, vol. 138, no. 2, 1991.
15. Phadke, A. G. Hidden Failures in Electric Power Systems. *International Journal of Critical Infrastructures*, vol. 1, no. 1, 2004.
16. Kovalev, B. I., and I. E. Naumkin. The Main Problems of Electromagnetic Compatibility of Secondary Circuits in High-Voltage Substations. In Collection of Reports of XV Scientific and Technical Conference "Relay Protection and Automatics of Power Systems" [in Russian]. Moscow, 2002.
17. U.S. Nuclear Regulatory Commission. Information Notice No. 94-20: Common-Cause Failures Due to Inadequate Design Control and Dedication. Washington, DC: U.S. Nuclear Regulatory Commission, 1994.
18. Matsuda, T., J. Kovayashi, H. Itah, T. Tanigushi, K. Seo, M. Hatata, and F. Andow. Experience with Maintenance and Improvement in Reliability of Microprocessor-Based Digital Protection Equipment for Power Transmission Systems, Report 34-104. Report presented at SIGRE, Session 30 August–5 September 1992, Paris.
19. He, S., L. Shen, and J. Lui. Analyzing Protective Relay Misoperation Data and Enhancing Its Correct Operation Rate. Paper presented at the IEEE/PES Transmission and Distribution Conference & Exhibition: Asia and Pacific, 2005, Dalian, China.
20. U.S.-Canada Power System Outage Task Force. Final Report on the August 14th Blackout in the United States and Canada. https://reports.energy.gov/BlackoutFinal-web.pdf
21. Shmuriev, V. Y. Digital Protective Relays [in Russian]. *Library of Electrical Engineering*, vol. 1, no. 4, 1999.

22. Kumm, J. J., E. O. Schweitzer, and D. Hou. Assessing the Effectiveness of Self-Test and Other Monitoring Means in Protective Relays. Paper presented at the 21st Annual Western Protective Relay Conference, 18–20 October 1994, Spokane, WA.
23. Terrazon Semiconductor. Soft Errors in Electronic Memory: A White Paper. Naperville, IL: Terrazon Semiconductor.
24. Baumann, R. C. Soft Errors in Advanced Semiconductor Devices: Part I: The Three Radiation Sources. *IEEE Transactions on Device and Material Reliability*, vol. 1, no. 1, 2001.
25. Dodd, P. E., M. R. Shaneyfelt, J. A. Felix, and J. R. Schwank. Production and Propagation of Single-Event Transient in High-Speed Digital Logic ICs. *IEEE Transactions on Nuclear Science*, vol. 51, no. 6, 2004.
26. Johnson, A. H., T. F. Miyahira, F. Irom, and L. D. Edmonds. Single-Event Transients in High-Speed Comparators. *IEEE Transactions on Nuclear Science*, vol. 49, no. 6, part 1, 2002.
27. Gurevich, V. Nonconformance in Electromechanical Output Relays of Microprocessor-Based Protection Devices under Actual Operation Conditions. *Electrical Engineering & Electromechanics*, no. 1, 2006.
28. Gurevich, V. Peculiarities of the Relays Intended for Operating Trip Coil of the High-Voltage Circuit Breakers. *Serbian Journal of Electrical Engineering*, vol. 4, no. 2, 2007.
29. Shneerson, E. M. Digital Relay Protection. *Energoatomizdat*, 2007.
30. Gurevich, V. Microprocessor Protection Relays: The Present and the Future. *Serbian Journal of Electrical Engineering*, vol. 5, no. 2, 2008.
31. Borisov, R. Insufficient Attention to the Problem of EMC May Turn into a Catastrophe [in Russian]. *News in Electrical Engineering*, vol. 6, no. 12, 2001.
32. Matveev, M. Electromagnetic Situation on Objects Determines EMC of the Digital Equipment. *News in Electrical Engineering*, vol. 1, no. 13, 2002.
33. Gurevich, V. Electromagnetic Terrorism: New Hazards. *Electrical Engineering & Electromechanics*, no. 4, 2005.
34. IEEE. IEEE Standard for Substation Intelligent Electronic Devices (IEDs) Cyber Security Capabilities: IEEE Std 1686-2007. Piscataway, NJ: IEEE, 2007.
35. Puliaev, V. I. Results of Usage of Relay Protection and Automation in Open Society "United Electrical Systems." In Collection of Reports of XV Scientific and Technical Conference "Relay Protection and Automatics of Power Systems" [in Russian]. Moscow, 2002.
36. Gurevich, V. Microprocessor Protection Relays: New Prospects or New Problems? *Electrical Engineering & Electromechanics*, no. 3, 2006, pp. 18–26.
37. Gurevich, V. I. Microprocessor Protective Relays: Alternative View. *Energo-Info*, vol. 4, no. 30, 2006, pp. 40–46.
38. Gurevich, V. Reliability of Microprocessor-Based Relay Protection Devices: Myths and Reality. *Serbian Journal of Electrical Engineering*, vol. 6, no. 1, 2009, pp. 167–186.
39. Gurevich, V. Reliability of Microprocessor-Based Protective Devices: Revisited. *JEEEC*, no. 5, 2009, pp. 295–300.
40. Gurevich, V. Hybrid Reed-Solid-State Devices Are a New Generation of Protective Relays. *Serbian Journal of Electrical Engineering*, vol. 4, no. 1, 2007, p. 85–94.
41. Gurevich, V. I. Hybrid Reed-Solid State Overcurrent Protection Relay with Harmonics Restraint. *Electronics-Info*, no. 1, 2009, pp. 26–28.

42. Shalin, A. I. About Efficiency of New Relay Protection Devices. Power and Industry of Russia (Selected Publications), vol. 203. Moscow: Power and Industry of Russia, 2007.

43. Shneerson, E. M. Operational Efficiency of Relay Protection Devices: A Reality and Abilities. *Energoexpert*, nos. 4–5, 2007, pp. 70–77.

44. Kudryashov, V. N., V. V. Balashov, A. G. Korolyov, and A. V. Sdobin. Experience with Using a Microprocessor Based Protection in Mosenergo. In Collection of Reports of XV Scientific and Technical Conference "Relay Protection and Automatics of Power Systems" [in Russian]. Moscow, 2002.

45. Ward, S., T. Dahlin, and W. Higinbotham. Improving Reliability for Power System Protection. Paper presented at the 58th Annual Protective Relay Conference, 28–30 April 2004, Atlanta, GA.

46. Gurevich, V. I. Cost of the Progress. *Components and Technologies*, no. 8, 2009.

47. Bittencourt, A., M. R. de Carvalho, and J. G. Rolim. Adaptive Strategies in Power Systems Protection Using Artificial Intelligence Techniques. Paper presented at the 15th International Conference on Intelligent System Applications to Power Systems, 8–12 November 2009, Curitiba, Brazil.

48. Laughton, M. A. Artificial Intelligence Techniques in Power Systems. In *Artificial Intelligence Techniques in Power Systems*. Herts, UK: Institution of Engineering and Technology, 1997, pp. 1–18.

49. Khosla, R., and T. Dillion. Neuro-Expert System Applications in Power Systems. In *Artificial Intelligence Techniques in Power Systems*. Herts, UK: Institution of Engineering and Technology, 1997, pp. 238–258.

50. Lyamets, U. Y., D. V. Kerzhaev, G. S. Nudelman, and U. V. Romanov. Multidimensional Relay Protection. In *Abstracts of the Second International Scientific-Technical Conference*, "Modern Trends in Development of Power System Protection and Automation," 7–10 September 2009, Moscow.

51. Kamel, T. S., M. A. Hassan, and A. El-Morshedi. Application of Artificial Intelligence in the Remote Power Line Protection. In *Abstracts of the Second International Scientific-Technical Conference*, "Modern Trends in Development of Power System Relay Protection and Automation," 7–10 September 2009, Moscow.

52. Bulychev, A., and G. Nudelman. Relay Protection. Improvement through Proactive Functions. *News of Electrotechnics*, vol. 4, no. 58, 2009.

53. Gurevich, V. I. Sensational "Discovery" in Relay Protection. *Energy and Industry of Russia*, vols. 23–24, nos. 139–140, December 2009.

54. CNews.ru. CIA: Hacking Electrical Grids Is Possible. CNews.ru: Newsline, 24 January 2008. http://www.cnews.ru/news/line/index.shtml?2008/01/24/285018

55. Gurevich, V. I. Electromagnetic Terrorism: The New Reality of the 21st Century. *The World of Techniques and Technologies*, no. 12, 2005, pp. 14–15.

56. Daamen, D. Avant-Garde Terrorism: Intentional Electromagnetic Interference. Twente: University of Twente, 2002.

57. Commission to Assess the Threat to the United States from Electromagnetic Pulse (EMP) Attack (EMP Commission). Report of the Commission to Assess the Threat to the United States from Electromagnetic Pulse (EMP) Attack. http://www.empcommission.org/docs/empc_exec_rpt.pdf./

58. Gannota, A. The Object of Defeat: Electronics. Independent Military Review, no. 13, 2001.

59. Loborev, V., U. Parfionov, and V. Fortov. Collapses of Noiseless Explosion. *Literary Gazette*, vol 5, no. 5865, 6–12 February, 2002.
60. Pokrovsky, V. Electromagnetic Factor. *Independent Gazette*, 8 October 2003.
61. Bäckström, M. Is Intentional EMI a Threat against the Civilian Society? Stockholm: SAAB Communication, 2006.
62. Gurevich, V. Reliability of Microprocessor-Based Relay Protection Devices: Myths and Reality. *Engineer IT*, part 1: no. 5, 2008, pp. 55–59; part 2: no. 7, 2008, pp. 56–60.
63. Gurevich, V. I. Reliability of Microprocessor-Based Protective Devices: Revisited. *Journal of Electrical Engineering*, vol. 60, no. 5, 2009.
64. Gurevich, V. I. Microprocessor Protective Relays: Searching for Optimality. *Energy and Industry of Russia*, vol. 21, no. 137, November 2009.

4

Logic Inputs in Digital Protective Relays

4.1 Reliability of Logic Inputs in DPRs

Problems with the reliability of microprocessor-based protective devices (MPDs) have arisen in connection with the worldwide moving away from electromechanical and static protective relays to digital protective relays (DPRs). As shown in References 1 and 2, the common misconception that DPRs have reliability considerably exceeding that of electromechanical and static protective relays is, in actual fact, no more than a myth fostered through the years by the advertising publications of DPR manufacturers. It is abundantly clear that such complex multipurpose structures as DPRs cannot, even theoretically, be free of problems, cannot be absolutely reliable, and cannot be devoid of failure statistics during 15–20 years of maintenance.

Previously, we have discussed the problems connected with insufficient reliability of the output miniature electromagnetic relays of DPRs intended for the direct tripping of high-voltage circuit breakers.[3] This chapter addresses a problem with the reliability of logic inputs in the DPR using the RE*316 series (REL, RET, REC, etc.) as an example. The RE*316 series has been widely used over the past 10–15 years.

Digital (logic) inputs in the DPRs of these types consist of a set of completely identical elements functionally representing the logical function *Prohibition* (see Figure 4.1). The direct input and prohibited input of each of these elements are connected respectively to the input circuit of the DPR and to the central processing unit (CPU) through optocouplers: Opt1 and Opt2. The starting signal from optocoupler Opt2 logically repeats the presence or absence of an input voltage. The functioning of this circuit can be blocked (prohibited) through a CPU internal logic. Thus, the blocking signal from the CPU through matching electronic circuits acts on the prohibiting input (an input of optocoupler Opt1).

The problem of this type of logic cell consists in the high level of an input signal (220–250 V DC) which needs to be attenuated up to a level of 1.5–2.0 V, at which point the Opt2 optocoupler operates. The total current consumed by the input circuit is split approximately 50/50 between resistors R1 and R2; therefore, both of them, basically, should be equal. However, there is no place

149

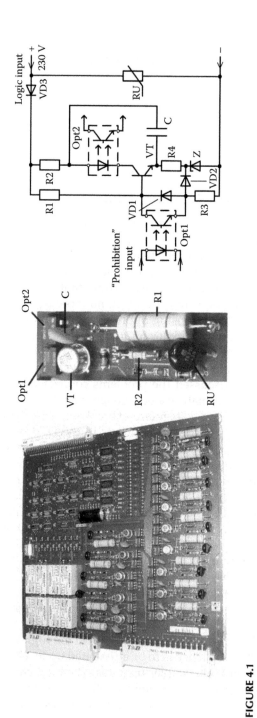

FIGURE 4.1

Logical inputs: printed circuit board (PCB) module of microprocessor-based protective device (MPD) RE*316 type, single-channel and IT circuit diagram.

on a printed circuit board (PCB) for two large resistors such as R1 and R2 (including the distance between them, which is necessary for cooling).

Therefore, to dissipate the superfluous power of this element, the DPR designers have assigned to a transistor, VT, working not in a switching mode, as is usual in such a circuit, but in an amplification mode (linear mode). It is natural that the resistor for R2 is therefore selected as being of low power and small size. In practice, such aspirations to miniaturization lead to serious problems: resistor R2 often completely burns out, leading sometimes to the burnout of sections of the PCB and even those components in its vicinity.

The first problem of this arrangement centers on the transistor, VT. In operating in amplification mode and continually dissipating the super-fluous power, it heats up to temperatures of 70–80°C. Unlike the usual resistor, the resistance of which slightly increases with increase of temperature, resistance of the *n-p-n* transition of the transistor with the increase of temperature essentially decreases owing to the drift of the operating point (referred to as the *quiescent point*) on the characteristic and increases in the coefficient of amplification. This leads to an increase of the collector current, that is, the current through resistor R2. With the simultaneous heating of many resistor (R1) and transistor (VT) combinations in 10–15 input circuits, the temperature inside of the input section (in the case that the DPR is divided into sections by internal diaphragms) can essentially increase. This leads to a further drift of a quiescent point of the transistor, further increasing its collector current and the overheating of resistor R1. Thus diode VD, intended for stabilizing the quiescent point of the transistor, appears ineffective because its temperature and the temperature of the transistor discriminate on 50–60 grades. Thus, the aspiration of the manufacturer to reduce MPD sizes, using a transistor instead of a powerful resistor for superfluous power dissipation, has led to a decrease of MPD reliability. The problem of insufficient reliability of this circuit connected with aspirations for miniaturization is not restricted only to this problem.

The second problem is the miniature disk ceramic capacitor, C, with a capacitance of 2.2 nF. Ceramic capacitors are considered to be among the most resistant to the effects of electric operational loads and are stable over long durations. However, in ceramic capacitors with badly isolated gaps between electrodes, there is the possibility of a decrease in the insulation resistance and even an electric breakdown due to the migration of metal ions of facings (especially silver) on the butt end of the capacitor, especially in conditions of a wet tropical climate. At an input voltage of the DPR below 100 V, capacitors such as these give no indication of the defects and at *procall*, a usual tester, show full serviceability. But already at voltages of 180–230 V, an outflow current through the capacitor increases (Figure 4.2) such that normal operation of the transistor and optocoupler becomes impossible. Moreover, there is a long-range influence of an input voltage of 220–230 V because of the

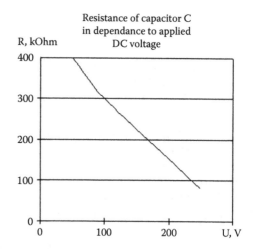

Resistance of capacitor C in dependance to applied DC voltage

FIGURE 4.2
Reason for R2 burning is increasing leakage current through capacitor C.

increased outflow current through this capacitor; the power dissipation on resistor R2 increases, and it simply burns out (Figure 4.2).

The third problem is the internal cutoff failures appearing in the course of time in the powerful R1 resistors (see Figure 4.1). This phenomenon presumably originates owing to a contact fault between metal cups at end faces of the resistor and a resistive layer on a surface of the ceramic cylinder because of its shelling in cheap resistors used in the DPR. The fourth problem is mass failures of the Opt2 optocouplers. For example, once we had to replace seven optocoupler units of type CNY17-2 on only a single PCB of a logic inputs unit of relay REC316! Optocouplers of this type have following parameters:

- Average forward current (a current through a light-emitting element, or I_F): 60 mA
- Maximum forward current (I_F): 100 mA
- Rated collector current of the output transistor: 50 mA
- The maximum collector current: 150 mA

It appears that these optocouplers operate in the DPR in a mode that is far from nominal; rather, they operate with very small input (forward) and output (collector) currents, practically on the bottom boundary line of these currents.

Besides, as a result of research carried out by us, it has been found that optocouplers, desoldered from nonworking inputs of the DPR, actually are

TABLE 4.1

Voltage Drop (V) on Output Transistors (Collector-Emitter) of CNY17-2 Optocouplers at Collector Current 0.5 mA and Input (Forward) Currents 2 and 2.5 mA

$I_F = 2.0$ mA	$I_F = 2.5$ mA
Optocouplers: Extracted from Nonworking Inputs of REC316	
1.471	0.992
1.027	0.344
0.574	0.188
1.452	0.967
1.315	0.769
0.734	0.201
0.634	0.178
New Optocouplers: Same Types, but Other Manufacturer	
0.134	0.120
0.143	0.127
0.132	0.119
0.144	0.127
0.139	0.123
0.139	0.124
0.134	0.120
0.143	0.125

not damaged. The difference between "working" and "nonworking" in DPR optocouplers has appeared only in their sensitivity to very small, lowest-limit forward currents (I_F); see Table 4.1.

Judging by the data in Table 4.1, the collector-emitter voltage drops of optocouplers from nonoperating input circuits of the DPR considerably exceed typical values (see Figure 4.3; 0.18 V at forward current $I_{IF} = 2$ mA and collector current 0.5 mA), while the same values for new optocouplers of other manufacturers are below the typical value.

Unfortunately, many optocoupler manufacturers (e.g., CNY17-2 type optocouplers are manufactured by the companies Agilent, Toshiba, QT Optoelectronics, Fairchild, Vishay, Liteon, Everlight, Isocom, and Opto Inc., among others) simply do not present these major characteristics of photocouplers in their datasheets; therefore, any claims regarding them are impossible to present. The claim can be made to designers of the electronic circuits (in our case, to the designers of RE*316 from company ABB) that the operation mode of the equipment (the DPR) is on the boundary line of the characteristic of electronic components (optocouplers); therefore, the usual slightest technological dispersions in parameters of these electronic components lead to full loss of the working capability of such crucial equipment as DPRs.

The fifth problem is the manufacturer selecting too wide a range of operating voltages (82–312 V) for logic inputs. This wide range of operating voltages

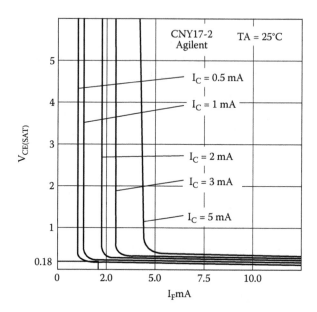

FIGURE 4.3

Voltage drop on the output transistor $V_{CE(SAT)}$ in dependence from the input current of the opto-coupler (I_F) for different collector current values.

is the source of two problems: low efficiency of overvoltage protection by means of the varistor and a problem of false operations.

The first problem is caused because by maintaining the working capability at a voltage of 312 V (which, needless to say, is absolutely not realistic in the practical condition with a DC network based on 220 V batteries), the clamping voltage of the varistor RU (7K391 type, in our case; see Figure 4.1) should be selected as not less than 650 V (because of the actual characteristics of varistors for a working voltage of 312 V DC). This means that all electronic circuit components will be exposed to the overvoltage reaching up to 650 V (for comparison, the maximum collector-emitter voltage allowed for transistor VT 2N3439 type does not exceed 350 V).

The second of these two problems is caused by too low a value of the bottom boundary line of an operating voltage range: 82 V. When one of the poles (positive or negative) in the DC power network, in which the DPR is connected, is grounded to earth, there is a voltage pulse equivalent to half the value of the rated battery voltage (that is, nearly 110 V) caused by the discharge of the capacity of wires of a DC network that is produced. As shown in Reference 4, at such conditions the spontaneous activation of logic inputs of DPRs occurs if an operating threshold of logic inputs is below half of the mains voltage of a DC network (that is, 110 V).

In just a fragment of the DPR circuit containing only ten electronic components, we have shown how many of the problems connected with DPR reliability can originate in actual conditions of operation. It confirms the thesis put forward by us that modern DPRs are not the model of perfection and reliability at all as advertising catalogs try to present them. Further, they represent a most complicated and not always reliable enough method to be employed. These problems should not be ignored as is done today; rather, they should be widely discussed.

4.2 Increasing Noise Immunity of the Logical Inputs in DPRs

Earlier, I offered solutions for solving the problem regarding the insufficient noise immunity of microprocessor-based protective relays.[2,5] For example, in 1996 the way of increasing the noise immunity for analog inputs of the microprocessor based on overcurrent protection using fast (0.8–1.5 ms) reed relays with high-voltage (3–5 kV) insulation was offered.[6] Fast relays such as these should short-circuit while constantly measuring inputs of the microprocessor relay until the input current reaches some threshold value. Besides analog (measuring) inputs, microprocessor-based protective relays are supplied with many logical inputs. Depending on the internal protection logic incorporated in the relay, in some cases activating some of these inputs may cause a false trip of the high-voltage circuit breaker, a high-voltage line, the transformer, or other important power equipment.

In the event that such an input activation is a consequence of interference affecting the relay, false disconnection of high-power equipment is a serious crash, accompanied by significant damage and requiring time-consuming analysis of the reasons for its cause. Unfortunately, in many cases it is not possible to reveal the true reason for a false operation of protective relay.

Investigation of several such situations in which I participated has shown that the frequent reason for false operations of microprocessor protection on power substations is grounding of wires (poles) of a DC network. Thus, nothing terrible happens with the network, as both its poles are in normal conditions and well isolated from earth, and the condition of the insulation is constantly monitored by special devices. However, this insulation causes significant capacity branching and long DC networks, especially in large substations, and is the origin of the pulses of a discharge current (and, accordingly, voltage) at the grounding of one of the wires (poles) of a network (see Figure 4.4). The maximal value of the voltage originating at the capacity of a discharging network can reach, basically, up to half of the value of the substation battery voltage, that is, 110 V. Actuating the logical inputs of the microprocessor relays of various types occurs at various levels of the initial voltage, since 50–60 V is the typical voltage for single-pole grounding

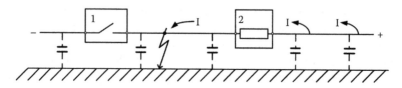

FIGURE 4.4
The circuit diagram for discharging a DC network capacity through logic input of the protective relay (2) with the open control contact (1) at single-pole grounding.

in a DC network. Moreover, power and pulse duration of a discharge current appear quite sufficient not only for actuating the logical inputs of microprocessor relays, but also even for energizing some types of auxiliary electromechanical relays with a rated voltage of 220 V and a minimal pickup voltage near 70–80 V.

Considering the limited pulse duration of a discharge current, technically it is possible to inject (through computer software) the additional program timer, providing a time delay of the input signal on 20–40 ms into a logical circuit of the microprocessor relay. A timer such as this will play the role of the filter, prohibiting pickups of the microprocessor relay of the troublesome short-time pulses of a current at the discharging of a DC network capacity. It is possible to use a resistor-capacitor (RC) circuit which is connected to the inputs of the relay and carries out the same function.

However, it is necessary to make the decision of, basically, whether or not doubling the time reaction of the protective relay to a fault mode (short circuit) is admissible. In our opinion, such delay of the protective relay actuation is inadmissible. In connection with what has been stated above, another solution for the problem is offered, that is, increasing the minimal threshold of logical input pickups to a level exceeding half the value of the DC network voltage. The selected level of the voltage is to be 150 V.

To implement this idea, we suggest using an elementary module as shown in Figure 4.5, consisting of two electronic components: a Zener diode with a rated voltage of 150 V and rated current of 5 mA, and a power thyristor for a maximum current of 7.5 A, maximum voltage of 800 V, and required gate current of 5 mA. At voltage levels below 150 V, the device is in the inactive condition; therefore, voltage pulses with magnitudes of up to 110 V, originating in a DC network ground, do not pass to logical inputs of the relay.

On the turning ON of the operating contact S, a voltage of 220 V will be abruptly applied to the Zener diode and it will instantly open, passing a current in the gate of the thyristor. The thyristor will open, and by the low forward resistance shunts the Zener diode and the gate circuit.

Now all operating current passes to logical input of the relay through the anode-cathode circuit of the thyristor. The voltage drop on the open thyristor does not exceed a fraction of a volt, which, in combination with the small current carried through it (15–50 mA), causes very insignificant dissipating power and does not heat the power thyristor.

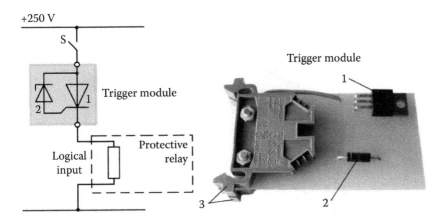

FIGURE 4.5

The circuit diagram and a design of the trigger module. 1: thyristor BT151-800L type; 2: Zener diode 1N5383 type; and 3: two terminal panels of the Wieland 9700A/6S35 type.

Both of the elements are placed on a small dielectric plate constructed with a glass-cloth-base laminate and measuring 70 × 40 mm. The plate is clamped by means of two screws between two panels of the standard terminal block, intended for installation on a standard Deutsches Institut für Normung (DIN, the German national organization for standardization) rail. After the installation of elements on a plate, they are covered by a layer of water-resistant varnish and are isolated by means of dielectric shrink tubing, providing mechanical protection of the elements on the plate. This device represents a small, easily constructed module (see Figure 4.6) having a very low cost (the thyristor together with the Zener diode cost less than $2.00).

The module is not connected to all the logical inputs, only to those that are especially critical. Upon activation it may affect pickups of a microprocessor relay and tripping high-voltage line or power electric equipment. For a single microprocessor relay, two to four modules may be required. The proposed device can also be used together with the usual auxiliary electromechanical relays when their real pickup voltage is less than half of the DC network voltage, that is, when there is a threat of their faulty operation.

It is necessary to emphasize that use of the proposed device does not reduce the reliability of relay protection, as electronic elements are completely deenergized at the opening of the operating contact S, and therefore cannot pickup spontaneously, for example owing to a breakdown or a spontaneous switching ON of the thyristor. Even allowing the probability (theoretically only) of a breakdown of the electronic elements of the module, the protective relay is simply returned to the mode of operation in which it was before using the module. The module can only have an effect on reliability of the relay in the breaking of the internal circuit. However, according to statistics, internal breakages in semiconductor devices originate only in the presence of very high carrying currents (when there is a burnout of the internal

FIGURE 4.6
External view of the single module mounted on a standard Deutsches Institut für Normung (DIN) rail.

semiconductor structure) and thus constitute no more than 5% of all faults. Under normal conditions of real currents carried through the thyristor in a real circuit of the module (15–50 mA), burnout of the internal thyristor structure is not possible. Some internal soldering joints and additional wire splices by means of the terminal block used in the module can really affect the reliability of relay protection. Therefore, the ensemble of these elements (in which latent defects are possible) is not used on the PCB, and leads of the thyristor and Zener diode are fixed before soldering by means of a thin copper tin-coated conductor that keeps the probability of failure of the module extremely low.

References

1. Gurevich, V. I. How to Equip a Relay Protection: Opinions of Russian Experts and a View from Outside. *Electric Power News,* no. 2, 2007, pp. 52–59.
2. Gurevich, V. Reliability of Microprocessor-Based Relay Protection Devices: Myths and Reality. *Engineer IT,* part 1: no. 5, 2008, pp. 55–59; part 2: no. 7, 2008, pp. 56–60.
3. Gurevich, V. Peculiarities of the Relays Intended for Operating Trip Coils of the High-Voltage Circuit Breakers. *Serbian Journal of Electrical Engineering,* vol. 4, no. 2, 2007, pp. 223–237.
4. Gurevich, V. I. Increasing Noise Immunity of the Logical Inputs in Microprocessor Based Protective Relays. *Electronics-Info,* no. 11, 2008, pp. 26–27.

5. Gurevich, V. I. Principles of Increase of Noise Immunity of Static Overcurrent Protective Relays. *Power Engineering and Electrification*, no. 2, 1992, pp.16–18.
6. Gurevich, V. I. Some Ways for Solving Problem of Electromagnetic Compatibility of Protective Relaying. *Industrial Power Engineering*, no. 3, 1996, pp. 25–27.

5

Problems with Output
Electromechanical Relays

5.1 Introduction

As is known, the switching capacity of relay contacts is determined by the area of contact surface, contact mass, contact force, and contact gap. The higher values that these parameters have, the higher the switching capacity of the contacts is. This is why powerful contacts differ from low-power ones, first of all in regard to their dimensions and second in regard to their gaps. A larger and more powerful coil is needed to create a large contact force and to move heavier contacts at a greater distance. Thus, one can state that for switching more powerful loads, a larger relay is needed (see Figure 5.1).

In old electromechanical protective relays as an element-switching trip coil of a high-voltage circuit breaker (CB), one used a special embedded auxiliary latching relay, with manual resetting and with an embed flag (target) indicating the relay condition. This relay is called an *auxiliary seal-in relay with target,* and it has powerful contacts with big gaps. They are especially meant for energizing up to 30 A with DC voltages of 250 V.

In next-generation protective relays—electronic analogues (or "static"), made up of integrated microcircuits and transistors—there is still a tendency to use large embedded output (trip) relays with powerful contacts meant for switching the CB trip coil (see Figure 5.2).

A new reality has appeared in the transition to the newest relay protections—microprocessor-based ones.[1-3] Hard competition in the market and a desire to maximally reduce the size of digital protective relays (DPRs) have resulted in the usage of subminiature electromagnetic relays as output elements (see Figure 5.3).

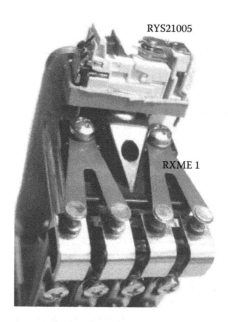

RYS21005

RXME 1

FIGURE 5.1
Subminiature relay RYS 21005 type located on V-shaped double-break high-power contacts of the auxiliary relay RXME-1 type intended for controlling a circuit breaker (CB) trip coil.

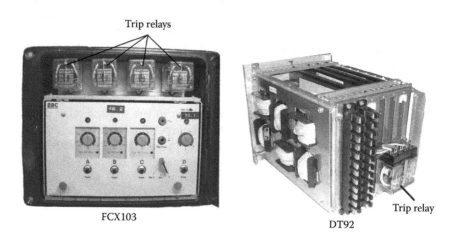

Trip relays

FCX103

DT92

Trip relay

FIGURE 5.2
Solid-state protective relays on discrete components with embedded power output relays.

ST2

RY611012

RTE24012

FIGURE 5.3
Printed circuit boards of the microprocessor-based protective relays with output electro-mechanical relays of different types.

5.2 The Analysis of Actual Operating Conditions of the Output Relays in Digital Protective Relays

According to manufacturer documentation, these relays are meant for applications in such systems as industrial automation, electronic power supplies, TV sets, domestic appliances, computers and communication systems, timers, and so on. In the technical characteristics of these relays, the switching capacity in DC is limited as a rule to 28–30 V and is to be used only for purely active (resistive) loads. At the same time, the maximum switching power in DC (sometimes it is curved lines of switching capacity in DC) enables calculating the maximum switching current at 250 V DC (see Table 5.1).

As is clear from Table 5.1, values of these currents, even with purely resistive loads, are 20–40 times less than in AC. Regarding the switching of

TABLE 5.1

Switching Capability of Miniature Electromechanical Relays Used in Microprocessor-Based Protection Devices

Relay Type (Manufacturer)	Maximal Switching Power (for Resistive Load)		Rated Current and Voltage (for Resistive Load)		
	AC	DC	AC	DC	For 250 V DC
ST series (Matsushita)	2000 VA	150 W	8 A; 380 V	5 A; 30 V	0.40 A
JS series (Fujitsu)	2000 VA	192 W	8 A; 250 V	8 A; 24 V	0.35 A
RT2 (Schrack)	2000 VA	240 W	8A; 250 V	8A; 30 V	0.25 A
RYII (Schrack)	2000 VA	224 W	8A; 240 V	8A; 28 V	0.28 A
G6RN (Omron)	2000 VA	150 W	8 A; 250 V	5 A; 30 V	—
G2RL-1E (Omron)	3000 VA	288 W	12 A; 250 V	12 A; 24 V	0.30 A

inductive loads in DC, this capability is not provided in technical specifications at all.

How did manufacturers of DPRs manage to use miniature (i.e., low-power) relays for direct switching of CB trip coils? Is it the case that requirements to control contacts of trip coils may have been reduced? By no means! In the technical specifications of all DPRs, manufacturers guarantee current switching of not less than 30 A at 250 V DC. Probably miniature relays themselves attained such perfection that now they are able to switch inductive loads (coils) with a current of 30 A at 250 V DC? Alas, technical specifications of subminiature relays used in DPRs do not say anything about such abilities of miniature relays. However, engineers of manufacturing companies that produce these relays to whom I addressed direct inquiries are categorical in rejecting such abilities of relays used in DPRs. Then is it clear that manufacturers of DPRs make such important and expensive ($10,000–$15,000) devices like DPRs with trivially improper elements? Reports about tests of output relay switching capacity which were submitted on our demand by the world's largest manufacturers of DPRs say that these relays have stood to the tests successfully and are acknowledged as valid for application. Then where is the logic? Perhaps manufacturers of DPRs conduct these tests improperly? On the contrary, DPRs with these miniature output relays have functioned successfully in many world power systems for many years. Then maybe the real operating conditions of these relays are much easier than the requirements mentioned in technical specifications? Let us try to sort out this situation. First of all, we will examine the real parameters of CB trip coils (see Figure 5.4).

Table 5.2 shows the results of measurements of general parameters of trip coils (L1, L3, and L4) of high-voltage CBs of several types, and also coils (L2) of special high-speed auxiliary latching relays with position fixing and

FIGURE 5.4
Trip coils of the CB on 160–170 kV from different manufacturers. 1: Hitachi Kokubo Works (GE-Hitachi, USA); and 2: AQ Trafo AB (Sweden).

TABLE 5.2

Main Parameters of Trip Coils to the High-Voltage CBs of Some Types and to the Lockout Relays

Parameter	Unit	L1	L2	L3	L4
Current, I	A	1.25	2.5	5	12
Inductance (coil on core), L	H	0.5	1.0	1.0	0.22
Resistance, R	Ω	200	100	50	22
Time constant, $\tau = L/R$	ms	2.5	10	20	10
Magnetic energy, E	J	0.40	3.12	12.50	15.80

manual release (lockout) which is sometimes included between the protective relay and CB.

The analysis of coil parameters given in Table 5.2 may lead to some interesting conclusions. First, a lockout relay is the load for contacts of miniature output relays not less than CB trip coils (see Figure 5.5). Experiments that I performed with the relay showed that even powerful contacts of the relay (with a contact diameter of 6 mm, and the gap between contacts was

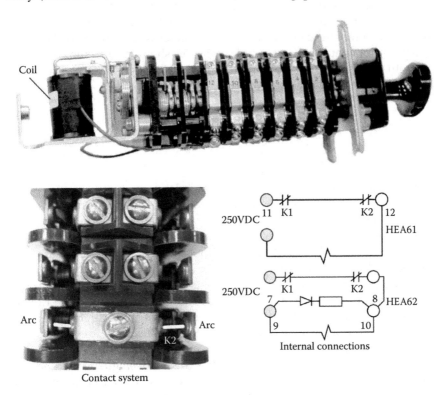

FIGURE 5.5
Latching hand reset auxiliary relay (lockout), 12HEA61 type.

TABLE 5.3

Switching Capabilities for Power Contacts of Lockout Relays

Load	Current (A) for Number of Contacts, Connected in Series		
	1	2	4
250 V DC, inductive	0.7	1.75	6.5
220 VAC, resistive	25	50	—
220 VAC, inductive	12	25	40

about 8 mm) are not able to break current (with arc) their own control coil series-connected with normally closed contacts at 250 V DC. Only two pairs of series-connected NC contacts (see the circuit in Figure 5.5) were able to break the arc appearing at disconnection. In the next modification of this relay (HEA62), even for two pairs of such powerful contacts one decided to make the switching process easier and to shunt the coil with a special arc-suppressing circuit composed of a diode and resistor. Manufacturer data given in Table 5.3[4] give a visual idea of the degree of load type influence on the switching capacity of even such powerful contacts.

Second, the time constant, $\tau = L/R$, which usually characterizes load type, is not a sufficiently informative value to allow conclusions of the real switching capacity of contacts. For example, Table 5.2 shows that in the L2 and L4 coils with the same L/R, considerably different energy is reserved which is evaluated with the following equation[5]:

$$E_L = \int_0^I id\psi = \int_0^I Lidi = \frac{LI^2}{2}$$

where L is load inductivity, and I is current in load.

Exactly this energy of the magnetic field is released on the contacts during the switching process. This means that the relay contacts will wear out differently during the switching of L2 and L4 coils with the same L/R value.

Third, the switched current value without indication of other parameters of inductive load (as, for example, in Table 5.3) is not a sufficient parameter for unambiguous evaluation of the contacts' switching capacity. For example, the current in the L2 coil is only two times stronger than in the L1 coil, whereas the energy reserved in L2 is almost eight times higher than the energy released during switching in L1. Experiments on these coils with fixing of arc power on contacts verified these conclusions.

Based on the above, one proposes using this load magnetic field energy as an index characterizing inductive load. In our concrete case for nominal voltages 250 V DC and 125 V DC, these values will be

$$E_{250} = 0.125I\tau,$$

$$E_{125} = 0.062I\tau$$

where I is current in load in amperes, and τ is load time constant in milliseconds. Thus, examination of real parameters of CB trip coils and high-speed lockout relays may lead to the conclusion that they are really serious inductive loads for contacts of protection devices' output relays.

In justifying the ability of miniature electromagnetic relays to control trip coils of high-voltage CBs, the manufacturers of DPRs usually refer to the fact that contacts of these relays just turn ON the trip coil of CBs. Turning Off of the coil is accompanied by intensive arcing and is implemented by auxiliary normally closed contacts of the CB itself, but not by the contacts of the miniature relay. This is why it is possible to turn on powerful trip coils of high-voltage CBs by means of low-power contacts of miniature relays. Is the statement unambiguous? It is well known that contact closure of electric appliances is accompanied by numerous contacts springing after first closure and further repeated closures (the process is called *bouncing*). This fact is reflected in technical literature and standards (see Figure 5.6).[6]

This means that there is no "pure closure" of contacts without numerous breakings in the process of relay actuation. Surely, the period of contacts being turned off (i.e., with arc burning) is minor during rebounds, but small distances between contacts in this period and the attendant compression make the risk of contacts sticking very real. That is why in existing standards, there are no great differences between the turning ON and turning OFF of circuits with inductive loads in direct current by the evaluation of contact switching capacity. For example, for application category DC-13 (control of electromagnets, coils of solenoids, and valves), according to the IEC 60947 standard,[7] the current of contacts both turning ON and turning OFF should

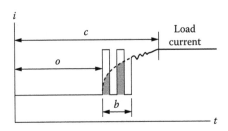

FIGURE 5.6
Oscillogram of relay-making process with contact bounce. *o*: time period from coil energizing up to first contact closing; *b*: bounce time; and *c*: time to stable closing. (*Source:* International Electrotechnical Commission, 2006.[6])

TABLE 5.4

Switching Capacity of Contacts Depending on the Type of Load for Control of Electromagnets, Valves, and Solenoid Actuators

Utilization Category IEC 60947-4	Type of Current	Switching Capacity of Contacts in the Mode of Normal Switching					
		Make (Switching ON)			Break (Switching OFF)		
		Current	Voltage	$\cos\varphi$	Current	Voltage	$\cos\varphi$
AC-15	AC	$10\,I_N$	U_N	0.3	$10\,I_N$	U_N	0.3
DC-13	DC	I_N	U_N	—	I_N	U_N	—
Switching Capacity of Contacts in the Mode of Infrequent Switching							
AC-15	AC	$10\,I_N$	$1.1\,U_N$	0.3	$10\,I_N$	$1.1\,U_N$	0.3
DC-13	DC	$1.1\,I_N$	$1.1\,U_N$	—	$1.1\,I_N$	$1.1\,U_N$	—

Note: I_N and U_N rated values of currents and voltages of electric loads switched by relay contacts.

not exceed a nominal (continuous) current, while for contacts functioning with AC, a tenfold value of turn-ON current is accepted (see Table 5.4).

However, the above is not the reason to conclude unambiguously that miniature relay contacts realizing activation of trip coils of CBs or coils of powerful high-speed auxiliary relays are really subjected to significant overloads. The issue is that during the activation of inductive loads, the current in it grows not linearly but exponentially. This means that the load circuit breaks during contacts bouncing, which happens when the current is less than nominal (see Figure 5.6). On the other hand, the fact that output contacts of miniature relays in protection devices do not fail when activated for the first time, but function for a long time in real operating conditions, also does not prove that these contacts function in the mode that is normal for them. It depends on the fact that even when contacts have visible arc-switching failure (i.e., nonclosure or nonbreaking of contacts), this does not take place immediately. Rather, a long process of defects accumulating on the contact surfaces takes place as a result of intensive evaporation of contact material from one contact and the transference of it to the other contact. Transient resistance of contacts increases, and so does their temperature. In miniature relays, this leads to the fusion of the plastic case near contacts, contamination of contacts, and further growth of transient resistance. After several thousands of these occurrences in simple relay, the final welding of contacts disintegrates or a break in one of the contacts takes place which appears as an absolute relay failure. As electromagnetic relays are usually meant for hundreds, thousands, or even millions of cycles, the operating regime in which the relay fails after several thousands of commutations instead of a million commutations is not acceptable and is not allowed by relay manufacturers. On the other hand, output relays in protection devices do not function with such intensity. The maximum number of actuations of these relays during their entire service life hardly exceeds several thousand. Then it is clear why relays functioning in the regime, which is abnormal for them, nevertheless provide serviceability of protection and even pass tests at the factories of DPR manufacturers. DPR manufacturing companies always submit these two facts in extenuation of using miniature relays for the direct breaking of high-voltage CBs. However, does it really mean there is no problem in the question? The process of relay contact failure in the operating mode is statistical, and the moment of failure depends on the number of accumulated defects and their size. This in turn is determined by concrete parameters of trip coils, frequency of DPR actuation, physical properties of the contact construction, and the dispersion of their parameters during relay assembling. The longer the relay functions, the higher the possibility of its failure and thereby the failure of protection of important power objects. Thus, the matter is not that miniature electromagnetic relays used in protection devices fail right after the first actuation or after a definite number of actuations, but that in the process of operation there is a steady erosion of their reliability and the strong possibility of failure grows.

5.3 Analysis of the International Standards and Technical Specifications

The foregoing discussion brings into question the methods of miniature relays testing in conditions not provided and not authorized officially by the manufacturer, brings into question validity criteria, and the like. Perhaps we will manage to answer these questions and clarify the situation with the help of international standards in this area. What are the standards? Judging by the names, for control relays of CB trip coils, two general standards are suitable: the IEC 60947-5-1 standard (Low-Voltage Switchgear and Controlgear Part 5-1: Control Circuit Devices and Switching Elements: Electromechanical Control Circuit Devices)[7] and the IEEE C37.90 standard (Relays and Relay Systems Associated with Electric Power Apparatus)[8]; see Table 5.5.

The definitions given in both standards are very close; however, the application area of the IEEE C37.90 standard is seen as part of the wider application area of the IEC 60947-5-1 standard. A rather strange limit of the C37.90 standard draws attention: the exclusion of industrial automation relays and other relays not meant especially for power device control. What serious fundamental differences exist between industrial automation relays meant for control of powerful contactors coils, coils of solenoids and valves of control systems of manufacturing processes, and relays meant for trip coils of CBs? They have the same voltage, the same current, and the same capacity! This limitation of C37.90 is not so harmless as it seems and has far-reaching consequences, as on one hand the standard describes the procedure of examination

TABLE 5.5

Scope and Object of the Standards IEC 60947-5-1 and IEEE C37.90

IEC 60947-5-1	IEEE C37.90
Control circuit devices and switching elements intended for controlling, signaling, interlocking, and so on of switchgear and controlgear.	Standard specifically for relay and relay systems used to protect and control power apparatus.
Also applies to specific types of switching elements associated with other devices (whose main circuits are covered by other standards).	It does not cover relays designed primarily for industrial control, for switching communication or other low-level signals, or for any other equipment not intended for control power apparatus.

of relays meant for the activation of trip coils of CBs, and on the other hand it excludes from the consideration relays of industrial purpose not meant specially for power device control. This means that this standard cannot be applied to DPRs in which miniature output relays (originally meant for industrial automation, communication equipment, or other similar equipment, but not for power device control) are used as elements directly controlling the trip coils of CBs. There are also strange differences in the methods of testing relays offered by these standards; see Table 5.6.

TABLE 5.6

Making and Breaking Capacities for the DC Load Test According to IEC 60947-5-1 and IEEE C37.90 Standards

	IEC 60947-5-1	IEEE C37.90	Conformance Grade for Discussed Usage
1	Load type: air-cored inductor in series with a resistor, $L/R < 300$ ms.	Load type: active resistor	—
2	Switching current for utilization category DC-13, designation N300: 1.1 A		
	Any other types of applications shall be based on agreement between the manufacturer and user	Making current: 30 A	—
3	Number of operation: 5000 at 10-second interval	Number of operation: 2000 in sequence: 0.2 s ON, 15 s OFF	—
4	Acceptance criteria: • No electrical or mechanical failures • No contact welding or prolonged arcing • Withstanding the power-frequency test voltage of $2U_{NOM}$, but not less than 1000 V	Acceptance criteria: not specified	—

Why does the C39.70 standard prescribe testing the switching capacity of contacts specially meant for switching CB trip coils (i.e., significant inductive loads) on purely resistive load? Why is the turn-on current strictly specified as 30 A in this standard, while the trip coils of modern breakers are meant for a much lower current? Why are the criteria for relay applicability not discussed during testing? Other questions that need answering are "Why does the switched current in the IEC 60947-5-1 standard not exceed 1.1 A during testing?" and "Why isn't there a separate mode of load make without z breaker (i.e., a typical mode of functioning of trip relay contacts)?"

An analysis of the standards' requirements for the testing of relay insulation and withstanding voltage also evokes bewilderment. For example, in IEC 60947-5-1 the list of relay parts is given with the test voltage attached. It appears, however, that this list contains no relay output contacts! IEC 60255-5[9] considers it possible to test these contacts, but it assumes the necessity of coordinating the test voltage between the manufacturer and customer.

As practice shows, in many cases the customer knows nothing about this article of the standard and does not coordinate anything with the manufacturer apart from the requirements to output relay contacts. This is why the manufacturer may indicate in technical documentation that the parameters of the protection device are in full compliance with requirements of the standard without any additional provisos.

C37.90 treats the testing of output contacts of protective relays completely differently. It supports such testing, but in processes of relay manufacturing it, in fact, does not allow customers to test this most important parameter of relay on their own. Why? Unfortunately, I did not manage to get clear answers to these questions even from the IEEE group responsible for this standard. Moreover, DPR manufacturers actively use these standards, refer to them in their documentation, and conduct their own testing based on them.

A very confused situation has emerged not only in the sphere of standards, but also in the sphere of the type of tests of protective relays, conducted by both relay manufacturers and the International Certifying Centre KEMA. I was permitted to get acquainted with type test protocols of different types of DPRs conducted by KEMA, Siemens, and ABB, and I discovered many strange things there, too. For example, at Siemens the switching capacity of output relays of DPRs is examined using AC (instead of DC!), and at KEMA this type of microprocessor-based protective device (MPD) testing is not conducted at all. At ABB the company tests of these contacts were conducted for protective relays of previous generations (the SPAD, SPAU, and SPAC series, for example) in which large auxiliary relays were used with rather powerful contacts, which were good in the switching of trip coils. For protection devices of the next generation (which we are discussing) in which miniature relays were first used, such tests are not even provided in the list of test types.

Lack of preciseness in international standards results in mistakes in the technical specifications of modern DPRs. I have analyzed many technical specifications of such devices, manufactured by world benchmark companies

TABLE 5.7

The Switching Parameters of the Output Relays Stated in the Specification of Microprocessor Relays Produced by AREVA

Protective Relay Types: Distance Line Protection (MiCOM P443) and Current Differential Protection (MiCOM P541 through P546)	Parameters Specified in Documentation	Our Comments
Standard general purpose contacts		
Rated voltage	300 V	
Continuous current	10 A	
Short-duration current	30 A for 3 s	1. For such making capacity at 300
Making capacity	250 A for 30 ms	V, rated voltage will obtain
Breaking capacity:		making power: 50A * 300V =
DC resistive	50 W	75.000 W (75 kW!). It is very
DC inductive (L/R = 50 ms)	62.5 W	difficult to give credence to such
AC resistive	2500 VA	capability for subminiature
AC inductive (PF = 0.7)	2500 VA	electromagnetic relays.
High break contacts for tripping		
Rated voltage	300 V	2. Higher breaking capacity for DC
Continuous current	10 A DC	inductive load (62.5 W) in
Short-duration current	30 A DC for 3 s	comparison to lower breaking
Making capacity	250 A DC for 30 ms	capacity for DC resistive load
Breaking capacity:		(50 W) contradicts known theory
DC resistive	7500 W	and practices.
DC inductive (L/R = 50 ms)	2500 W	

in the field, on the switching capacity of output relays. These include the 7SD61 and 7SA522 (Siemens); MiCOM P541 and P546 (AREVA); T60, D60, and L90 (General Electric); REL561 and REL670 (ABB); and BEI-GPS100 and BEI-CDS240 (Basler), among others. All of them contained mistakes and inaccuracies, or just lacked the most important parameters to avoid unambiguous conclusions about such relays' applicability. As an example, one may consider the set of parameters given in the relay specification of AREVA; see Table 5.7.[10] I repeatedly addressed AREVA and asked for explanations of these strange parameters. The first time I got different information having no relation to the questions, and later I got no answers at all.

5.4 Improvement of DPR Output Circuits

As is known, the switching capacity and electrical endurance of contacts in electromechanical relays at voltages above 30 V largely depend on the kind

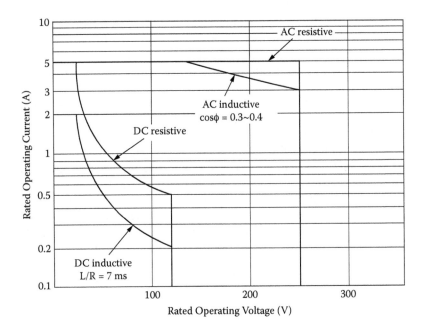

FIGURE 5.7

Typical switching ability for contacts of the miniature electromagnetic relays applied in micro-processor-based protective devices (MPDs) at various kinds of loads.

of switching voltage and the kind of load. It is known that in DC switching at voltages above 30 V, the parameters are much lower than those for AC switching, and for inductive loads these parameters are even lower (see Figure 5.7). And, because of contact bouncing during the make process, the DC switching ability of contacts at a make does not exceed the switching ability at a break.

In the previous section, it has been shown that the parameters of the miniature electromechanical relays used as output elements of DPRs are inappropriate for the actual operating conditions, that is, the switching of an inductive load (trip coils of the CBs or coils of the auxiliary relays) at 220 V DC. The contact systems of these relays simply are not intended for switching loads such as this owing to the fact that the contacts rapidly wear out due to electric erosion, sharply reducing the functioning reliability of the DPRs (see Figure 5.7).

The wide usage of such nonsuitable electromechanical relays in practically all types of DPRs, and also in the microprocessor-based devices providing communication between the protective relays and in other analogous devices which are available in the market, has come about due to the aspiration of manufacturers to reduce the mass and the dimensions of these devices. In a large number of such output relays, sometimes reaching up to several tens, the sizes of the DPRs need to be doubled—even tripled if one is using big output relays with the suitable parameters. It is clear that DPR manufacturers cannot adopt such a step as DPR sizes are one of the important advertising criteria in the struggle for the potential buyer. But who pays for this? The DPR user pays. The question arises, "Is it possible to solve the problem?" Yes, I answer, and I offer alternatives for solving the problem.

The general solution offered here consists of rejecting the installation of output electromechanical relays inside the DPR. Such relays should not be part of DPRs at all, and should be chosen and received by a user (or be delivered complete with the DPRs) separately depending on the concrete conditions. These conditions can vary considerably, placing different demands on the output relays, that is, which contacts are to be used in the circuits. For example, for switching low-voltage low-current signals (so-called dry circuits) in electronic circuits, it is necessary to use a relay with bifurcated gold or gilt contacts, whereas for switching current of a few amperes at 220 V AC, silver-based alloy contacts are necessary. For switching inductive loads at 220 V DC, in general, it was observed that special relays are necessary; practical implementation for them can be effected in two principal ways:

1. Installing in the DPR a special solid-state element intended for switching coils of external electromechanical relays
2. Making each output channel in the form of a voltage source for power, sufficient for operating external electromagnetic relays or semiconductor relays

We shall examine each of these ways in the course of this section.

Today in the market of semiconductor devices, there is a wide assortment of miniature high-voltage metal-oxide semiconductor field effect transistors (MOSFETs) and insulated gate bipolar transistors (IGBTs) for voltages of 1000–1500 V and currents of 5–50 A, and also matching drivers with optical decoupling for them. For instance, it is possible to specify transistors of the following types: IXYS05N100, STFV4N150, STP4N150, STP5NB100, STP5NK100Z, and so on associated with the TO-220 case type; and also transistors STPW4N150, STW11NK100Z, APT15GN120BDQ1, APT35GN120B, IRG4PH20K, IXDH20N120, IXGH25N160, and so on associated with the TO-247 case type.[11]

As is known, some problems may arise when using IGBT transistors represented by the correct organization of their control circuits; however, today these problems are successfully solved and there are numerous drivers for controls of IGBT transistors on the market which are made as small modules (see Figure 5.8). In such drivers, all necessary elements are contained inside for reliable switch-ON and switch-OFF IGBT transistors. Single modules of this type and two IGBT transistors form analogs of high-quality changeover contacts, and are galvanically isolated from the internal control circuits of the DPR.

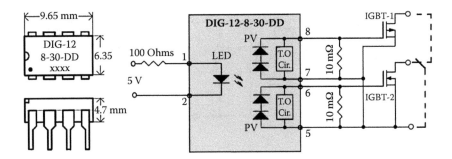

FIGURE 5.8
Modern galvanic isolated driver with a pair of insulated gate bipolar transistors (IGBTs), formatives single changeover contact.

Installed on the DPR printed circuit board (PCB) without a heatsink and supplied with the overvoltage protection element (varistors of sufficient power), these transistors are capable of making and breaking coils of the external auxiliary relays and small contactors directly with currents of up to 1 A at the rated voltage of 220 V DC. Recently, the well-known firm CP Clare has mastered mass production of high-voltage MOSFETs in the ISOPLUS264 case type with built-in drivers having optical decoupling (see Figure 5.9). This device can be controlled directly by a current of 10 mA from the internal electronic circuits of the DPR, and can carry out switching the coils of external electromagnetic relays with currents of up to 1 A at the rated voltage of 220 V DC. In this way, DPR developers have pretty much exhausted the first method, mentioned above, for solving the problem.

We now turn our attention to the second way for solving the problem, and we will see that there are many more possible alternatives. First, there are some alternatives in using the CPC1788 type component presented above as a structural element of the external module, for example for switching-ON

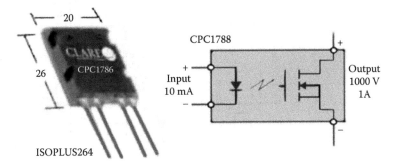

FIGURE 5.9
High-voltage metal-oxide semiconductor field effect transistor (MOSFET) CPC1788 type in an ISOPLUS264 case with a built-in optical decoupling driver made by CP Clare.

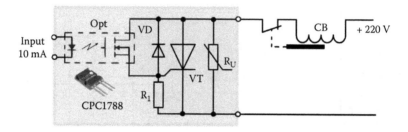

FIGURE 5.10

The circuit diagram of the external module on the basis of the CPC1788 element, connected to the digital protective relay (DPR) output for switching the trip coil of the CB with a mains supply of 220 V DC. VT: thyristor; and R_U: varistor.

the trip coil of the CB (see Figure 5.10) or switching the auxiliary electromagnetic relay (see Figure 5.12).

CPC1788 has considerable sensitivity that allows connecting directly in the circuit of the existing DPR stage without any changes of operating modes, instead of the coils of miniature electromechanical relays used today. When the single-output DPR contact must switch-ON a group of trip coils belonging to different CBs, it is possible to use a power demultiplexor on thyristors (see Figure 5.11) connected to the output of the above-mentioned amplifier.

In the market, there are also other types of solid-state relays suitable for use at 220 V DC, for example the SSC1000-25 with a maximum voltage of 1000 V and a maximum current of 25 A (see Figure 5.13). There are, however, a couple of characteristics of this relay that must be taken into account. First, it has a minimum switching current (20 mA), and, second, it has a minimum control voltage (12 V). The first characteristic means that the device can only be used in applications for loads that are powerful enough, such as trip coils

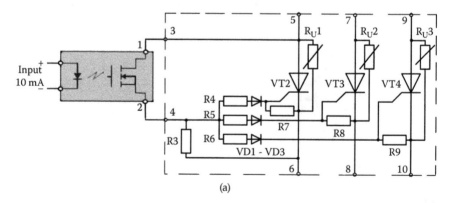

(a)

FIGURE 5.11

(a) Power demultiplexor on thyristors for switch-ON of the group of trip coils. (b) Circuit diagram for external connection of a power demultiplexor (DU). CB1, CB2, and CB3: three-phase circuit breakers.

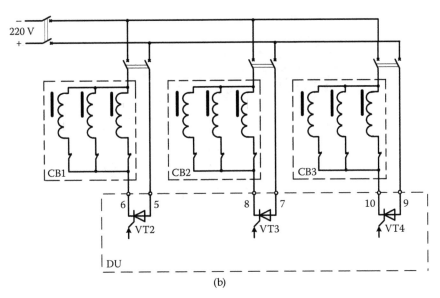

(b)

FIGURE 5.11 (continued).

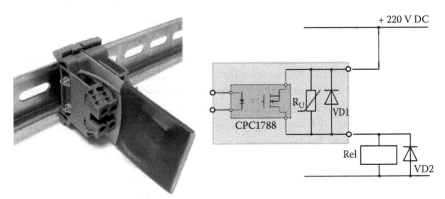

FIGURE 5.12
The external view and circuit diagram of the external module of an elementary construction based on the CPC1788 element connected to the DPR output for switching auxiliary relays.

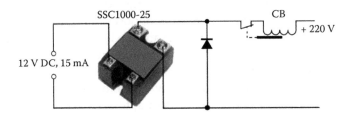

FIGURE 5.13
The connection diagram of the powerful solid-state relay SSC1000-25 type for switching the CB trip coil.

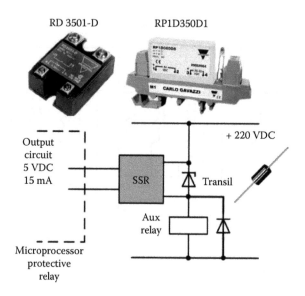

FIGURE 5.14
External output elements of the DPR based on low-power solid-state relays (SSRs) for switching coils of auxiliary electromechanical relays at 220 V DC. Transil®: the semiconductor element intended for overvoltage protection of SSRs.

of the CBs, powerful contactors, and so on. The second characteristic engenders changes that must be made in the internal electronic circuit DPR, since ordinarily the coils of the built-in output relays are operated by voltages of 5 V. In order to use SSC1000-25 as an external output relay, the development engineers of a new DPR have to provide an increase in the level of the operating voltage, up to 12 V.

Less powerful solid-state switching devices intended for a current no more than 1.0–1.5 A can be used for controlling low-power auxiliary electromechanical relays widely used in relay protective systems (see Figure 5.14). These solid-state relays have no lower limits of switching currents and can be operated by voltages of 5 V; this frees the development engineers from the necessity of changing the internal output circuits of the DPR. On the other hand, if a new DPR is developed to increase the power of a control signal up to values of 12 V and 100 mA, there is an opportunity of using the external powerful electromagnetic relays without any auxiliary solid-state elements.

It is necessary to note that many loads used on installations of the electric power industry in a 220 V DC network have an inductive character. This imposes special requirements on auxiliary electromechanical relays, the contacts of which should be capable of switching to such loads. The analysis of specifications of the widest used types of electromagnetic relays shows that the majority of them are not intended for switching inductive loads at 220 V DC.

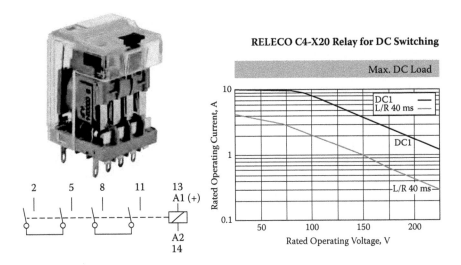

FIGURE 5.15
The C4-X20 (RELECO) relay type (with the cover partially removed) with two double make contacts and its switching ability on a DC.

For this purpose, relays with special designs providing a double make of the load are used (Figure 5.15) or those containing a permanent magnet (blowout magnet) near the contacts intended for repulsing an electric arc from the contact gap (see Figure 5.16).

On the market, there are also smaller sized multicontact relays with blowout magnets, even intended for mounting on the PCB (see Figure 5.17). Relays with triple make contacts are available as well (see Figure 5.18), enabling one to operate a sufficiently powerful inductive load at 220 V DC, such as trip coils of CBs, especially the old types. There are also good prospects in the use of reed switch–based relays, especially with the new small-sized, high-power reed switches made by Yaskawa under the BESTACT® trade mark (see Figure 5.19). Reed switches of this type have double make contacts (basic and arc quenching), with sequence switching allowing the making and breaking of high inductive loads (L/R = 100 ms) with a current of 0.2 A and a voltage of 220 V DC. Control coils of relays based on this reed switch have parameters analogous to those observed above for electromechanical relays.

When implementing the modernization of DPRs intended for external relay connections based on changing their output channels in the form of 5 V or 12 V sources, as detailed above, it is necessary to take into account the necessity of protecting the MPD internal circuits against overloads and faults due to incorrect connections of these outputs. One way of constructing such outputs is presented in Figure 5.20.

In our opinion, the ideal solution is using built-in powerful high-voltage optocouplers (like CPC1788) in the new MPD types, instead of the internal

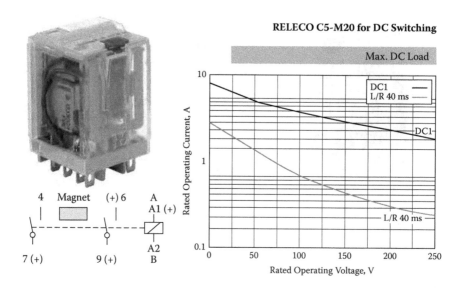

FIGURE 5.16

The C5-M20 (RELECO) relay type with two make contacts and a blowout magnet, and the switching ability of its contacts for an inductive load.

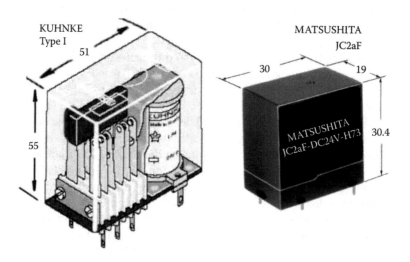

FIGURE 5.17

Small-sized relays with blowout magnets.

miniature electromechanical relays used today, for control of any type of external auxiliary relays chosen by a user. Another solution is delivery by the manufacturer of a separate module with the output relays of various types (upon request of a user) together with DPRs provided with a special connector for coupling with the DPRs. Destination and redestination (address

FIGURE 5.18
The RMEA-FT-1 (RELEQUICK S. A.) relay type with one triple make contact, capable of switching currents up to 3 A in an inductive load at 220 V DC.

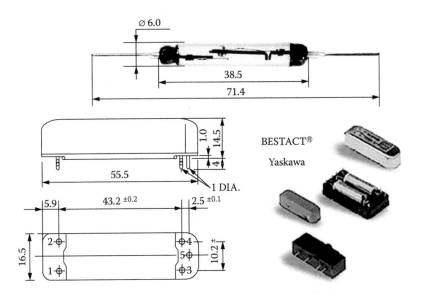

FIGURE 5.19
Powerful R14U (R15U) reed switch type with double make contact and the relays as its basis, manufactured by Yaskawa.

modification) of these or other DPR outputs for using these or other kinds of output relays can be effected through software.

The designs observed above are intended for use in newly developed DPRs. The question arises, "What should users do with the thousands of devices already in use and those continuing to be released to the market?" It is obvious that modification of such devices can be carried out only with external modules, without making any changes in the internal DPR circuits.

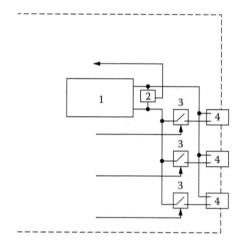

FIGURE 5.20
Alternative construction of DPR outputs made in the form of sources of an operating voltage. 1: the separate power supply of 5 V or 12 V with current limiting and protection against short circuits; 2: the voltage transducer for monitoring the serviceability of the power supply; 3: low-voltage optocouplers; and 4: output connectors for connecting external relays.

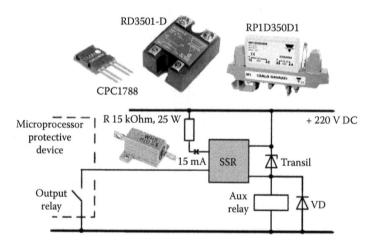

FIGURE 5.21
Diagram of the connection of the external solid-state relay (SSR) module to a DPR output for switching an inductive load (a winding of the auxiliary relay) at 220 V DC.

One possibility of such modules is shown in Figure 5.21. Any of the components discussed above can be used in it. As for control circuits of solid-state modules at voltages as low as 12–36 V, the resistor R (in the circuit diagram) is required for dissipation of the main part of the network voltage 220 V and must be included.

Another solution is that the arc protective modules of the passive type, connected in parallel to contacts of the relay, can be used, for example a

FIGURE 5.22
Passive arc-protective module containing series-connected RC elements (produced by RIFA).

resistor-capacitor (RC) circuit of self-made or industrial types (Figure 5.22), manufactured by many companies. These are specifically for switching the inductive load by output contacts of existing DPRs which are currently available on the market.

More effective protection of DPR output relay contacts against an electric arc is provided by protective modules of the active type, containing semiconductor elements such as transistors (Figure 5.23). Naturally, modules of this type are much more complex and expensive than modules of the passive type. Even a more simplified version of such a module (U.S. Patent 5,703,743) contains two transistors (IGBT and FET types), one triac, three diodes, and three Zener diodes. A more sophisticated updating (U.S. Patent 6,956,725) consists of the current transformer, a rectifier bridge, and some capacitors and resistors in addition to the above-listed elements. Such modules are sold on the open market by Schweitzer Engineering Laboratories and can be successfully used by any consumer-used DPR. The choice of type of protective module depends on the concrete parameters of the switching load. At "light" loads, with the time constant not exceeding 7–10 ms, elementary

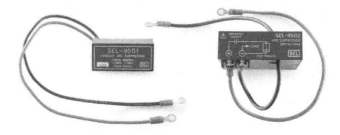

FIGURE 5.23
Smart (active-type) arc-quenching modules SEL-9501 and SEL-9502 types (manufactured by Schweitzer Engineering Laboratories).

FIGURE 5.24
ST-REL7-HG220/4X21 (Phoenix Contact) auxiliary relay type with the additional electronic components installed on the PCB inside of the relay case.

RC modules can be used, and for heavy loads with L/R = 30–50 ms, active-type modules are more suitable.

In the case where, as the load, the coil of small low-power auxiliary relay is used, this can be simply replaced by a relay of the same type, but with coils with a voltage of 12–48 V connected in series with damping resistors of matching power. In addition, the coil of the relay should be shunted with a diode, connected in the opposite direction (as in Figure 5.20). Use of the coil of the auxiliary relay on smaller voltages (i.e., consisting of a considerably smaller number of coils connected in series with the resistor) essentially decreases the time constant of the coil $\tau = L/R$ and makes an easy switching process. As a matter of fact, the same principle is also used in some types of serial production relays, for example in the small auxiliary relay of ST-REL7-HG220/4X21 type (with the contacts that are not intended for switching an inductive load at 220 V DC), manufactured by Phoenix Contact. The relay contains the small PCB with installed diodes, resistors, and a light-emitting diode (LED) inside of its case (see Figure 5.24). The input circuit of the relay has a very high resistance and can be switched absolutely freely by contacts of output miniature relays of any type at a voltage of 220 V DC.

5.5 Conclusions

1. Currently, there are no unambiguous proofs that contacts of miniature electromagnetic relays widely used in MPD devices function in the modes acceptable for them and ensure necessary reliability when switching trip coils of CBs. Therefore, it is necessary to relate

to this as a real problem and to take measures for the prevention of failures in the output relays.

2. Manufacturers of miniature electromagnetic relays used in MPDs should include the following parameter in technical specifications of their relays: make without break of inductive loads at 125 V DC and 250 V DC in the infrequent switching mode. International standards dealing with the switching capacity of relay contacts should be amended with a parameter specifying the following: make without break of inductive loads, which is in compliance with real parameters of trip coils of CBs or powerful auxiliary relays. These standards should be mutually coordinated.

3. I offer for consideration the possibility of replacing the $\tau = L/R$ parameter with the parameter of characterizing switching energy E for different rated voltages. One should create a standard (or add a separate paragraph to existing standards) for typical symbols of the most important parameters of relay contacts' switching capacity, which should be obligatory for inclusion in technical documentation, and examples of such parameters should be recorded in the technical specification.

4. Requirements for the testing of relay contacts specially meant for energizing the trip coil of CBs (IEEE St. C37-90) should be brought into conformity with real service conditions. Contacts applicability criteria in the testing process should include analysis of their condition to ensure the necessary reliability of switching.

5. Manufacturers of MPDs should revise technical specifications concerning the parameters of output relays and bring them into conformity with reality.

6. Consumers of MPDs should more carefully analyze specifications of the equipment bought and demand from manufacturers test record sheets of compliance with standards requirements.

References

1. Gurevich, V. *Electric Relays: Principles and Applications*. London: CRC Press, 2005.
2. Gurevich, V. Nonconformance in Electromechanical Output Relays of Microprocessor-Based Protection Devices under Actual Operation Conditions. *Electrical Engineering and Electromechanics*, vol. 1, 2006, pp. 12–16.
3. Gurevich, V. I. How to Equip a Relay Protection: Opinions of Russian Experts and a View from Outside. *Electric Power News*, no. 2, 2007, pp. 52–59.
4. HEA Multicontact Auxiliary. Fairfield, CT: GE Industrial Multilin, n.d.

5. Tayev, I. S. *Electrical Appararus: Basic Theory* [in Russian]. Moscow: Energiya, 1977.
6. International Electrotechnical Commission (IEC). International Standard IEC 61810-7: Electromechanical Elementary Relays—Part 7: Test and Measurement Procedures. Geneva: IEC, 2006.
7. International Electrotechnical Commission (IEC). International Standard IEC 60947-5-1: Low-Voltage Switchgear and Controlgear Part 5-1: Control Circuit Devices and Switching Elements: Electromechanical Control Circuit Devices. Geneva: IEC, 2003.
8. ANSI/IEEE C37.90-1989. IEEE Standard for Relays and Relay Systems Associated with Electric Power Apparatus. New York: *IEEE*, 1989.
9. International Electrotechnical Commission (IEC). International Standard IEC 60255-5: Electrical Relays—Part 5: Insulation Coordination for Measuring Relays and Protection Equipment: Requirements and Tests. Geneva: IEC, 1999.
10. AREVA. MiCOM P543/4/5/6 Current Differential Relay: Technical Data Sheet P54x/EN TDS/A22. Paris: AREVA.
11. Gurevich, V. *Electronic Devices on Discrete Components for Industrial and Power Engineering*. Boca Raton, FL: Taylor & Francis Group, 2007.

6

Problems with External Power Supplies

6.1 Electromagnetic Disturbances in the Power Network

Many cases of malfunctions and even damage of microprocessors are caused as a result of the impact of electromagnetic disturbances (e.g., blackouts, noise, sags, spikes, and surges) from the power supply network in the operation of digital protective relays (DPRs).

6.1.1 Blackout

A blackout results in total loss of utility power.

- *Cause*: Blackouts are caused by excessive demand on the power network, lightning storms, ice on power lines, car accidents, construction equipment, earthquakes, and other catastrophes.
- *Effect*: Current work in random access memory (RAM) or the cache is lost. Total loss of data stored on read-only memory (ROM).

6.1.2 Noise

More technically referred to as *electromagnetic interference* (EMI) and *radio-frequency interference* (RFI), electrical noise disrupts the smooth sine wave one expects from utility power.

- *Cause*: Electrical noise is caused by many factors and phenomena, including lightning, load switching, generators, radio transmitters, and industrial equipment. It may be intermittent or chronic.
- *Effect*: Noise introduces malfunctions and errors into executable programs and data files.

6.1.3 Sag

Also known as *brownouts,* sags are short-term decreases in voltage levels. This is the most common power problem, accounting for 87% of all power disturbances according to a study by Bell Labs.

- *Cause*: Sags are usually caused by the startup power demands of many electrical devices (including motors, compressors, elevators, and shop tools). Electric companies use sags to cope with extraordinary power demands. In a procedure known as *rolling brownouts,* the utility company will systematically lower voltage levels in certain areas for hours or days at a time. Hot summer days, when air-conditioning requirements are at their peak, will often prompt rolling brownouts.

- *Effect*: A sag can starve a microprocessor of the power it needs to function and can cause frozen keyboards and unexpected system crashes, which result in both lost and corrupted data. Sags also reduce the efficiency and life span of electrical equipment.

6.1.4 Spike

Also referred to as an impulse, a spike is an instantaneous, dramatic increase in voltage. A spike can enter electronic equipment through AC, network, serial, or communication lines and damage or destroy components.

- *Cause*: Spikes are typically caused by a nearby lightning strike. Spikes can also occur when utility power comes back online after having been knocked out in a storm or as the result of a car accident.

- *Effect*: Catastrophic damage to hardware occurs. Data will be lost.

6.1.5 Surge

A surge is a short-term increase in voltage, typically lasting at least 1/120 of a second.

- *Cause*: Surges result from the presence of high-powered electrical motors, such as air conditioners. When this equipment is switched off, the extra voltage is dissipated through the power line.

- *Effect*: Microprocessors and similar sensitive electronic devices are designed to receive power within a certain voltage range. Anything outside of expected peak and root mean square (RMS; considered the average voltage) levels will stress delicate components and cause premature failure.

Malfunctions of and damage to DPRs owing to electromagnetic disturbances are described in the literature. For example, mass malfunctions of microprocessor-based time relays occurred in nuclear power plants in the United States.[1] A review of these events indicated that the DPRs failed as a result of voltage spikes that were generated by the auxiliary relay coil controlled by the DPR. The voltage spikes, also referred to as *inductive kicks*, were generated when the DPR contacts interrupted the current to the auxiliary relay coil. These spikes then arced across the contacts of the output relay of the DPR. This arcing, in conjunction with the inductance and wiring capacitance, generated fast electrical noise transients called *arc showering* (electromagnetic interference). The peak voltage noise transient changed as a function of the breakdown voltage of the contact gap, which changed as the contacts moved apart and/or bounced. These noise transients caused the microprocessor in the DPR to fail.

6.2 Applying Uninterrupted Power Supply

The organization of the supply system of relay protection is very important. Power units are supplied by powerful accumulator batteries with a constantly connected charger, or by an *uninterrupted power supply* (UPS) cushioning the negative impact of the factors listed above; however, the same system supplies driving gears of power switches and many other devices, causing spikes. Besides, investigations of UPS systems[2] have shown that at certain conditions, noise spikes and high harmonics can get into microprocessors through grounded circuits, and neither a UPS nor filters can prevent this. In addition, UPS devices have their own changeover times. Usually specifications for a UPS indicate a switch delay time of 3–5 ms, but in fact under certain conditions this time may increase by a factor of more than 10. In a normal mode, the load in some types of UPS devices (offline types) is usually fed through a thyristor switch (bypass), which must become enabled when voltage diverts below 190 V or above 240 V. After that, the load is switched to the output of the inverter supplied from the accumulator battery.

The time of disabling of the thyristor switch is summed up from the time of decrease of current flowing through the thyristor to the zero value t_0 and turn-OFF time t_q, after which the thyristor is capable of withstanding the voltage in the closed (OFF) position. For different types of thyristors $t_q = 30$–500 microseconds, and t_0 depends on the proportion of induction and pure resistance of the circuit of the current flowing through the disabled thyristor (if the circuits breaks). As a result of switching OFF of the input circuit breaker or pulling out of the UPS supply cable from the terminal block, $t_0 = 0$–5 milliseconds, which will provide very high performance; however, at actual interruption of supply, a break usually occurs in circuits of a higher voltage level, and not in the circuit for 120/220/380/400 V. The thyristor switch is then shorted to the second winding of the network transformer and the load connected to it. If the beginning of the supply interruption coincides with the conductivity interval of the thyristor, the duration of the process of current decaying may exceed 400 milliseconds, and changeover to reserve power supply (inverter) may be delayed for inadmissibly long times. If the beginning of supply interruption coincides with the no-current interval, switching requirements are similar to those at input cable break. It follows that at interruptions of supply and voltage falls-through coinciding with the interval of current flowing, the working source will change over to the reserve supply for inadmissibly long times.[3]

The user who sets up and checks the UPS usually reaches conclusions about its serviceability by switching the input circuit breaker OFF. As shown above, this doesn't always correspond to conditions of actual transient processes (short-circuiting of the input to the induction shunt), which creates the possibility that after installation and successful check of the UPS by switching OFF the input circuit breaker, some voltage falls-through (coinciding with intervals of current flowing) may lead to disabling ("suspension") of the microprocessors. That is why in some types of UPS of the OFF-LINE category, in order to provide high performance during changeover in the short-circuit mode, high-speed electromechanical relays instead of thyristor ones are used.

To determine the actual time of changeover of the UPS, one should make an experiment imitating short circuits of the working source, to inductively activate a shunt while a special measuring system records the transient processes. But who runs such experiments, and where?

Thereupon, another aspect of the problem gains our attention: suspensions and malfunctions of operation of the microprocessor of the UPS in emergency modes on high-voltage circuits. When the control microprocessor malfunctions, alternation of switching-ON and switching-OFF of power semiconductor elements of the inventor may be disturbed and short circuit loop making, followed by automatic switching-OFF of the input circuit breaker of the UPS. This same phenomenon can happen to automatic chargers whose microprocessors are supplied from an external auxiliary UPS. Such incidents quite often occur in practice, but nobody yet has concerned him- or herself with a serious analysis of the reasons. It is quite possible that the reasons for such emergency switching of a UPS, and of the chargers, are similar to those for the case considered in this section.

6.3 Other Problems in the AC Network

According to IEC 61000-4[4,5] voltage sags (sometimes referred to as *dips*) are brief reductions in voltage (below 0.8 U_N), lasting from tens of milliseconds

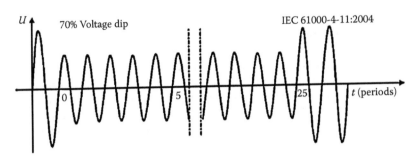

FIGURE 6.1
Example (from IEC 61000-4-11) of the 70% voltage sag during 25 cycles (0.5 sec).

to 15 seconds (see Figure 6.1). As is known, the main reason for voltage sags in the 0.4 kV network of the substation auxiliary supplies is short circuits in external high-voltage grids. In manufacturing plants, such voltage sags are frequently associated with the working modes of the powerful electrical equipment, for example with starting power motors. Voltage sags are an important criterion of the power quality.

6.3.1 Voltage Sags in Manufacturing Plants' 0.4 kV Networks

Sags in the 0.4 kV networks of manufacturing plants may seriously disturb the production cycle due to the mass disconnection of electrical motors through the releasing of the contactors followed by the auto-starting of a number of motors which may cause additional serious decreases in the voltage level and exacerbate the problem.[6,7]

As shown in Reference 8, upon voltage cessation for an electromotor during 0.4–0.8 seconds, the vectors of the residual electromotive force of the motor can occur in antiphase with the vector of the network voltage. As a result, upon network voltage restoration, the high-magnitude current pulse will flow through the motor, causing the tripping of the protective circuit breaker and the disconnection of the motor.

On the other hand, short voltage sags with durations of less than 200–300 ms (most frequent in a 0.4 kV network) do not harm the motors. For these reasons, the means for ameliorating sags in networks in manufacturing plants have included, as usual, some technical solutions for keeping the contactors closed during a sag: special dynamic voltage sag compensators, a UPS, and the like. Because such compensators and high-power UPS are very expensive, different electronic devices[9,10] have been developed that guarantee feeding the contactor's coil from a DC power supply during short sags.

As is known, in the pickup process of an AC contactor there is a considerable variation of the current consumed by its coil, thereby leading to considerable variation in the core's attractive forces needed for contactor pickups. When feeding a contactor's coil from a DC supply, such current variations (and, as a result, also the attractive force variations) will not be present and the contactor does not work properly.

These above-mentioned electronic devices use four levels of DC voltages for feeding the contactor coil, which simulates the natural attraction force characteristics at contactor pickups on AC. These electronic devices with integrated circuits (microchips) are not intended, however, for use with powerful contactors with low resistance coils (10–15 Ohm) and high inrush currents. For example, the power consumed by the coil of the 3NTF54 contactor at pickup is 1.6 kVA on AC and 1.2 kW on DC (with a special starting coil).

For large contactors with powerful coils, special devices have been developed which work on another principle (see Figure 6.2). The device consists of an undervoltage relay KU, a timer K1 for an "Impulse-ON" standard function, and a simple DC power supply consisting of transformer T, rectifier bridge VD2, and low-voltage, high-capacity capacitor C. When control switch S1 is closed, the 230 V AC voltage is applied to the voltage relay KU, which picks up if the applied voltage is more than the minimum allowed value (180 V in our case) and closes its contact in the feeding circuit of the KT timer. The timer instantly picks up and by its contact connects the contactor's coil to the 230 V AC network through rectifier bridge VD1 and limiting resistor R1.

The direct current near 5A will be carried through the contactor's coil. Such current produces an electromagnetic force equivalent to that of the

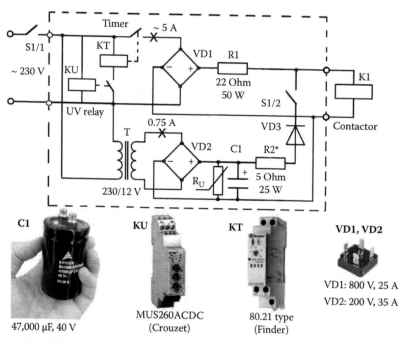

FIGURE 6.2
Circuit diagram of the control device for a powerful AC contactor.

natural starting current leading to the usual connection of the contactor to the 230 V AC network. Simultaneously, capacitor C1 is quickly charged. Due to the presence of diode VD3, capacitor C1 is charged from a low-voltage (12 V) DC power supply only and is completely isolated from the high voltage (near to 70 V) applied to the contactor's coil during the starting process. After 2–3 seconds following the contactor pickups (the time is determined by the timer's internal setting), the timer's contact breaks off a high starting current in the contactor's coil. Thus, diode VD3 will instantly unlock and the low-voltage power supply with the charged capacitor C1 will connect to the contactor's coil. At this moment of time, the contactor's coil is fed by the lowered DC current, limited, in addition, by the low-resistance resistor R2. Selection of the value of this resistor depends on the specific contactor type. For instance, for a 3TF54 type contactor, the resistance of this resistor has to be as low as 5 Ohm. At this resistance, the reliable holding of the contactor in the closed position is provided over a long period of the AC voltage decreasing (up to 140–130 V) and, at the same time, the allowable temperature of heating of the coil (not to exceed 50–60 degrees) is provided.

The research that has been done has shown that when feeding the contactor's coil with the lowered DC current, its sensitivity to decreasing power supply voltage level is sharply reduced. For example, the contactor that was tested was held in the closed position at a voltage reduction on the coil from 12 V up to 2–3 V, that is, in 4–6 times. This positive property is used in this device for holding the contactor for short-term downturns of the voltage level in AC networks. For very deep voltage sags or even full voltage loss, holding the contactor is effected by the energy of capacitor C1. From the results of the tests, it appears that small capacitors with capacities in the range of 47,000 microfarads on voltage 40 V are capable of holding the large contactor (in our case, such as 3TF54) for periods of 1.3–1.5 seconds. This is quite sufficient for short-term voltage sags in real life in the AC networks.

The rectifier bridge VD needs to be selected with considerable reserve in connection with current, because high-current pulses flow through it during capacitor charging

At the fall of the voltage in AC network to a level lower than 160 V, the undervoltage relay KU opens its contact and disconnects the feeding circuit of the timer KT. However, the position of the output contact of the timer does not change, and the capacitor continues feeding the contactor's coil from the low-voltage DC power supply until the restoration of the proper voltage level in the AC network or until the capacitor C1 energy is fully exhausted (which will occur if the voltage sag lasts over a long time interval), at which time the contactor will be disconnected. With the restoration of a voltage in a network up to a level not less than 180 V, the undervoltage relay KU again will pick up and the timer feeding will appear, and the working cycle of the device as described above will repeat.

The Impulse-ON function, which is sometimes called *interval, fleeting, single shot, power ON, single-shot leading edge*, or *rising-edge pulse*, is not something exotic, and it represents the standard function (designated, sometimes, as function number 21). The timers actualizing such functions are widely available in the market. These are, for example, timers of series PBO (Meander, Saint Petersburg, Russia); CT-VWE, CT-WBS, and CT-VWD (ABB); BC7931 (Dold & Soehne); MICV and NMICV (General Electric); KRDI (ABB); 3RP15 (Tyco Electronics); DIL-ET-11-30-A (Moeller); DDT and TZ (Tempatron); 87.21 and 81.01 (Finder); MURc3 (Crouzet); RE7-PR11 (Telemecanique); M1SMT (Broyce Control); 3RP1505 (Siemens); TDRPRO-5100 (Magnecraft); and many others. Unfortunately, only some of them, such as 81.01, 80.01, and 80.21 (Finder); 821 (Magnecraft); 4604 (Artizan); and some others equipped with powerful output contact are suitable for switching coils of large contactors. Using timers of the other types will necessitate using an addition auxiliary relay with powerful contacts inserted in the circuit instead of the timer's contacts.

Voltage relay KU can be used as either an undervoltage relay with an adjustable trip level or hysteresis, which does not require a separate power supply. Such relays as SUA145 (Bender), EUS (EID Electronics), MUS260ACDC (Crouzet), M200-V1U (Multitek), RM4-UB3 (Telemecanique), PVE (Entrelec), UAWA (Thiim A/S), BQP1202 (Midland Jay, a division of Midland Automation, United Kingdom), PKH-1-1-15 (Meander), and others suit these requirements.

The device is assembled in a closed plastic container with the dimensions $210 \times 160 \times 90$ mm. It is abundantly clear that the device can be used with medium-sized contactors as well as with small contactors. In both these cases, the capacitance of the capacitor and the power of the transformer (therefore, its cost and dimensions) will dramatically decrease.

It needs to be noted that some manufacturers (including Siemens, manufacturer of the powerful 3TF series of contactors) provide the possibility for feeding AC contactors from a DC power supply (a DC network with a power battery). In this case, the contactor's condition is fully independent of sags in the AC network. It offers one more way of solving the problem, but, on the other hand, realizing this solution is not simple because of special starting characteristics which need to be simulated as mentioned above. Siemens offers two special windings for 3TF5 series contactors for use: a powerful pickup winding (PW) and a low-power holding winding (HW) (see Figure 6.3).

Changeover from one winding to another after contactor K1 pickups is effected with the help of an additional contactor, K2, connected in series with powerful contacts (for disconnection of the high inductive load at 230 V DC) and an additional auxiliary contact block (95, 96) on the main contactor.

With the presence of a powerful battery and the possibility of setting the DC voltage to the contactor's location, the problem can be solved by means of a more intelligent method. As has been already mentioned, timer 81.01 (Finder) type and a small switching type supply power for output voltage 12 V and current 1.2 A (see Figure 6.4). Only two not very expensive off-the-shelf devices are needed for realizing the solution. The cost of these set elements is only about £80.

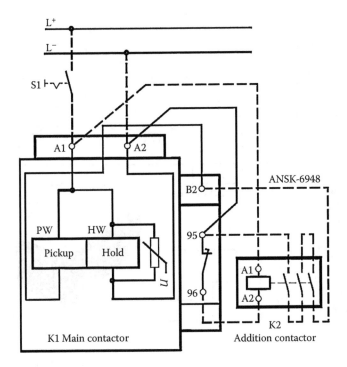

FIGURE 6.3
Solution offered by Siemens.

6.3.2 Voltage Sags in 0.4 kV Auxiliary AC Network

The peculiarity of the low-voltage auxiliary AC network in power substations is that it does not contain devices that allow for short pauses in the power supply, and almost all of the critical power consumers (relay protection, emergency modes recorders, communication system, and signaling and remote control) are fed, as usual, from a power substation battery. At the same time, the power electronic systems with microprocessor controllers such as invertors, battery chargers, and power supplies are fed from the auxiliary AC network. Practical experience has shown that such devices do not "love" short voltage interruptions (50–200 ms) with subsequent voltage restoration. Sometimes such devices have time to hang through an automatic change-over from the main to spare power supply (transformer). Another problem of the power battery chargers with a powerful input transformer is the high inrush current at sudden interruption and subsequent input voltage restoration that causes full charger disconnection by the electromagnetic releaser of the input circuit breaker. This state of affairs is considerably aggravated in some cases when even single voltage sags with durations of 100–200 ms provoke multiple pickups and releases of the electromagnetic contactor during the sags.

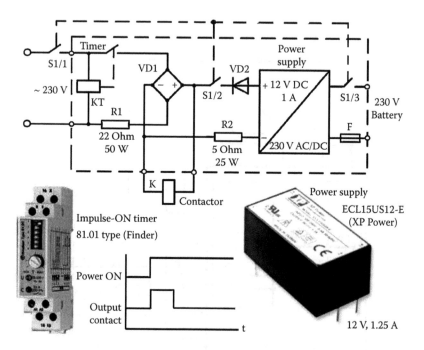

FIGURE 6.4
Device for feeding a large AC contactor from a DC network with a powerful battery.

6.3.3 Problematic Action of the Powerful Contactors as a Changeover from the Main to Reserved Auxiliary AC Power Supply on Substations

For increasing the reliability of an AC network 0.4 kV in power substations, ordinarily two auxiliary transformers, feedings from different lines, are used. One of them connects to the 0.4 kV network permanently, and the other one automatically connects at voltage disappearance on the first transformer. Connecting and disconnecting AC network 0.4 kV to these transformers are affected by means of two powerful electromagnetic contactors on currents 200–400 A with AC control coils. These contactors are the major elements of the auxiliary network on which, in many respects, reliable work of all power substations depends.

As an object of research, the electromagnetic contactor 3TF54 type (Siemens) with switching capability 300 A has been taken (see Figure 6.5), which is used at changeover in the auxiliary AC network 0.4 kV in power substations. During the research, the oscillograms of pickups and releases of the contactor have been recorded at the feeding of the control coil from the AC supply (Figures 6.5 and 6.6). The oscillogram in Figure 6.5 shows the presence of a high starting (inrush) current caused by small impedance of the coil until the moment of the closing of the contactor's magnetic circuit. Oscillograms

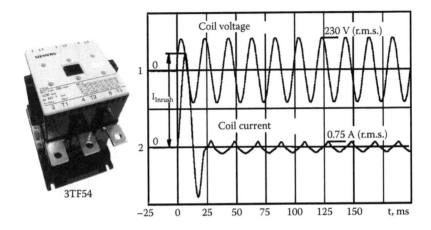

FIGURE 6.5
Electromagnetic AC contactor 3TF54 type (Siemens) and oscillograms for current and voltage on its coil at switch-ON.

shown in Figure 6.6 allow determining the time pickups and time release of the contactor (i.e., the reaction time of the contactor for voltage sags).

Analysis of the oscillograms has shown that full switch-ON time (i.e., the time interval between the moment that the voltage is applied to coil and the moment of main contacts closings) is about 20 ms. And the full switch-OFF time (i.e., the time interval between the disconnection of voltage on the coil and the moment of main contacts openings) is 15–18 ms.

According to the data sheet, 10–30 ms for nominal voltage is applied before disconnection, and 10–15 ms for voltage 0.8 of the nominal value. Such small time delays for such large contactors mean that during typical voltage disturbances with alternate voltage level sags and restorations, the contactor will have time to connect and disconnect the main power circuit several times. Moreover, as shown in Reference 11, the reaction of the contactor during a 75% voltage sag is more serious than 100%, since the releasing time for the first case is shorter on 40–50% than in the second case and may be 10 ms even for large apparatus.

One more problem of the power AC contactor was discovered during the research stage. It was found that reducing voltage across the coil up to 150–135 V causes high vibration of the contactor magnetic system with magnitude that is sufficient for closing and opening its main contacts. The same phenomenon arises when the AC voltage across the coil increases from 0 up to 160–185 V. The possibility of working in such a mode with such speed means that even at a single voltage sag (to level 135–150 V during 100–200 ms; see Figure 6.7), the power generator transforms multiple interruptions of the main voltage in a substation auxiliary AC network. The same result appears when trying to connect the contactor's coil to a power supply with a voltage of 150–170 V.

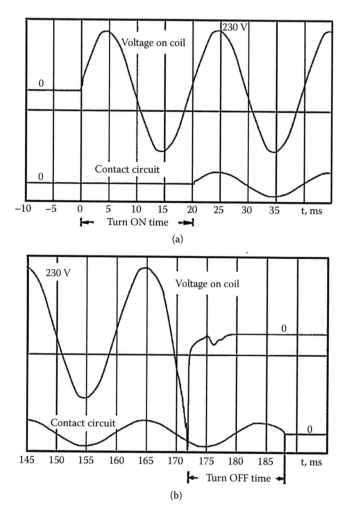

FIGURE 6.6
Oscillograms of contactor, type 3TF54, switching-ON and switching-OFF.

6.3.4 Offered Solution for the Problem

In view of character of the loads fed from the auxiliary AC network in substations (power electronic equipment sensitive to short sags), the technical solution offered for contactors intended for use in manufacturing plants (retention of the contactor in a closed position at short sags) is not the correct technical solution for the substation 0.4 kV network, because, through closed contactor's contacts, the short sags will affect sensitive equipment and provoke its disturbances.

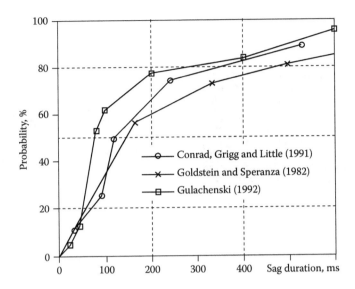

FIGURE 6.7
Sag durations in a power AC network according to some research.

In our opinion the problem must be solved not by means of retention of the contactor in the closed position at voltage sags, but by means of rapidly (during 10–12 ms) disconnecting the contactor's coil at voltage levels dropping below 160 V and returning it to its initial condition at voltage level restoration up to 185 V with a time delay of 5–10 s. A single interruption of 5–10 s duration in the auxiliary 0.4 kV AC substation power network does not cause serious disturbances in the substation equipment due to a power battery feeding most of the important substation consumers. At the same time, such an algorithm of the contactor's work may prevent serious disturbances and failures in AC power electronic equipment.

For fast contactor disconnection at voltage network drops, most electronic relays available in the market are not suitable because their minimal reaction time is, usually, 100 ms. During a time interval of this duration, the contactor will make several disconnections and connections of the main power.

As a result of our research, only a few types of devices suitable for contactor's control were found (see Figure 6.8). One of them is an undervoltage relay combined with a timer: the Brown-Out Timer GBP2150 type, manufactured by Midland Jay. The reaction time of this device for voltage drops of 30% is only 5 ms. The release time after voltage restoration up to 80% can be adjusted to intervals of 1–10 seconds. Another good example is the Russian undervoltage relay PKH-1-3-15 type, manufactured by ZAO Meander. Such devices are ideal solutions for our purposes. For decreasing loads on the output contacts in these devices, an additional auxiliary fast electromagnetic

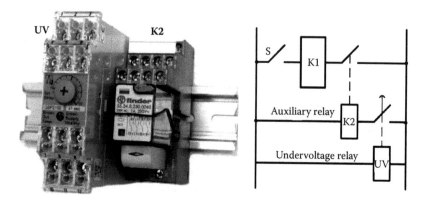

FIGURE 6.8
Device for fast forceful disconnection of the main contactor at voltage sag.

relay, type 58.32.8.230 (Finder), with powerful output contacts and a releasing time of 3 ms, is employed.

6.3.5 Conclusion

For manufacturing plants with electrical motors as dominant consumers and for power substations with power electronic equipment as dominant consumers in a 0.4 kV AC network, different methods must be used in contending with voltage sags. In the first case, the devices with a capacitor or compact switching power supply for detention, the main contactor in closed position during short voltage sags, as described above, can be used. In the second case, it is recommended using a simple device for fast forceful disconnection of the main contactor at voltage sags below 25–30%.

6.4 Problems in DC Networks

6.4.1 Problems of Power Supplies of DPRs in Emergency Mode

As is known, both auxiliary AC and DC voltages are used at power substations. Use of DC auxiliary voltage increases the essential reliability of relay protection due to use of a powerful battery, capable of supporting the required voltage level on the crucial elements of the substation at emergency mode with the AC power network disconnected. However, this increase of reliability comes at the cost of an essential rise in price of the substation and its maintenance. On the other hand, electromechanical relays of all types do not demand an external auxiliary power supply for proper operation, as their operation requires input signals only. There may be some problem when it is necessary to energize the trip coil of the high-voltage circuit breaker at loss of auxiliary voltage in the emergency mode, but this problem has been solved for a long time and simply enough through the use of a storage capacitor. It is constantly charged at the normal operating mode from the AC auxiliary power supply through a rectifier and provides a power current pulse to the trip coil on operation of the protective relay in the emergency mode.

A modern capacitor trip unit contains, in addition, little nickel-cadmium cells and a low-power solid-state inverter for an output voltage of 250 V, through which the main capacitor is constantly recharged from a battery while the auxiliary voltage is disconnected.

The power capacity of the inverter makes mill watts, which are spent only for compensation of self-discharging of the capacitor. Such compact devices (Figure 6.9) are issued by many companies and allow keeping the capacitor charged for several days. Clearly, in such conditions, sufficient reliability of relay protection, even on an operative alternating current, is provided. For this reason, the operative alternating current is applied very widely.

The situation began to change with the introduction of microprocessor-based relays and the mass replacement of electromechanical relays by them. To the many problems caused by this transition,[12–14] one more problem was added. As is known, the internal switching-mode power supply, admitting use as auxiliary AC and DC voltages, has an overwhelming majority of DPRs. Therefore, at first sight, there should be no reasons to interfere with the use

FIGURE 6.9
One of the modern capacitor trip units providing accumulation and long storage of energy for a feed of trip coil of a circuit breaker in the absence of an auxiliary voltage.

of an auxiliary AC voltage on substations with DPRs. The problem arises when there is not enough power for the normal operation of an overwhelming majority of DPRs and only the presence of corresponding input signals (as for electromechanical relays) and also requires a feed from an auxiliary supply. How will the DPR behave at loss of this feed at failure mode when the hard work of the microprocessor and other internal elements is required? How will the complex relay protection (containing some DPRs, incorporated in the common system by means of the network communication when there are also losses of auxiliary feed) function? How will the DPR behave during voltage sags (brief reductions in voltage, typically lasting from a cycle to a second or so, or tens of milliseconds to hundreds of milliseconds) during failure? We shall try to understand these questions.

The internal switching-mode power supply of the DPR contains, as a rule, a smoothing capacitor of rather large capacity, capable of supporting the function of the relay during a short time period. According to research which has been led by General Electric[15] for various types of DPRs, this time interval takes 30–100 ms. In view of that time of reaction, the DPR for emergency operation lays in the same interval and depends on that type of emergency mode; it is impossible to tell definitely whether protection will have sufficient time to work properly. At any rate, it is not possible to guarantee its reliable work. It is a specially problematic functioning of protection relays with the time delay, for example the distance protection with several zones (steps of time delay, reaching up to 0.5–1.0 s and more). Also, it is possible to only guess what will take place with the differential protection containing two remote complete sets of the relay, at a loss of a feed to only one of them.

Voltage sags are the most common power disturbance. At a typical industrial site, it is not unusual to see several sags per year at the service entrance, and far more at equipment terminals. These voltage sags can have many causes, among which may be peaks of magnetization currents, most often at inclusion of power transformers. Recessions and the rises of voltage arising sometimes at failures and in transient modes are especially dangerous when coming successively with small intervals of time. The level and duration of sags depend on a number of external factors, such as the capacity of the transformer, impedance of a power line, remoteness of the relay from the substation transformer, size of a cable through which feed circuits are executed, and so on. DPRs also have a wide interval of characteristics on allowable voltage reductions. As mentioned in Reference 14, various types of DPRs keep the working capability at auxiliary voltage reduction from the rated value of up to 70–180 V. Thus, DPRs with a rated voltage of 240 V suppose a greater (in percentage terms) voltage reduction than devices with a rated voltage of 120 V. It is also known that any microprocessor device demands a long time from the moment of applying a feed (auxiliary voltage) to full activation at normal mode. For a modern DPR with a built-in system of self-checking, this time can reach up to 30 seconds. It means that even after a short-term failure with auxiliary voltage (voltage sag) and subsequent restoration of voltage level, relay protection still will not function for a long time.

What is the solution to the problem offered by the experts[14] from General Electric? Fairly marking that existing capacitor trip devices obviously are not sufficient to feed DPRs, as reserved energy in them has enough only for the creation of a short-duration pulse of a current and absolutely not enough to feed DPRs, I have come to the conclusion that it is necessary to use a UPS for feeding the DPR in an emergency mode. My second recommendation is to add an additional blocking element (a timer, e.g., or the internal logic of a DPR), which will prevent closing of the circuit breaker before the DPR becomes completely activated. Both recommendations are quite legitimate. Here, only usage of UPS with a built-in battery is well known as a solution for maintenance of a feed of crucial consumers in an emergency mode. This solution has obvious foibles and restrictions (both economic and technical). Use of blocking for switching-ON of the circuit breakers can be a very useful idea which should be undoubtedly used; however, it does not always solve the problem as failures of voltage feeding connected to the operation of the circuit breaker are always a possibility.

In our opinion, a more simple and reliable solution of the problem is use of a special capacitor with a large capacity connected in parallel to the feed circuit of every DPR instead of UPS usage. High-quality capacitors with large capacities and rated voltage of 450–500 V are sold today by many companies for approximately €150–200 and are not deficient; see Table 6.1.

Elementary calculations show that when charged up to 250 V, one 5000 μF capacitor is capable of feeding a load with consumption power 30–70 VA up

TABLE 6.1

Parameters of Capacitors with Large Capacity and Rated Voltage of 450–500 V

Capacity (μF)	Rated Voltage (V)	Dimensions (Diameter, Height; mm)	Manufacturer and Capacitor Type
6000	450	75 × 220	EVOX-RIFA PEH200YX460BQ
4700	450	90 × 146	BHC AEROVOX ALS30A472QP450
10,000	450	90 × 220	EVOX-RIFA PEH200YZ510TM
4000	500	76.2 × 142	Mallory Dur-Cap 002-3052
4000	450	76.2 × 142	CST-ARWIN HES402G450X5L
6900	500	76.2 × 220	CST-ARWIN CGH692T500X8L

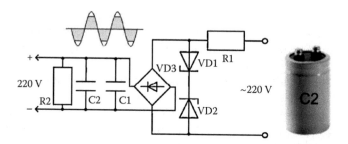

FIGURE 6.10
The device for a reserve feed of DPRs at an emergency mode with AC auxiliary voltage.

to voltage decreasing to a minimum level of 150 V during 3–5 s, which is quite enough for operation of the DPR in the emergency mode.

Use of such capacitors for auxiliary voltage of 220 V AC requires, naturally, a rectifier and some more auxiliary elements (Figure 6.10). In this device, a capacitor of large capacity is designated, such as C2. The C1 auxiliary—not electrolytic—capacitor with capacity in some microfarads serves for smoothing pulsations on the electrolytic capacitor C2. It is possible to also include, in parallel to C1, one more ceramic capacitor with a capacity of some thousand picofarads, for protection of C2 against the high-frequency harmonics contained in mains AC voltage. A R1 (200–250 Ohm) resistor limits the charging current of C2 at a level near 1A. The same resistor also limits pulse currents proceeding through back-to-back connected Zener diodes VD1 and VD2. Resistor R2 has high resistance and serves to accelerate the discharging capacitor up to a safe voltage at switching-OFF of the auxiliary voltage. Zener diodes are intended for the maximal value voltage limits of capacitor C2 at a level of 240 V. Without such limitations on the device, output voltage would reach a value of more than 300 V due to the difference between RMS and peak values of voltage. That is undesirable both for microprocessor-based protective device (MPDs) and for C2.

The Zener diodes slice part of a voltage sinusoid in which amplitude exceeds 240 V, forming a voltage trapeze before rectifying. As powerful Zeners for rating voltage above 200 V are not at present on the market, it is necessary to use two series-connected Zeners with dissipation power of 10 W and rating voltage of 120 V each (such as VD1 or DD2—e.g., types 1N1810, 1N3008B, 1N2010, or 5223A).

As further research of this type of situation clarified, the problem of maintenance of reliable feed DPR is relevant not only for substations with AC auxiliary voltage, but also for substations with DC voltage. Many situations where the main substation battery becomes switched-OFF from the DC busbars are known. In this case, nothing terrible occurs, as the voltage on the busbar is supported by the charger. However, if during this period an emergency mode occurs in a power network, the situation appears to be no better, since use of an AC auxiliary voltage as a charger feeds from the same AC network. Usually an electrolytic capacitor with some hundreds of microfarads for smoothing voltage pulsations is included on the charger output. Since not only many DPRs but also sets of other consumers are connected to the charger output, it is abundantly clear that this capacity is not capable of supporting the necessary voltage level on the busbars during the time required for proper operation of the DPRs. Our research has shown that such high capacitance as 15,000 μF does not provide proper functioning of DPRs as consumption from the charger reaches up to 5–10 A.

For maintenance of the working capability of DPRs in these conditions, it is possible to use the same technical solution with the individual storage capacitor connected in parallel to each DPR feeding circuit. Now the design of the device will be much easier, due to a cutout from the circuit diagram of Zeners and the rectifier bridge (Figure 6.11). The resistor R (100 Ohm) is necessary for limiting of the charging current of the capacitor at switching-ON auxiliary voltage with a fully discharged capacitor. Diode VD1 should be for a rated current of not less than 10A. A high-capability quick blow fuse F (5 A/1500 A, 500 V) is intended for protecting both the feeding circuit of the DPR and the external DC circuit at damaging of the capacitor.

The prototype of such a device with the capacitor 3700 μF (Figure 6.12) has shown excellent results at tests, with the various loadings which simulate DPRs of various types with different power consumption at the nominal voltage 240 V.

One more variant of the solution of this problem for substations with DC auxiliary voltage is to not use an individual capacitor for each DPR, but rather a special "supercapacitor" capable of feeding a complete relay protection system set together with conjugate electronic equipment within several seconds. Such supercapacitors can already be found on the market under brand names such as *supercapacitors, ultracapacitors, double-layer capacitors,*

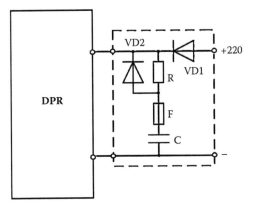

FIGURE 6.11
The device for a reserve feed of DPRs at an emergency mode with DC auxiliary voltage.

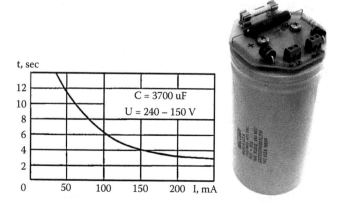

FIGURE 6.12
The prototype of the device for a reserve feed of DPRs.

and also *ionistors* (for Russian-speaking technical literature). There are electrochemical components intended for storage of electric energy. On specific capacity and speed of access to the reserved energy, they occupy an intermediate position between large electrolytic capacitors and standard accumulator batteries, differing from them in their principle of action, based on their redistribution of charges in electrolytes and their concentration on the border between the electrode and electrolyte.

The capacity of modern supercapacitors reaches hundreds and even thousand of farads. Today, supercapacitors are produced by many Western companies (e.g., Maxwell Technologies, NessCap, Cooper Bussmann, and Epcos) and also some Russian enterprises (e.g., ESMA and ELIT); however, the rated voltage of one element does not exceed, as a rule, 2.3–2.7 V. For higher voltage, separate elements connect among themselves in parallel and series as consistent units (Figure 6.13).

Unfortunately, supercapacitors are not so simply incorporated among themselves as ordinary capacitors; they demand leveling resistors at series cells connection and special electronic circuits for alignment of currents at parallel cells connection. As a result, such units turn out to be rather "weighty," expensive, and not so reliable (there could be enough damage to one of the internal auxiliary elements to cause failure of the entire unit). For example, a combined supercapacitor manufactured by the NessCap firm, with a capacity of 51 F and voltage of 340 V, weighs 384 kg! One unique company known to us which produces individual modules (i.e., not containing too many low-voltage cells inside) for high voltage (Figure 6.14) is the Canadian firm Tavrima. Its ESCap90/300 type supercapacitor (see Table 6.2) approaches our purposes quite well.

Another example is the supercapacitor module Sitras® series from Maxwell Technologies (Figure 6.15). At use of supercapacitor SC, the feeding circuit of

FIGURE 6.13
Internal design of a high-voltage (ten voltages) supercapacitor, assembled from a number of low-voltage elements.

ESCap 90/300

FIGURE 6.14
High-voltage supercapacitors made as a single module and main parameters of the ESCap90/300 type capacitor.

the protective relays should be allocated into a separate line connected to the DC busbar through diode D (Figure 6.16).

Due to the large capacity of the supercapacitor, the voltage reduction on feeding input of DPRs at emergency mode (with loss of an external auxiliary voltage) will occur very slowly, even after passage of the bottom allowable limit of the feeding voltage.

From my personal experience, cases of false operation of the microprocessor systems have been known to occur at slow feeding voltage reduction,

TABLE 6.2

Main Parameters of Tavrima ESCap90/300
Type Supercapacitor

Rated voltage (V)	300
Capacitance (F)	2.0
Maximum power (kW)	75
Maximum energy (kJ; at 300 V)	90
Internal resistance (Ohm)	0.3
Dimensions (mm)	Diameter 230 × 560
Weight (kg)	35
Temperature (°C)	−40 to +55
Price per unit (for 2006; $)	1000.00

Sitras®

Nominal voltage – DC 750 V

Number of ultracapacitors in single
module – 1344

Energy stored – 2,3 kWh

Energy saving per h – 65 kWh/h

Max. power – 1 MW

Capacitor efficiency – 0,95

Temperature domain – 20 to 40°C

FIGURE 6.15
High-voltage supercapacitor module Sitras® series from Maxwell Technologies.

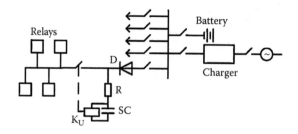

FIGURE 6.16
Example of usage of the supercapacitor as a group power supply for protective relays at emergency mode with DC auxiliary voltage. K_U: voltage relay; and SC: the supercapacitor.

below allowable levels. This can be explained by the existence of different electronic components of a high degree of integration serving the microprocessor, having different allowable levels of voltage feeding reduction and stopping the process of voltage reduction serially, breaking the internal logic of the DPR operation. If such equipment is found in the DPR, used on the given substation in parallel to the supercapacitor, it should be connected to a simple voltage-monitoring relay K_U, which disconnects the supercapacitor at a voltage reduction below the lowest allowable level, for example lower than 150–170 V.

6.4.2 System for Supervision Substation Battery Connectivity

An auxiliary DC power supply substation system, shown in Figure 6.17, includes main and reserve auxiliary transformers, power battery, chargers, DC busbars, and a distribution cabinet. It is an important substation system upon which the reliability of relay protection, automatic system, control, and communication depends. According to Reference 16, disturbances in this system may lead even to a full power system collapse.

A modern charger provides many different internal protective and signaling systems connected to emergency modes, while battery protection boils down, usually, to using a fuse. At the same time, there is always present the risk of failures in the contacts between the battery and the busbar; failures in the links connecting series of separate battery banks; failures in the internal structure of the accumulators; and failures of batteries due to natural disasters, such as earthquakes and the like. It is enough to take into account that a 230 V voltage substation battery contains some 106 separate accumulators, connected together in series by means of more than 200 links, the interruption of any one of which can lead to complete battery malfunction.

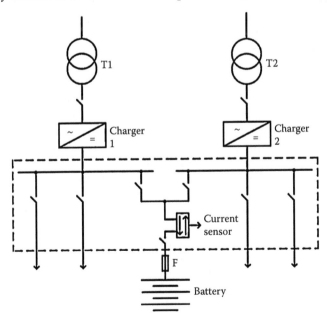

FIGURE 6.17
Typical single-line circuit diagram of a substation DC system.

6.4.2.1 Existing Methods for Supervising Substation Battery Connectivity

The Bender Company manufactures a device which can be used for monitoring the harmonics level in the DC network.[17] For a serviceable battery connected to a DC busbar, the harmonics level is very low. It was assumed that the disconnection of the battery from the DC busbar will dramatically increase the harmonics level produced by the charger and pickups output relay in the Bender monitoring device. Unfortunately, the harmonics sum generated by the charger is high depending on its output current (i.e., from the external load on the DC busbar). More to the point, in reality modern chargers provided with large filter capacitors with a total capacitance in the range of 5000–15,000 microfarads in output circuits determine a very low harmonics level on a DC busbar even with a disconnected battery. We conclude from this that using a harmonics level as the criterion for monitoring the substation battery connectivity is not applicable.

In Reference 18, a device for supervising substation battery connectivity based on a periodically pulsed increased voltage level on battery terminals and the measurement of current pulses passed through a battery is described. In Reference 19, a method for supervising the substation battery connectivity by injection of an audio-frequency signal into the supervised circuit and the measurement of a voltage drop on the circuit terminals on this frequency is described. In References 20 and 21, various devices for measuring the battery impedance as a criterion for supervision of battery circuit connectivity are offered, but in Reference 22 it is shown that the conventional methods for measuring battery impedance are ineffective because of very low value of the impedance in the power substation battery and, therefore, very low value of AC voltage which needs to be measured. Measuring such low values of AC voltages in real substation conditions is problematic, as noted in Reference 22. Nevertheless, the AREVA Company (formerly Alstom) offers a special device, the Battery Alarm 300 (BA300),[23] which is specifically intended for measuring battery impedance (Figure 6.18). This device in parallel with the battery periodically connects a resistor by means of a semiconductor switch for short periods of time (50 µs). The resistor produces a short current pulse with magnitude 1 A. This current produces a small voltage drop across the battery terminals that is used for calculating the battery impedance.

Insofar as the battery charger contains large filtering capacitors in its output circuits, it is abundantly clear that the current pulse through the measurement resistor in the BA300 device will be formed not only by battery, but also by the discharging of these filtering capacitors. In connection with this, AREVA suggests inserting a chock, intended on a full charger current, in the output circuit of each charger. Regarding where a consumer obtains such chocks on currents 30 A or 100 A and how much they will cost, AREVA is somewhat reticent. Even so, their device itself costs not a little (approximately €750).

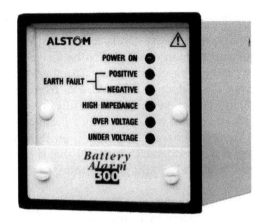

FIGURE 6.18
Device BA300 type (AREVA) for constantly monitoring substation battery impedance, voltage, and ground insulation levels.

For the sake of justice, it is necessary to note that constantly monitoring the impedance of the battery can reveal not only the fact of a full break of the battery circuit, but also the deterioration of the general condition of this circuit even before its full break. Besides, the BA300 enables supervising additional parameters of the battery, such as voltage and the ground insulation level. However, for me the specific goal is limited only to the supervision of substation battery connectivity, and for this, the suggested solutions should be the most simple, reliable, and inexpensive ones possible so that it is possible to use a large number of them to cover all the batteries available in a power system.

6.4.2.2 Suggested Methods for Supervising Substation Battery Connectivity

In distinction from the complexity (and, consequently, costliness) of the known methods for supervision of substation battery connectivity, we are suggesting another method, based on the measurement current which permanently passes from the battery to the busbar or from the chargers to the battery (Figure 6.17) in a typical system. Even a fully charged battery continues to consume a small current (referred to as a *floating current*) in a range of about 0.5–3.0 A from the charger, depending on the battery power and condition. Therefore, it is reasonable to consider that if the current in a battery circuit is reduced to a level less than 0.1 A, this unequivocally testifies to breakage of this circuit.

It is necessary to pick a controller capable of giving the appropriate output signal when decreasing a current in a supervised circuit lower than 100 mA. The problem with the selection of such a controller consists of, first, the direction in which the current in a supervised circuit can change when reversing, and, second, the change of value of a current in the circuit occurs in very wide limits: from 0.1 A up to 100 A, that is, by a factor 1000. Therefore, a high-sensitivity controller should be reliably protected from the influence of the high current value and should supervise a current in both directions.

According to these requirements, we constructed various units and tested some different systems for supervising the DC current in the battery circuit, based on different principles.

6.4.2.3 Device for a Supervision Battery Circuit Based on a Nonlinear Shunt

Using a nonlinear shunt allows considerably simplified supervision of current that changed in wide limits. The shunt in the device employs two back-to-back connected Schottky diodes. Forward voltage drop on one of the diodes (depending on the current direction) is changed according to the graph shown in Figure 6.19.

As can be seen from the graph in Figure 6.19, due to nonlinear diodes' characteristics, the forward voltage drop is changed from 0.20 to 0.65 V (i.e., by a factor of three while the current changed by a factor of 100). This feature of the diodes provides proper protection of sensitive input of the controller at high currents. On the other hand, enough high-voltage drops on diodes at small currents reduce the requirement of the sensitivity of the controller.

To test this idea, a circuit was put together modeled on the controller shown in Figure 6.20, using two standard devices produced by Conlab (Israel): DCT-3 and DCM-1. First the insulated transducer with high-voltage insulation input the circuit from output, and also the converter input voltage ±100 mV into the standard output signal 4–20 mA was connected to them. Then the controller itself with two programmed relay outputs was connected. The controller operated with input signals varying between 4 and 20 mA. For the two

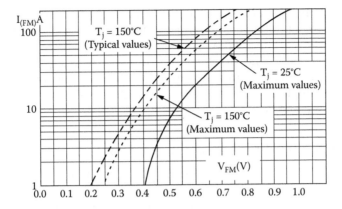

FIGURE 6.19
Dependence forward voltage drops on the Schottky diode STPS200170TV1 type from direct current (for different temperatures of a semiconductor structure).

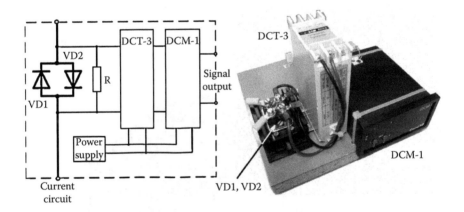

FIGURE 6.20
Model of the system based on a nonlinear shunt for a supervision battery circuit. DCT-3 and DCM-1: electronic transducers (Conlab, Israel); and VD1 and VD2: unit with back-to-back connected Schottky diodes STPS200170TV1 type (STMicroelectronics).

back-to-back connected Schottky diodes, STPS200170TV1, manufactured by STMicroelectronics, was used.

Experimental examination of this system indicated that it was thoroughly passable. The output relay picks up at the reduction of the current in the supervised circuit below 50–60 mA and releases at current increases up to 130–140 mA. The presence of a high hysteresis in this case is a positive feature which increases the stability of system operation.

At the same time, an important deficiency was noted in connection with strong heating of the diodes unit at high currents. Installing a small heatsink

(Figure 6.20), the temperature of the diodes unit achieved about 70°C at 25 A and carried for a period of 15–20 minutes. From this, it is clear that for operating at currents around 100 A, the diodes unit has to be combined with a large heatsink, or a ventilator for forced heatsink blowing has to be used. Another disadvantage of this system is the necessity for using a separate 24 V power supply for feeding the DCT-3 and DCM-1 devices. The cost of this system is about $600.

6.4.2.4 Using a Standard Shunt as a Current Sensor

Using a standard linear shunt, for example 100 A/60 mV or 100 A/100 mV and a current pickup level of 0.1 A, the sensitivity of the controller has to be 10 microvolts (unlike the previous example, where the sensitivity of the controller can be as low as hundreds of millivolts). It is unlikely that a controller available in the market has so high a sensitivity.

At our request, Conlab gave us another system for testing (shown in Figure 6.21) which met the requirements mentioned above for using a standard shunt. During the test this system exhibited the following results: pickup current levels of 60 to 80 mA (through the shunt), and release current levels of 120 to 160 mA. The same hysteresis is a positive property of such a system because its presence increases the system stability. One of the

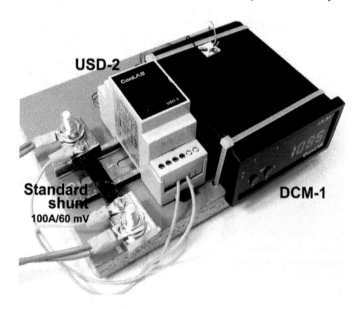

FIGURE 6.21
Model of the system based on a standard shunt for a supervision battery circuit. USD-2: electronic transducer (input: ± 200 µV, output: 4–20 mA); and DCM-1: programmed electronic transducer for 4–20 mA with relay output (Conlab, Israel).

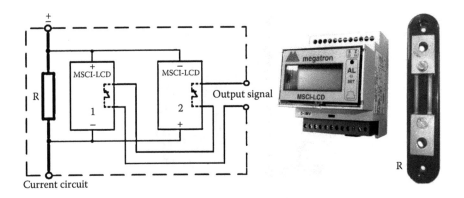

FIGURE 6.22
System for supervision battery circuit based on a standard shunt and two universal controllers with relay output MSCI-LCD-S type (Megatron, Israel).

disadvantages of this system is the necessity of using an external 24 V power supply for feeding the transducers. The cost of this system is about $600.

The MSCI-LCD-S type serial line controller, manufactured by Megatron (Israel) with some slight modifications to meet our requirements, appeared as a very successful variant and demonstrated very stable operation with high sensitivity (Figure 6.22). This controller practically did not react with the AC component when measuring the input and therefore had a high noise stability, and could feed from the 230 V network.

It was noticed that the MSCI-LCD-S type controller lacked the ability to detect the polarity of the input signal. In the modified variant of the controller, the measuring input is protected against the high voltage applied at high currents carried through the shunt, and also protected against voltage polarity reversals applied to measuring the input during changes of the current direction in the shunt.

A single problem is the necessity of using two identical controllers connected in opposite polarity to the shunt for measuring the currents, proceeding in both directions, and connecting the normally closed contacts of its output relays in series. Thus the output signal will appear only in the case that the current will be lower than 0.1 A in both directions. Considering the small cost of a single controller (about $130), it turns out that when having to use two such controllers, the total cost is much lower than that of the second system provided by Conlab.

6.4.2.5 Use of the Hall-Effect Sensor in Systems for Supervision Battery Circuits

The monitoring systems considered above demand the insertion of additional elements (such as diodes or a shunt) in a power circuit (cable) connecting the

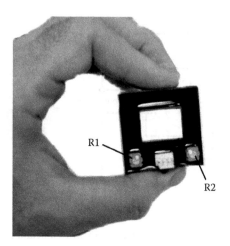

FIGURE 6.23
Hall-effect transducer HAL50-S type with built-in electronic amplifier for measurement of DC current. R1 and R2: potentiometers for amplifier adjusting.

battery with a DC busbar. However, there is a variation in which there is no necessity to cut the circuit (cable) and no necessity for inserting additional elements. This variation is based on using a Hall-effect sensor (transducer) in the form of a framework through which the power cable connecting the battery and busbar is passed (Figure 6.23).

Some companies, for example CR Magnetics,[24] have proposed DC current relays with built-in Hall sensors. However, from the company's answers we received, it seems that such devices cannot provide carrying currents varying between 0.1A and 100A and pickups at currents below 0.1 A. We therefore undertook the attempt to develop the current relay of our own design based on separate Hall transducers, such as HAL50-S, manufactured by the Japanese branch of the LEM company and designed for currents up to 150 A, and the AM22D type controller, manufactured by the Israeli company Amdar Electronics & Controls.

This type of Hall transducer consists of a built-in electronic amplifier and two potentiometers which deduce outside for adjusting the amplifier characteristics. Due to the amplifier, the output signal of the transducer is as high as ±4 V at nominal current ±50 A. We assumed that such a high output signal at nominal current would provide enough high-level signal, also at a current of 0.1 A.

Unfortunately, even at zero current the output signal level drifted and its instability exceeded the level of the functional output signal at a current of 0.1A. Thus, we would not be successful in receiving a comprehensible signal from such a system that would be applicable for use in our supervision system.

6.4.2.6 The Newest Developments and Prospects for Their Application

More recently, in many cases after I had made inquiries addressing the problems noted above to the manufacturers, some companies improved the devices and now offer new products that are more suitable for the parameters for monitoring DC systems. For example, transducers type CR5211-2 (advertised by CT Magnetics) are ostensibly specially intended for monitoring DC system circuits without the external shunt and are ideally suited to our needs.

These transducers are advertised as being mounted directly on the cable. The sensing element of this device is not the Hall element (which is poorly suited for monitoring small currents), and a high-frequency transformer the magnetic condition of which varies depending on the magnitude of the direct current flowing through the monitored cable. According to the manufacturer's explanation, "[T]he ranges 2 to 10 Amp utilize an advanced Magnetic Modulator technology." This principle of such a sensing transducer reminds one of the magnetic amplifiers that were widely used in automatics in the 1950s and 1960s. The list price of this device is $125. Another device of an analogous principle is offered by Powertek (United Kingdom): the Current Sensor CTH Type 2 for a current range of 0–5 A, costing about $360. It is only a transducer, with a transformed input signal in an output voltage. If we add to it a voltage relay, for example SM125 024VDC 4V (Carlo Gavazzi); a timer H3DE-S1 (Omron Industrial Automation); and a miniature DIN-rail power supply 24 V DC, for example type DSP10-24-PSU (TDK-Lambda), the cost of this complete set will increase from $300 for the first example of the current sensor to $520 for the second one (Figure 6.24).

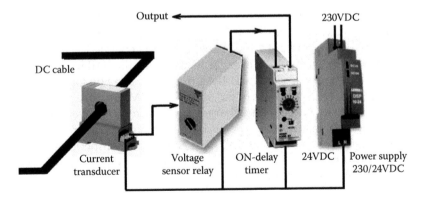

FIGURE 6.24
Structure of the DC circuit-monitoring system based on a novel type of small current transducer.

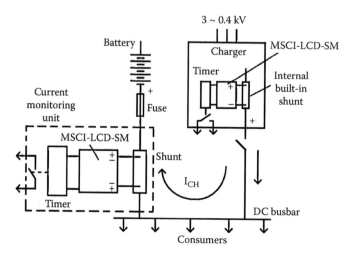

FIGURE 6.25
Circuit diagram for monitoring a DC system.

Megatron, the Israeli company mentioned above, upon my request has also modified its MSCI-LCD-S controller and has made it insensitive to the polarity of current flowing through the shunt. It enables using only one such modified controller (MSCI-LCD-SM), at a cost of about $130, instead of two; this is depicted in Figure 6.22. Considering the increase of the device cost, there was an idea of adding the additional controller for monitoring the charger condition of the DC-monitoring system (Figure 6.25).

In all chargers, there is the built-in shunt intended for the measurement of an output current. The controller MSCI-LCD-SM can also be connected to this shunt. This controller watches the serviceability of the charger.

As has been shown in our experience, there is an assemblage of types of internal faults of chargers, at which no signals are generated.

Thus the battery without the charger feeds the loads until its voltage does not decrease to the certain level and the supervision relay (voltage relay) generates a signal about a fault. Use of the MSCI-LCD-SM controller would allow increasing the reliability of the charger monitoring and even completely cancel the voltage monitoring. The addition of a simple timer, for example the H3DE-S1 (which costs about $30) with an ON-delay function and a time delay of 10–15 s to the circuit, has allowed preventing output from spurious signals at switching in DC systems.

The main current-monitoring unit is placed in two standard plastic cases. In one of them the shunt, and in the other the controller with the timer, are mounted (Figure 6.26). The additional complete set of the controller with the timer is mounted directly in the charger (Figure 6.27). The sensitive input of the controller is connected directly to the regular internal shunt of the charger.

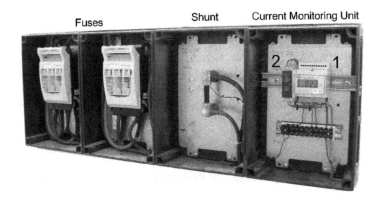

FIGURE 6.26
Main DC-monitoring unit mounted in a substation.

FIGURE 6.27
Additional current-monitoring unit mounted in charger. 1: controller; and 2: timer.

6.4.2.7 Conclusion

On the basis of a comparative estimation of parameters and test results of some variations of systems for supervision circuits of the substation battery 230 V that were discussed above, we have come to the conclusion that from the standpoint of the greatest stability, greatest reliability, and least cost, the system of choice is that based on a standard shunt and modified MSCI-LCD-SM type controllers, manufactured by Megatron. This is the system that we recommend for wide use in substations and power stations.

6.4.3 Measures for Improving the Reliability of DC Battery Chargers

Powerful battery chargers, which are generally three-phase, generating output currents of 30–250 A at rated voltages of 24–230 V, are used extensively at electric power substations and power plants, gas compressor and gas-distributing stations, and so on for powering critical relay protection and automatic equipment, control and regime-monitoring systems, signaling and telecommunications under the normal operating conditions, as well as charging storage batteries powering these systems when AC power fails.

Normally, the modern battery charger is a combination of powerful electrical assemblies, like a supply transformer, choke and high-current thyristors, and a sophisticated electronic system that often contains microprocessor-based controllers. Such chargers are equipped with numerous systems protecting from overvoltage or undervoltage in the three-phase AC mains and complicated output current and voltage regulators generating commands for high-current thyristors fixed to phase angles of the input AC voltage. Therefore, operating regimes of the battery charger depend heavily on the parameters of the voltage in AC mains.

Our study of several new battery chargers, such as AEG type D400G24/250BWrug-CFpux (Germany) and ELCO (Israel), found that the devices were very sensitive even to the smallest voltage dips in the AC mains that appeared when turning on high-powered consumers. For example, turning on outdoor Hg-lamps caused quick response of the AEG devices, while voltage dips were 20–50% of the rated value and lasted some 2.5–2.9 ms. Each switching on was followed by the turning off of the battery charger and subsequent turn on in 1.0–1.5 seconds. See Figure 6.28.

A follow-up oscilloscope study of two new battery chargers of that type showed that the responsiveness to voltage dips does not depend on the phase angle or the parameters of the dip; each dip resulted in short interruptions of battery chargers.

When the battery charger turns on, a sophisticated electronic control system provides the stepwise start-up of separate charger systems within 5–7 seconds under a defined operating algorithm. Often the mains transient processes, actuating protective systems at this stage, cause malfunctions of the start-up algorithm and charger "hang-up." Generally, such malfunctions

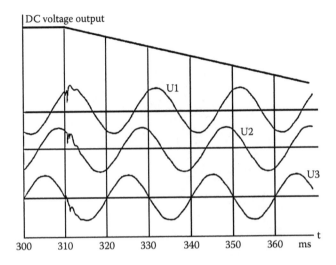

FIGURE 6.28
Disconnection of DC output in modern chargers due to small dips in the main power.

occur due to voltage dips and are rectified within a short period of time in 0.4 kV mains (in 0.4 kV mains, it results from emergency transient modes occurring in high-voltage lines). Therefore, to avoid battery charger emergency modes, it is recommended using a special, fast-speed, minimum-voltage relay (referred to as a Brown-Out Timer) of type GBP4450 (manufactured by Midland Jay) that reacts within 0.5–4.0 ms upon voltage reduction to 80% of the rated value and returns back to the normal mode at 85% voltage after a specified delay, adjustable within 6–60 seconds. See Figure 6.29.

Such a relay as this must cut out the battery charger very fast upon the first short voltage dip and return it to the initial condition after a specified delay (i.e., after termination of the transient process). This should prevent battery charger hang-up caused by multiple interleaved voltage dips. However, unfortunately, this is not the only problem with battery charger powering. Even if after a short interruption, the device automatically returns to the operating mode without hanging up, such modes are usually rather problematic. The problem is that the charger returns to the normal mode stepwise according to the defined algorithm imbedded in the charger control system.

In many types of chargers, the first step is the automatic enabling of a powerful idle-running supply transformer through the contactor. As is known, the enabling of powerful idle-running supply transformers results in a magnetizing current inrush which can reach hundreds of amperes. For example, according to Benning Elektrotechnik und Elektronik manufacturer data, magnetizing current inrush of a type D400G216/30 BWrug-TDG battery charger may reach 320 A. Magnetizing current inrush upon startup of an ELCO battery charger reached 600–900 A according to our research. The exact value of the magnetizing current amplitude depends heavily on dozens of factors

FIGURE 6.29
Extra-fast-speed, minimal voltage relay with delayed return (Brown-Out Timer), type GBP4450.

and may even vary within the same device. Differences in the magnetizing current can be explained by the different distances between the chargers and the auxiliary supply transformer, that is, by different resistances of the power circuits limiting the maximum current, auxiliary transformer power, excitation phase, transformer core magnetic state at the moment of start-up, design of transformer, and so on. Most types of battery charges are protected from overloads and internal short circuits with automatic breakers or quick-break fuses installed on the power line side. Parameters of these breakers or fuses are chosen based on the overload capability of the internal power elements (thyristors) such as to provide maximum protection. High-magnetizing inrush currents into the internal supply transformer actuate these switches, while quick-break fuses blow, switching the battery charger off. If there are attending personnel near such chargers, they just return the automatic breaker to the initial (actuated) condition. Problems occur when faulty breaker operations happen at unattended substations, especially at night.

Simple current limiters can protect automatic breakers from magnetizing current inrush. Certain types of battery chargers are equipped with delayed contactors (the delay can be up to several seconds) connected in series with the automatic breaker and required for the setup of an internal controller. Current limiters for such devices have the simplest design (Figure 6.30) consisting of three resistors $R1$–$R3$ connected in parallel (for each phase) with contacts of K contactor.

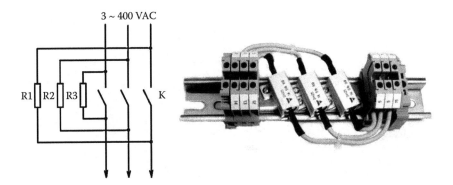

FIGURE 6.30
Circuit diagram and layout of inrush current limiter.

When the breaker is switched on and contactor is switched off, the small current (0.1–0.2 A) limited by the resistors goes through the supply transformer, providing its premagnetization within the contactor delay time. After the contractor actuation, resistors are bridged by the contactor contacts and do not affect device operation. Studies on real devices have shown that resistance of resistors could reach 500–1000 Ohm, while the power did not exceed 25–50 W. However, as transformer magnetizing current inrush is not the only reason for false activation of automatic breakers, the inrush current limiters are not always effective. Short voltage dips, phase displacement, and phase voltage asymmetry often result in the failure of electronic control systems of semiconductor devices, causing conductivity on semiconductor devices which do not have to be conductive simultaneously as well as short-circuit-actuating automatic breakers. It is evident that inrush current limiters are not able to fix the problem in these situations.

There is a more effective solution of the problem regardless of the reason for automatic breakers' faulty disconnection: automatic battery charger reclosure. This can be done by an automatic breaker with the same parameters as the basic one because overload and short-circuit protection specified by the manufacturer should be maintained in all regimes. In the simplest case, it is possible to use existing automatic breakers which should be equipped with special drives returning them to the initial (enabled) condition. Such drives of FW7 type (shown separately in Figure 6.31a, and shown together with automatic breaker in Figure 6.31b) are available from Moeller Electric (Germany). It has a pusher interlocking with an automatic breaker handle to lift it up (or advance) under a certain time delay after the automatic disconnection of the breaker. It is obvious that the device of this type is not universal, as its dimensions should exactly correspond to the dimensions of the breaker. Besides, this type of drive is not able to push button-type automatic breakers. The cost of such a device (around $500) also prevents it from wide use.

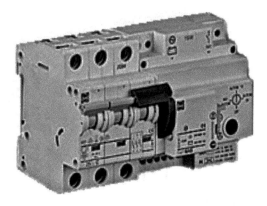

FIGURE 6.31
Automatic remote-driven breaker type FW7 (Moeller Electric).

In view of the above, we have developed a simple device (Figure 6.32) compatible with any type of battery charger. This device consists of an additional automatic breaker CB1′ of the same type as the standard CB1, contactor K2, auxiliary relay K1, and turn-on delay timer. When CB1 is activated, relay coil K1, being powered through timer break contact, stays continuously enabled while its break contact of K1 is open. Upon accidental disconnection of the automatic breaker CB1, relay K1 becomes deenergized and its contact closes, thereby powering the timer. After several minutes, required for the completion of all transient processes in the mains, the timer becomes active and its contact provides an additional break of the relay coil K1 supply circuit, which switches and generates supply circuit for the coil of contactor K2. Then, the consumer is switched on, automatically bypassing the disabled automatic breaker CB1 through the contacts of the K2 contactor and the additional automatic breaker CB1′.

In order to return the device to the initial condition, it is required to disconnect the battery charger supply circuit for a short time from the outer control panel and return the main automatic breaker CB1 to its initial (enabled) position. This device may be equipped with switching elements of any type with AC 220 V coils. The contactor capacity must correspond to the battery charger's power.

It is recommended to be very careful when repairing battery chargers equipped with automatic reclosure devices. If you are going to repair the battery charger, it is required to ensure there are proper visual notices that the external power source must be disconnected.

About 30 sets of inrush current limiters with automatic reclosure devices were produced and installed into powerful battery chargers manufactured by ELCO with 30 A and 100 A output current at 220 V.

Combined implementation of the solutions described above will help significantly improve the reliability of powering such critical systems as relay protection and automatic control.

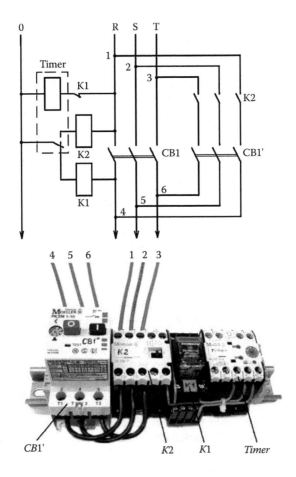

FIGURE 6.32
Battery charger automatic reclosure device reacting on the accidental actuation of an automatic breaker.

References

1. U.S. Nuclear Regulatory Commission. Information Notice No. 94-20: Common-Cause Failures Due to Inadequate Design Control and Dedication, March 17. Washington, DC: U.S. Nuclear Regulatory Commission, 1994.
2. American Power Conversion (APC). *The Power Protection Handbook*. West Kingston, RI: APC, 1994.
3. Dshochov, B. D. Features of Power Supply of Computer Network Elements [in Russian]. *Industrial Power Engineering*, no. 2, 1996, pp.17–24.

4. *International Electrotechnical Commission* (IEC). IEC 61000-4-11 Ed. 2.0 b:2004: Electromagnetic Compatibility (EMC): Part 4-11: Testing and Measurement Techniques: Voltage Dips, Short Interruptions and Voltage Variations Immunity Tests. Geneva: IEC.

5. *International Electrotechnical Commission* (IEC). IEC 61000-4-34 Ed. 1.0 b:2005: Electromagnetic Compatibility (EMC): Part 4-34: Testing and Measurement Techniques: Voltage Dips, Short Interruptions and Voltage Variations Immunity Tests for Equipment with Input Current More than 16 A per Phase. Geneva: IEC.

6. Melhorn, C. J., T. D. Davis, and G. E. Beam. Voltage Sags: Their Impact on the Utility and Industrial Customers. *IEEE Transactions on Industry Applications*, vol. 34, no. 3, 1988, pp. 549–558.

7. McGranaghan, M. F., D. R. Mueller, and M. J. Samotyj. Voltage Sags in Industrial Systems. *IEEE Transactions on Industry Applications*, vol. 29, no. 2, 1993, pp. 397–404.

8. Fishman, V. Voltage Sags in Industrial Networks [in Russian]. *Electrical Engineering News*, vol. 5, no. 29, and vol. 6, no. 30, 2004.

9. Kelley, A., J. Cavaroc, J. Ledford, and L. Vassalli. Voltage Regulator for Contactor Ridethough. *IEEE Transactions on Industry Applications*, vol. 36, no. 2, 2000, pp. 697–703.

10. Andgara, P., G. Navarro, and J. I. Perat. A New Power Supply System for AC Contactor Ride-Through. Paper presented at the International Conference on Electric Power Quality and Utilisation, 9–11 October 2007, Barcelona.

11. Iyoda, I., M. Hirata, N. Shigei, S. Pounyakhet, and K. Ota. Effect of Voltage Sags on Electro-magnetic Contactor. Paper presented at the International Conference on Electric Power Quality and Utilisation, 9–11 October 2007, Barcelona.

12. Gurevich, V. *Electrical Relays: Principles and Applications*. London: Taylor & Francis, 2005.

13. Gurevich, V. Nonconformance in Electromechanical Output Relays of Microprocessor-Based Protection Devices under Actual Operation Conditions. *Electrical Engineering & Electromechanics*, vol. 1, 2006, pp. 12–16.

14. Gurevich, V. Microprocessor Protective Relays: New Perspectives or New Problems? *Electrical Systems and Networks*, no. 1, 2006, pp. 49–60.

15. Fox, G. H. Applying Microprocessor-Based Protective Relays in Switchgear with AC Control Power. *IEEE Transaction on Industry Applications*, vol. 41, no. 6, 2005, pp. 1436–1443.

16. Skok, S., S. Tesnjak, and M. Filipovic. Risk of Power System Blackout Caused by Auxiliary DC Installation Failure. In Proceedings of the IASTED International: PowerCON 2003: Special Theme: Blackout conference, December 2003, New York.

17. Bender. Ripple Detector RUG1002Z: Insulation and Voltage Monitoring of DC System RGG804: Operation Manual. Grünberg, Germany: Bender GmbH.

18. Phansalkar, B. J., P. N. Tolakanahalli, N. R. Pradeep, and S. K. Saxena. Method and System for Testing Battery Connectivity. U.S. Patent No. 6,931,332, August 16, 2005.

19. Russo, F. J. High Sensitivity Battery Resistance Monitor and Method Therefore. U.S. Patent No. 5,969,625, G01R31/36, October 19, 1999.

20. Wurst, J. W., T. Chester, and C. Morris. On-Line Battery Impedance Measurement. U.S. Patent No. 5,281,920, G01R31/36, January 25, 1994.

21. Burkum, M. E., and C. M. Gabriel. Apparatus and Method for Measuring Battery Condition. U.S. Patent No. 4,697,134, G01R31/36, September 29, 1987.

22. Kozlowski, J. D. A Novel Online Measurement Technique for AC Impedance of Batteries and Other Electrochemical Systems. Paper presented at the Sixteenth Annual Battery Conference on Applications and Advances, January 2001, Long Beach, CA.
23. AREVA. Battery Alarm 300: User Manual BA300. Paris: AREVA.
24. CR Magnetics. Direct Current Sensing Relay CR5395 Series: Data Sheet. St. Louis, MO: CR Magnetics.

7

Testing Digital Protective Relays

7.1 Problems with DPR Testing

Relay protection constitutes a major part of any power system that provides for continuous control of the main operation modes of power system elements and generates tripping commands for the failed parts or elements of the system. Faulty operation of relay protection owing to internal malfunctions can lead to the development of massive failures and even to the collapse of the power system with huge attendant financial losses. For this reason, the performance of the relay protection has to be periodically tested. There exists a vast variety of relay protection devices with different operating principles and construction.[1] Lately, all but digital protective relays (DPRs) have been completely driven out of the market. The choice of DPRs has been driven by various reasons and not at all by their absolute advantage over electromechanical or analog electronic devices. DPRs based on various principles of operation have their advantages and disadvantages.[2-4] However, one of the problems is the complexity of the procedures for testing their operation.

Usually the operational condition of relay protection devices is checked with specific settings used for the relay operation in a certain network point. Any change of the settings during the normal relay operation requires repeating the working condition test with these new specific settings. When electromechanical relay protection devices were used, this procedure was quite reasonable since any change of settings was affected by the mechanical shifting of the internal relay elements, switching the taps of the built-in transformers, or the like. Following a change of settings, a failure in the internal relay circuits connected to a new tap of the transformer (e.g., rupture of a wire, contact failure, or insulation damage), relay imbalance caused by the change of the mechanical parts position, relay "grinding," and other problems might occur. Therefore, an electromechanical relay normally used with a fixed setting did not necessarily ensure its normal operation with other settings.

For DPRs, a change of settings is not accompanied by physical changes in their internal structure. The input and output circuits in a DPR, as well as the logic elements, central processing unit (CPU) or power supply, and so forth, operate regardless of the selected settings or operation modes. Moreover, activation or deactivation of specific relay functions is not related to the physical condition of its circuits. So, checking the appropriateness of the selected protection logic and the correctness of the setting calculation for specific conditions of a certain circuit is a totally different task, not at all connected to testing the working condition of the relay. Furthermore, these tests are performed by engineering services responsible for setting the calculations and selecting the internal logic of the relay operation rather than by the relay operation personnel who is responsible for its proper operation. In fact, when an operational test is performed, simulation of all the real situations and all the possible combinations of factors acting in a real circuit becomes for all practical purposes impossible. Detecting such situations is not a part of the protective relay operational test. Moreover, it can be demonstrated that the results of the relay tests with the nominal settings, only, are a positive measure because of a so-called human factor (which causes about 50% of relay malfunctions). The fact is that the settings selected for a specific operational regime in multifunctional DPRs make the testing of a specific relay function possible only by the desensitization or complete disabling of another competing function. Often a failure to reenable such desensitized or disabled functions to their initial condition causes wrong protective functions in emergency modes. A similar approach to a problem of DPR tests is used in Reference 5. In this document (having the status of a standard), all tests of the protective relay are divided into two kinds: calibration tests (setting and configurations of the relay) and functional tests. For functional tests, such periodicity as once in 4 years for all types of relays (including electromechanical and microprocessor-based ones) is required, but for calibration tests the periodicity of once in 4 years only for electromechanical relays is established. *Routine (periodic) calibration (i.e., checking of relay settings) for DPRs is not required at all.* The authors of Reference 6 have come to similar conclusions: at routine tests of the DPR, there is no necessity testing of internal setting preservation.

Experts of one of the leading manufacturers of relay test equipment, Omicron,[7] distinguish the three kinds of the DPR tests: type tests, commissioning tests, and routine tests. Type tests are understood as functional tests and possess the basic characteristics of DPR testing. At commissioning tests, such additional test procedures as calibration must be completed also (i.e., checking the actual setting and the internal logic programmed for concrete application is made also). And at routine (periodic) tests, the serviceability of the DPR is actually checked only once. Thus, according to Omicron experts' opinion, the testing of DPRs with an actual setting is required only once at the commissioning of a new protection and should not be repeated with routine and type tests.

7.2 New View of the Problem

On the basis of the discussion in Section 7.1, several principles applicable for testing the DPR may be formulated:

1. In order to verify the proper operation of complex multifunctional DPRs at their inspection, at their start-up after repairs, or during periodic tests, there is no need to use the actual settings at which the relay is to be operated in a certain network's point.

2. In order to test the working condition of a DPR, it should be tested for proper operation at several of its most critical *preset* character-istic points as well as in several *preset* characteristics constituting its most complicated (combined) operation modes, including the dynamic operation modes with *preset* transition processes specific for standard power networks (not necessarily for a specific point). Such tests should cover all the physical outputs and inputs of the relay. At the end of tests and verification of its proper operation, all the test settings should be automatically replaced with a set (file) of actual settings prepared in advance.

3. To the best of our knowledge, such testing of DPRs in the most complicated modes of operation will allow for more comprehen-sive testing of the DPR than employing a very limited test at strictly limited specific settings that will be further used during the relays operation.

4. Integrated testing of the DPR during the stage in which it is put into operation under the most severe operating conditions enables ruling out additional tests of the relay operation at every change of setting during routine operation.

The principles formulated above provide us with a different insight on the problem of testing DPRs. One can assume that the first devices for testing protection relays appeared almost simultaneously with the protection relays themselves. Of course, they were as primitive as the protection relays. At first, they were simple calibrated inductance coils (see Figure 7.1) and rheostats.

As relays have been improved, test units for them also have became more complex. Test benches have appeared (Figure 7.2) containing inductances and active resistance sets, by means of which one could set angles between current and voltage within a wide band and examine rather complex elec-tromechanical relays. In different power systems, one could establish dif-ferent times for periodic testing of relay protection, for example once every 2–3 years, but usually they were observed more stringently.

FIGURE 7.1
Set of inductances from General Electric for testing of electromechanical protective relays.

As microprocessor-based protective relays appeared on the market, the situation changed radically. Producers of these devices claimed that the microprocessor-based relays did not need periodic examinations, because they had powerful embedded self-test systems. This claim for DPRs appeared in advertising literature almost as their main advantage over electromechanical and analog electronic relays. Large advertising campaigns launched by DPR producers played this role as well.

Many specialists in the field of relay protection believed unreservedly in this advertising gimmick, as they did not have opportunities to verify in practice if this statement were true, even though it was patently obvious that it was impossible to create a test system on the basis of DPR inner microprocessors which could examine the physical repair of many thousands of electronic components. Functionally, it is also impossible to test the repair of, for example, an input unit or an output unit without turning it on and examining the relay's reaction in supplying the signal to them. In practice, it has turned out that most DPRs simply do not sense the substitution of the whole printed board of one kind for the board of another kind which is not compatible with relay current settings. I have already discussed the advertising gimmicks connected with DPRs, those of "self-diagnosis," in numerous other publications.[8,9]

Unlike DPR manufacturers, relay protection test system (RPTS) manufacturers have always affirmed that all protection relays should be periodically

FIGURE 7.2
Advanced test set TURH-20 type (ASEA) for testing of electromechanical protective relays containing sets of inductances and resistances.

tested, including DPRs, as so-called self-diagnosis covers not more than 15% of the software and hardware. Though DPR manufacturers have asserted that periodic examinations of protection devices were unreasonable, RPTS manufacturers have been intensively developing and putting on the market ever newer test systems.

7.3 Modern Test Systems for Digital Protective Relays

As today's DPR construction principles are general for most manufacturing companies, it is natural that test systems that are offered nowadays by different companies are rather alike, and not only in their appearance (see Figure 7.3) but also in their specifications. Modern RPTSs are completely computerized devices without any physical controls on the cover (control is via a computer connected to the device's RS232 connector). All that shows are the sockets for external wires and the RS232 connector for the computer. RPTSs such as these cost tens of thousands of dollars.

These systems are designed for running three types of tests: steady state, dynamic, and transient. Steady-state tests are used for examining the basic settings of relay actuation and as such are referred to as *preliminary*

FIGURE 7.3
The modern computerized test systems of the last generation for testing of multipurpose microprocessor-based protective relays.

examinations of relay. Dynamic tests are used, in general, to inspect complex protection behavior, such as distance or differential protection, in different areas of characteristics and protection zones based on input parameter (current, voltage, and angle) changes with time.

Transient tests are based on the injection into input circuits of relays of current and voltage which correspond with the COMTRADE transient files retrieved from recording devices, recorded real short-circuit transient processes, or simulated short-circuit transient processes based on files with the same format created artificially by means of special software. The test results are entered into a database realized, as a rule, on the basis of Sybase SQL Anywhere and automatically structured as a standard protocol that can be sent to a printer. RPTS manufacturers usually offer sets of test procedures (libraries) in the form of macros for different types of tests and even for some widespread relay types.

7.4 Modern RPTS Problems

Modern RPTSs are truly super-flexible and have many functional possibilities. These RPTSs allow simulating almost any working condition of protection relays which can occur in practice, including the creation of artificial COMTRADE files. Through these files, one can create artificial distortion of current waveforms, harmonic simulations, shifts of current sinusoid relative to the axis (simulation of the DC component), simulations of the circuit breaker response, automatic modeling of the most complex polygonal characteristics of distance protections, synchronization of differential protections by means of satellites, and more. One of the drawbacks of these modern complex RPTSs is the necessity to enter hundreds of parameters in many tables for every single relay test. However, embedded libraries of test procedures give little help in practice as they do not free one from having to fill in countless tables. To this, one should add the considerable flexibility and universality of the tested object (DPR), which also requires entering a large number of parameters from many dropdown menus and tables. The smallest discrepancy of DPR and RPTS settings leads to wrong results. And given this, it is often impossible to know that the results one has received are wrong. Even in cases when the mistake is obvious (e.g., an obtained relay characteristic does not correspond with the theoretic one), it is very difficult to identify where exactly the mistake was made: in the DPR or RPTS settings. I can confirm from my own experience that searching for a mistake is extremely difficult and requires much effort and time. Working with the *power system model* applied in RPTS for distance protection tests is not less difficult. In order to adjust the RPTS parameters in this mode, one has to know numerous parameters of real electrical networks that have to be entered with special indices in many tables. The technician and even the engineer of relay protection services often do not know many of these real mains parameters and applied indices, and this is why engineers from other power system services have to take part in relay test procedures.

7.5 Offered Solutions

Psychologists established long ago that the more buttons and levers (real and virtual, i.e., software) an operator has to manipulate, the lower the efficiency of a person's cooperation with such technique. Human perception simply does not grasp many functions and possibilities of such sophisticated techniques. How can one combine the universality and broadest functional possibilities of an RPTS with the abilities of an average technician or engineer of a relay protection service who needs quick and accurate examination of limited relay types? Does one overcome the great difficulties by developing and adjusting one's own procedures and create one's own library of macros as his or her basis, as was foreseen by the RPTS manufacturers? We are ready to offer more a radical solution to this problem:

1. It is unreasonable from a technical and economical point of view to use modern microprocessor-based RPTSs for testing even the simplest electromechanical relays, such as a current and voltage relay (e.g., type PT-40 or PH-54, as was attempted by the producers of Russian RPTS of RETOM-51 type). One can use simpler test devices

with much more effectiveness. There is no point in developing test procedures for automatically testing these relays, if the procedures do not involve testing hundreds of identical relays in their production process.

2. One can consider the reasonable usage of modern embedded libraries of test procedures requiring the entering of a huge number of parameters and exact knowledge of numerous indices only for complex electromechanical protection devices of the old type (e.g., distance protection LZ-31).

3. In order to test a modern, complex multifunctional DPR, one has to develop one for all types of RPTS software platforms, the requirements of which should be fixed by international standards. An example of such a general software platform is the well-known Sybase SQL Anywhere, which is widely used for the creation of databases in different data collection and processing devices, simulators, and test units of different producers. Another example is the universal format COMTRADE, which is used in all types of microprocessor-based fault recorders and, for that matter, in all types of RPTSs for the simulation of transient modes.

4. Application programs for working with RPTSs of different types can have absolutely different interfaces, but all of them should be implemented on the basis of a general standard program platform.

5. DPR producers should provide their protection devices with two CDs. One of them should contain full settings for specific modes of DPR protection functioning or for characteristic points of the characteristics for typical examples. The other CD should contain full sets of settings for the RPTS (each one under the number corresponding with the number of protection settings) and the circuit diagram of DPR external connections to RPTS inputs and outputs.

6. In our opinion, effective usage of modern RPTSs for testing modern, multifunctional DPRs is ensured only in the case when the whole test procedure comes to downloading settings XX.1 in the DPR, downloading settings YY.1 in RPTS, connecting the DPR to the RPTS, and … making a cup of coffee.

7.6 Digital Rate of Change of Frequency Relays and Problems in Testing It

7.6.1 What Is the Rate of Change of Frequency Relays?

The frequency of alternating current in electric power networks is the paramount parameter of the operating mode of a network. Even small frequency deviations can bring about serious malfunctions in the networks and demands for urgent corrective action. In many cases, a parameter for an emergency mode in a network is not just the absolute value of frequency; rather, it includes the frequency's change in time. Such a parameter as rate of change of frequency (ROCOF) is today the principal parameter that is monitored by numerous specialized digital relays (the ANSI code for relays of this type is 81RL), which are available on the market, for example UFD34, MRF2, G59, PPR10, LMR-122D, FCN950, KCG593, MFR 3, MFR 11, LS 4, VAMP 210, SPCF 1D15, 256-ROCL, and many other types.

There are two cases in which ROCOF relay protection is used in the core:

1. For automatic load shedding, that is, for the switching-OFF part of the loading upon detection of sweeping changes of frequency. It is necessary to note that in the emergency mode in a high-voltage power network, changes of frequency can be distinctive in different sections of a network depending on the power of the separate substations which are available in this network. In addition, at sweeping decreases of frequency in a branch network, there are overflows of power between the energy sources powering this network, accompanied by frequency fluctuations in the network as a whole. Thus, the absolute value of the downgraded frequency is not constant and therefore cannot serve as the criterion for pickup adjustment of the frequency relay and switching-OFF of a part of the loading. A much more reliable criterion for load shedding is the ROCOF function which is used as an additional criterion upon detection of the decrease of an absolute value of frequency below the set level.

2. For an instantaneous interdiction of repeated connection of the generator to a power network if it has been disconnected before even for a short term (isolated from a network). In the latter case, this protection is referred to as *loss of mains, loss of grid*, or *islanding protection*. A high-voltage circuit breaker tripping and the separation of a section of a network with the generator (that is, the formation of an isolated "island") from the main network (that is, "loss of mains") lead to infringement in the balance of power in the isolated section of the network and the origination of its oscillations, accompanied by frequency fluctuations. Very quickly, however, the frequency can return to normal under the action of an automatic voltage excitation controller in the generator or if the load of the generator is insignificant. However, the situation remains potentially dangerous as the frequency of the generator can change at any moment during the change of its load, thus the automatic reclosing of the circuit breaker will lead once more to the emergency mode of operation. For this reason, ordinary frequency relays are not applied in this situation. The ROCOF relays are capable of detecting the frequency fluctuations of

fractions of a second at once after the circuit breaker trips and stopping its automatic reclosing.

The set point ROCOF for the protective relays is calculated according to concrete parameters of a network, the generator, and loads,[10] and can be essentially different for various networks. For example, in networks of England this set point is a constant, 0.125 Hz/sec, but in nearby Northern Ireland it ranges between 0.45 and 0.50 Hz/sec. In Reference 11, it is shown that a wrong choice of the set point of the relay for this parameter results in false pickups, or insufficient sensitivity. This imposes certain requirements as to the accuracy of the protective relay.

7.6.2 The Algorithm of Frequency and ROCOF Measurements

The algorithm of frequency measurements in digital protective relays is related to the allocation of transition points of a sine wave through the zero value that allows eliminating the effect of sinusoid distortion on the accuracy of the frequency measurement (see Figure 7.4).

The sinusoid signal, as a rule, is first converted to a square wave and filtrated, and then short pulses are formed in transition points of a sine wave through the zero value. Intervals between these pulses are filled with the high-frequency pulses produced by a highly stable quartz resonator with a fixed oscillation frequency (usually 100 kHz). The pulse counter, with a very high degree of accuracy, counts the number of these pulses located and disposed within the interval determined by the transitions through zero value of the sinusoid (that is, from period T of an input signal). The error in measurement of frequency by the modern digital relays with this algorithm does not exceed, as a rule, ± 0.01 to 0.005 Hz.

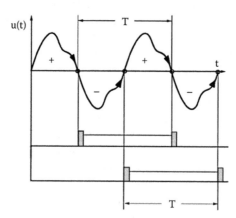

FIGURE 7.4
Principle for measurement frequency of the distorted periodical signal.

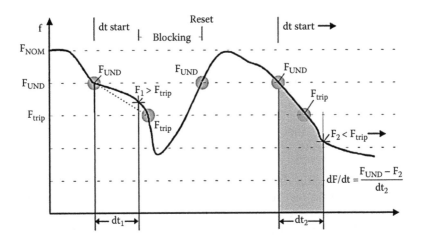

FIGURE 7.5
Principle of functioning of the ROCOF relay for load sheding. F_{NOM}: rated network frequency; F_{und}: lower frequency; F_{trip}: critical value of the frequency to starting ROCOF function; dt start: time interval for which relay remains in an active mode after the lower frequency (F_{und}) detecting; blocking: deactivation relay at expiration determined time interval (dt_1), even if frequency remains lower; reset: returning the relay to the initial position at increasing frequency up to and greater than the F_{und} value; and F_2: the lowest value of critical frequency in the time interval dt_2 for which ROCOF value is calculated.

The algorithm of measurement of the ROCOF is different. We shall observe the operation of this algorithm of such relay in an example of automatic load shedding (see Figure 7.5). As shown on Figure 7.5, the ROCOF function is started in the relay only if the value of the controlled frequency drops below the critical level F_{trip}. If this drop does not occur, the relay will deactivate after a specified time interval (dt_1), even if the frequency remains downgraded. The ROCOF function is put into effect only when the frequency in a network drops below the critical level F_{trip}. Therefore, the frequency is measured at two points: F_{und} and F_2 with a time interval dt_2 between these two measurements. If the value dF/dt for these measurements is more than the set point, the relay is activated, disconnects a part of the load, and reestablishes the power balance in the network.

A more complex algorithm, which is realized via auxiliary elements (Table 7.1), also determines a higher error of the relay in the ROCOF mode in distinction to the usual function of just monitoring the frequency. Whether we are talking about this algorithm or the one previously discussed, one thing is quite certain: the accuracy that is required when testing such a crucial protective relay as the ROCOF. Testing such relays is only possible in the presence of a special simulator that implements the ROCOF function. With the progress made in the field of microprocessor protective relays, many companies today make simulators for testing these relays. These simulators are even supplied with the capability of simulating the ROCOF function.

TABLE 7.1

Parameters of Some Common Types of the Frequency Protective Relays

Type	Manufacturer	Error of Pickup	
		On Frequency (Hz)	ROCOF (df/dt) (Hz/sec)
Module SPCF 1D15 to relay SPAF 340C	ABB	0.01	0.15
FCN950	ABB	0.005	0.05
MRF2	Woodward SEG	0.03	0.1

However, when testing the very precise FCN950 relay type in the ROCOF mode, an interesting fact has been detected: the relay behaved differently, depending on what type of simulator was used, thus the maximal error in the threshold of the relay pickups exceeded 10%. This is absolutely inadmissible, in our opinion. The analysis of the technical specifications of the simulators of the various types manufactured by the leading companies of the world—EPOCH-III (Multi-Amp), ORTS (Relay Engineering Service), F-2250 and F-6150 (DOBLE), PTE-300-V (EuroSMC), DVS3 mk2 (T&R Test Equipment), CMC256 (Omron), T-1000 and DRST-6 (ISA), PTR233/133 (Francelog Electronique), FREJA 300 (Programma), and MPRT (Megger), among others—has shown that there is no mention in the specifications of accuracy in the mode of ROCOF generation for any of these devices.

7.6.3 A Suggested Method for Precise ROCOF Measurement and Calibration

All manufacturers determine only the precision in the mode of continuous frequency generation. But as we have seen in Table 7.1, for protective relays the error in the ROCOF mode is approximately one exponent higher than in an operating mode monitoring an absolute value of frequency.

It is obvious that the same should be expected from the simulators which are implementing the ROCOF function simulation. The accuracy of the reproduction of an absolute value of frequency, declared by manufacturers, does not mean the same accuracy of ROCOF reproduction. What is this accuracy? And how are we to calibrate the simulator used for testing such crucial devices as the protective relays?

To solve this problem, we have developed the following simple technique. The output signal generated by a simulator in the ROCOF mode (with parameters close to the set point of the protective relay) is recorded with the high resolution by means of a digital recorder in a memory-recording mode (we used a multichannel Hioki-8842 digital recorder). Further, a time axis of the plotted sinusoidal signal was marked by means of cursors at a fixed time interval T (between 0.5 and 1 s), and by means of the cursors the

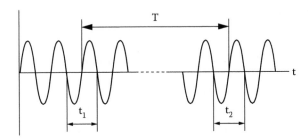

FIGURE 7.6
Suggested method for precise ROCOF measurement.

period for two single sinusoids was measured: beginning (at t_1) and ending (at t_2) in the fixed time interval (see Figure 7.6). The frequencies of the beginning ($f_1 = 1/t_1$) and the ending ($f_2 = 1/t_2$) sinusoids within the time interval can be calculated. And from this, it is possible to calculate the ROCOF = $(f_1 - f_2)/T$.

The digital recorders (manufactured by Hioki, Yokogawa, and others) with high-resolution capabilities in a memory-recording mode enable determining the time intervals of the low-frequency (45–60 Hz) signals with accuracies of parts of a millisecond. This technique was used for calibration of simulator types F-2253 (Doble) and T-1000 (ISA) in a function of ROCOF simulation. The results of the calibration are presented in Table 7.2.

Using this technique, it is possible not only to make periodic calibrations of the simulators of any type in a mode of ROCOF generation, but also to check the usability of concrete simulators for testing of relays. For example, from the results presented above, it is possible to consider that simulator T-1000 is not suitable for testing the relay type FCN950, but it can be applied for testing the relay type SPAF 340C.

TABLE 7.2

Result of Calibration of Simulators in ROCOF Mode
According to a Suggested Method

	ROCOF, Hz/sec		
		Error	
Simulator's Set Point	Value Measured and Calculated by Suggested Method	Hz/sec	%
Simulator F-2253 Type			
0.4	0.395	0.005	1.37
Simulator T-1000 Type			
0.4	0.449	0.049	+12.15

References

1. Gurevich, V. *Electrical Relays: Principles and Applications*. Boca Raton, FL: CRC Press, 2005.
2. Gurevich, V. Microprocessor Protection Relays: New Prospects or New Problems? *Electrical Engineering & Electromechanics*, no. 3, 2006.
3. Gurevich, V. A Problem of Power Supply of Microprocessor-Based Protective Relays in Emergency Mode, *Electricity Today*. Transmission & Distribution, no. 8, 2006.
4. Gurevich, V. Peculiarities of the Relays Intended for Operating Trip Coils of the High-Voltage Circuit Breakers. *Serbian Journal of Electrical Engineering*, vol. 4, no. 2, 2007.
5. PJM. PJM Relay Testing and Maintenance Practices: PJM Interconnection: Relay Subcommittee. Norristown, PA: PJM, 2004.
6. Nelson, J. P., P. K. Sen, and A. Leoni. The New Science of Protective Relaying in the Petro-Chemical Industry. Industry Applications Magazine (*IEEE*), vol. 6, no. 2, 1998.
7. Testing Multifunctional Relays: An Updated Approach, vol. 4. Carrollton, TX: Omicron Electronics, 2010.
8. Gurevich, V. Reliability of microprocessor-based relay protection devices—myths and reality. Part I. *Engineer IT*, May 2008, pp. 55–59.
9. Gurevich, V. Reliability of microprocessor-based relay protection devices—myths and reality. Part II. *Engineer IT*, July 2008, pp. 56–60.
10. Vieira, J. C., W. Freitas, Z. Huang, W. Xu, and A. Morelato. Formulas for Predicting the Dynamic Performance of ROCOF Relays for Embedded Generation Applications: *IEEE Proceedings*. Generation, Transmission and Distribution, vol. 153, no. 4, 2006, pp. 399–406.
11. Ding, X., and P. A. Crossley. Islanding Detection for Distributed Generation. Paper presented at the IEEE International Conference Powertech 2005, 17–30 June 2005, St. Petersburg, Russia.

8

Electromagnetic Intrusions on Digital Protective Relays

8.1 Electromagnetic Vulnerability of DPRs

The problem of electromagnetic compatibility (EMC) of electronic equipment has arisen together with the advent of this kind of equipment, as it receives and transmits electromagnetic radiation. The problems resulted from both internal interferences of assemblies and external emission of various origins. For decades, the problems of EMC have been the prerogative of specialists in electronics, radio engineering, and communications. Suddenly, over the last 10–15 years, this problem has become critical for the power industry. Of course, rather high electromagnetic fields have always existed in electric power facilities. However, electromechanical devices which have been applied for decades in automation, control, and relay protection were not addressed to electromagnetic fields too much as there were no significant EMC problems encountered. The last two decades has showed a sharp change-over from electromechanical to digital protective relays (DPRs) and automation. Moreover, the change-over included both the construction of new substations and power plants and the replacement of old electromechanical protective relays (EMPRs) at the old substations, built in those days when nobody assumed using microprocessor technologies, with the up-to-date DPRs. The latter have proved to be very sensitive to electromagnetic interference coming "out of thin air," penetrating through operating power circuits, voltage circuits, and current transformers. Some malfunctions of DPR were caused by mobile phones[1] and similar types of equipment. There have been other cases, such as malfunctions of DPRs at operating capacities of the Mosenergo, Ochakovskaya, and Zubovskaya substations (all in Russia). The operating algorithm of protection was affected by lightning, excavation work nearby, electric welding, and other types of interference. The Lipetsk substation start-up was postponed for 6 months due to faults of DPRs while they spent nearly $1.5 million for the DPRs. As a result, the substation was commissioned using a set of conventional defenses.[2] In practice, the shorting on the 110 kV side can cause protection failures on the 330 kV side, and

interference during switching of the same voltage rating penetrated inputs
(through the auxiliary circuits) of the relay protection apparatus operating
under the other voltage rating.[3]

According to Mosenergo, faults due to improper operation of relay protection amount to 10% of the total number of malfunctions and basically relate
only to microelectronic- and microprocessor-based relays.[4] Such a high percentage of malfunctions due to insufficient EMC results from the fact that
the DPR interference sensitivity is much higher than that of traditional electromechanical protection. For example, according to Reference 4, when
electromechanical relay operation can be affected by the energy of 10^{-3} joules,
the energy of only 10^{-7} joules causes the malfunctions of the microchips. The
difference is about 4 orders of magnitude, or 10,000 times.

The level of damage depends on the insensitivity of each circuit component and the energy of the powerful interference as a whole, which can be
absorbed into the circuit without the appearance of any defect or failure. For
example, although the switching noise caused by the inductive load with an
amplitude of 500 V is a twofold voltage surge, it is unlikely to lead to a failure
of an electromagnetic relay with a 230 V AC coil due to its insensitivity to
this kind of interference and its short duration (it lasts only several microseconds). The situation is different if the chip is powered from a 5 V DC source.
Impulse interference with an amplitude of 500 V is a hundredfold higher
than the supply voltage of the electronic component and leads to the inevitable failure and the subsequent destruction of the device. Surge resistance
of the chips is several orders of magnitude lower than that of the electromagnetic relays.[5] Surge resulting from lightning and switching in power plants
can damage or destroy both electronic devices and complete systems. Long-
term statistics confirm that the number of such incidents of damage doubles
every 3 to 4 years.[5] This statistic is in agreement with the so-called Moore's
law.[6] In 1965 Moore showed that the number of semiconductor components
in microchips doubles roughly every 2 years, and this trend has remained
valid for many years. If, some 10 years ago, the so-called transistor-transistor
logic (TTL) chip contained 10–20 elements per square millimeter, and had
a typical supply voltage of 5 V, now the popular chip can contain nearly
100 complementary metal-oxide semiconductor (CMOS) transistors on every
square millimeter of the surface and has a supply voltage of only 1.2 V.

The up-to-date solid-state technologies, for example silicon-on-sapphire
(SOS), raise the number up to 500 elements per square millimeter of the
surface.[7] It is obvious that such chips require even lower supply voltages,
and it is even more obvious that such improved microelectronics integrity
reduces insensitivity of its components to a high-voltage surge due to the
diminishing thickness of insulating layers and reduced operating voltage
of semiconductor elements. This means that a faulty substation DPR would
work properly during laboratory testing, making it impossible to determine
the reason of the fault at the substation (Figure 8.1). Statistics gathered by
the representatives of major Japanese manufacturers of DPRs point clearly

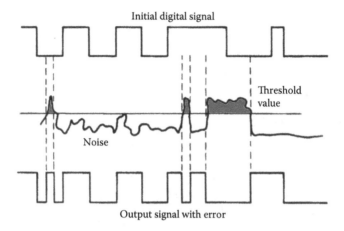

FIGURE 8.1
Impact of the low-energy interference onto digital devices.

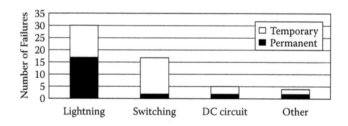

FIGURE 8.2
Statistics from Japanese manufacturers about the destruction of digital protective relay (DPR) due to electromagnetic impact.

to this feature of the DPR (Figure 8.2).[8] As seen from Figure 8.2, short-term nonrepeated malfunctions (faults) of DPRs prevail in most cases. This was also confirmed by another group of researchers.[9-14] According to their data, such faults account for almost 70% of the total number of DPR damage, while up to 80% of the faults occur in microchips.

According to Reference 4 and practical experience of OJSC Mosenergo, there have been many cases of DPR failures caused by electromagnetic interference. The most obvious was the inclusion of Siemens relay protection devices at TPP-12 of OJSC Mosenergo as per the design developed by the Atomenergoproekt Institute. The design didn't take the EMC requirements into account. As a result, within the period of August–December 1999 alone, there were more than 400 false information signals registered on discrete and analog inputs of DPR[4] due to interference impact. It should be considered that the cost of each DPR failure is ten times higher than that of the electromechanical relay due to the high number of functions provided by each DPR.

8.2 Lightning Strikes

Lightning strikes are the most powerful sources of impulse effects on the equipment of power stations and substations. Nearly 50 lightning strikes hit the earth's surface every second, and on average each square kilometer of the earth's surface is stricken by lightning six times per year.

The lightning voltage can be up to 100 million volts. Usually, construction standards for lightning rods assume that the lightning current is about 200 kA and its duration is about 1 ms; however, in reality the lightning current rarely exceeds 20–30 kA. The temperature of the channel during the main discharge can exceed 25,000°C. The length of the lightning channel can be from 1 to 10 km, and its diameter can be several centimeters. When lightning strikes the lightning rod, the current (in the form of bell-shaped pulse; see Figure 8.3) flows into the ground and spreads in all directions up to dozens and even hundreds of meters away, resulting in voltage drops due to ground resistance. Since the soil layers closest to the current entrance point have the largest resistance, they also show the highest voltage. The farther away from the point, the lower the current resistance and voltage (Figure 8.3).

In order to reduce the potential induced upon the flow of lightning current into the ground, residential and industrial buildings should be equipped with a large enough metal grid embedded under the foundations. However, the resistance of such grounding systems is still far from zero (Figure 8.3) and even the residual pulse potentials induced in the grounding system and penetrating through the cables to the inputs of electronic equipment are dangerous. Interference of this kind is called *conductive*. Besides this interference, high-current impulses flowing through the lightning rod generate disturbing electromagnetic fields that affect all conductors located nearby. This is the so-called *inductive effect*. There are also capacitive pickups, when short (i.e., high-frequency) surge voltage pulses from high-voltage power lines enter the low-voltage circuits through capacitive couplings of the transformer windings.

During the propagation, the interference transforms many times from one kind into another, so such classification is rather tentative, especially when it comes to high-frequency processes (lightning discharge current impulse with rather steep fronts of 8 microseconds and 20 microseconds [Figure 8.3] can be considered as such a high-frequency process). Therefore, in order to make an accurate analysis of the current spreading into the ground through the grounding device, it is required to consider both of these components.

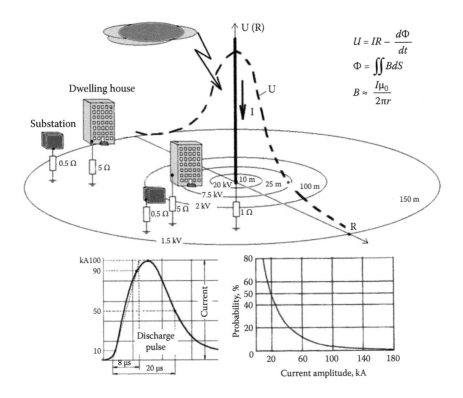

$$U = IR - \frac{d\Phi}{dt}$$

$$\Phi = \iint B dS$$

$$B \approx \frac{I\mu_0}{2\pi r}$$

FIGURE 8.3
Processes occurring in the lightning rod upon a lightning strike.

Moreover, the interference, which entered electronic equipment through electromagnetic fields or conductors, undergoes multiple transformations immediately inside the equipment due to parasitic capacitive and inductive couplings between elements or assemblies of the equipment. Thus, the high-frequency component of interference can propagate into hardware, bypassing the filters and protective elements. There is also another way for the lightning discharge interference penetration: the current flowing through the grounded metal DPR casing and grounded shields of numerous connecting cables. Therefore, it is very difficult to provide a proper level of electromagnetic interference protection. It is particularly difficult to realize at old substations where grounding systems were designed for electromechanical protection, which is much more stable to electromagnetic impact than microprocessor-based ones. Considering that dangerous potential peaks in grounding circuits occur not only upon lightning strikes but also upon emergency shorting in power mains, the problem becomes even more complicated. In some cases, grounding rings of power equipment are separated from those of electronic equipment to prevent such potential peaks in electronic equipment circuits. However, such separation is not possible to arrange at existing substations. It is our opinion that the only way to avoid the influence of high electromagnetic interference on DPRs is to develop a complex solution.

8.3 Switching Processes and Electromagnetic Fields Generated by Operating Equipment

Switching processes and electromagnetic fields generated by operating electrical equipment are the second most influential source of impulse interference affecting the DPR under normal operating conditions.

The usual sources of switching noise in the power industry are high-voltage switches and breakers, low-voltage relays and contactors, and controlled capacitor banks. Powerful electric drive frequency converters, corona discharge, and electrospark technologies generate electromagnetic emissions, which are dangerous for electronic equipment. Thus, the routes of interference penetration into DPRs may be very different: from direct induced pickups at low-voltage conductors and substation secondary circuit cables (Figure 8.4), to surge and high-frequency overvoltage occurring in the secondary windings of current and voltage transformers (Figures 8.5–8.7).

The less the duration of an arc's burning upon disconnection of the high-voltage circuit with switching apparatus, the greater the amplitude of an induced voltage surge in the secondary circuits. Therefore, the highest overvoltage is generated by vacuum circuit breakers, then goes to the SF_6 switch and oil switch, and then to a series of air circuit breakers. This explains the

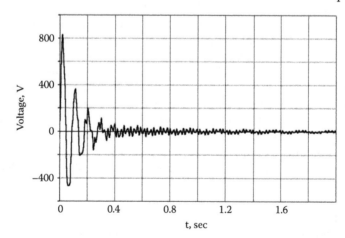

FIGURE 8.4

Voltage relative to the ground induced in a low-voltage control cable upon the switching process in a high-voltage circuit.[10]

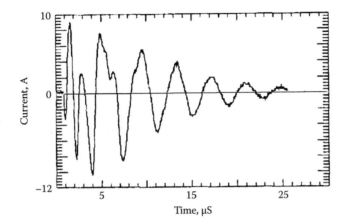

FIGURE 8.5
Current induced in the secondary winding of the current transformer upon switching the air circuit breaker with 500 kV voltage.[11]

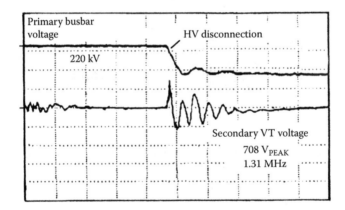

FIGURE 8.6
Pulse-switching overvoltage in the secondary circuit of a voltage transformer upon operating 220 kV air isolators.[12]

different amounts of DPR damage resulting from the operation of circuit breakers and disconnectors insulated with SF_6 gas and air (Figure 8.8). It should be noted that the high-voltage interference can be induced in the control cables by switching low-voltage circuits, particularly those that contain inductance. The nature of the switching transient process depends on many factors, and therefore the induced voltage can vary significantly even at the same substation. Since it is rather difficult to make theoretical calculations of such voltage surges, the easiest way to define them is to make direct measurements.

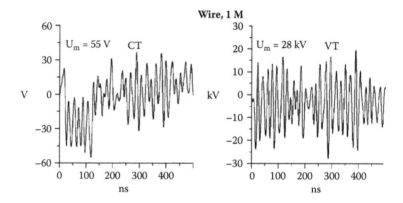

FIGURE 8.7

High-frequency noise in a 1-m-long piece of conductor, connected to secondary circuits of current and voltage transformers upon switching processes in SF$_6$ gas at 245 kV.[13]

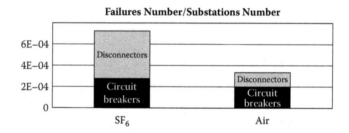

FIGURE 8.8

A comparative amount of DPR damages resulting from overvoltage upon the operation of circuit breakers and disconnectors with SF$_6$ gas and air insulation.[8]

Significant voltage surges, transformed into the secondary circuits, occur also on capacitor bank switching (Figure 8.9). An effective measure to avoid the induced voltage surge at the inputs of electronic equipment and its supply terminals is the wide use of elements with a nonlinear characteristic: gas-discharge arresters, varistors, special semiconductor elements based on voltage-regulator diode, and so on switched in parallel with the protected object (e.g., in parallel to the DPR input) and between each terminal of the object and the ground. So far, the best results are obtained from resistors with a nonlinear characteristic, made from compacted ZnO powder (at least from silicon carbide, barium titanate, and other materials), such as the widely spread varistors.

Today there are quite a few of them produced: without casing, in various casings, and with different auxiliary elements (fuses, alarm flags, etc.). Varistors must be properly selected. Unfortunately, wrong varistors are often installed even in equipment manufactured by leading world producers;

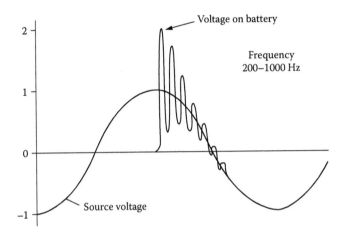

FIGURE 8.9
Overvoltage occurring upon capacitors switching.

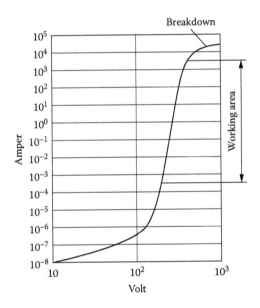

FIGURE 8.10
Typical current-voltage characteristic (IVC) of zinc oxide varistors (ZOVs).

thus, they provide no protection. Since the varistor current-voltage characteristic (IVC) is far from ideal (Figure 8.10), how to choose it properly can be a problem.

On one hand, the varistor should not transmit current of more than 1 mA (the default value for modern varistors of Western production) at maximum operating voltage (otherwise, it just overheats and burns). On the other hand, its "actuation" voltage, referred to as *clamping voltage,* should be much lower than the voltage that the protected equipment's electronic components withstand (otherwise, these components will "protect" the varistor). Since characteristics of IVC varistors do not allow them to fulfill these conditions, the maximum voltage that electronic components designed to operate in 220 V mains can withstand should not be less than 1000 V. However, such electronic components are more expensive than low-voltage ones, and their other characteristics are worse. For example, 1000–1200 V transistors have a significantly lower gain factor and much higher open-state voltage drops than the 400–500 V transistors of the same type. Therefore, the transistors with a maximum voltage of 500 V that they can withstand are often used in 220–250 V circuits of DPR power supply sources, emergency recorders, and other electronic equipment of the world's leading manufacturers. Varistors cannot protect the electronic components at such ratios of the operating and maximum withstanding voltage.

8.4 Issues with Control Cable Shielding

In order to protect control cables from induced voltages, they should be shielded and properly installed as far from the lightning rods and power cables as possible, or in cable trays. There are several types of such trays: plastic trays with aluminum inserts, metalized plastic trays, and aluminum trays.

In general, the efficiency of the metal screen (i.e., the degree of electromagnetic field attenuation) results from two properties: the absorption of energy during the passage of electromagnetic waves through a conducting medium, and the reflection of the wave at the interface between two mediums. Both of these phenomena depend on the electromagnetic wave frequency and the shielding material. The best electromagnetic absorption is provided by ferromagnetic materials (e.g., iron, Permendur, and Permalloy), and the best electromagnetic-wave reflection is provided by diamagnetic materials (e.g., copper and aluminum). The effectiveness of ferromagnetic material shielding decreases subsequent to the increase in field density due to saturation, while the effectiveness of diamagnetic shields decreases in proportion to frequency increase due to the increase in resistance. For a number of technical and economic reasons, copper mesh (braided) screens and aluminum profiles became the most popular.

Since the depth of electromagnetic wave penetration into metal is in an inverse ratio to the wave frequency, it is obvious that the thicker the metal shielding is, the better it will attenuate the electromagnetic field within a wide frequency range. For example, if a 0.6-mm copper screen provides effective shielding at 500 kHz, you'd need a copper screen of about 6 cm thick for 50 Hz (for a ferromagnetic shielding, the thickness can be 5 mm).

Based on the foregoing, it is obvious that plastic metalized trays have the lowest shielding effect, while they are widely used for shielding control cables. Such construction becomes effective only at frequencies exceeding 600 MHz. At frequencies below 200 MHz, it does not work at all. Usually control cable pickups at substations have much lower frequency than the specified 200 MHz, so the use of plastic metalized trays is fully senseless. However, aluminum trays and copper braid on the cables are still able to reduce induced voltage most of the time; that's why they are more widely used. The greatest attenuation of the pickups in a wide frequency range can be provided by installing control cables in steel water pipes.

For successful operation of shielding, it is necessary to ensure that the induced charge flows down into the ground. Ideally, the potential along the entire length of the shielding should be equal to the ground potential, so sometimes shielding of very sensitive high-frequency electronic circuits is equipped with multiple grounding per every 0.2 λ (λ – length of the electromagnetic field wave). At substations, shielded cables can be installed in parallel to the potential equalizing copper bus, grounded on both sides. However, the shielding is grounded more frequently on one or both sides (Figure 8.11).

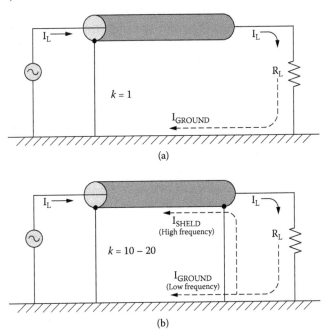

(a)

(b)

FIGURE 8.11
Operation of a shield grounded on one side (a) and on both sides (b).

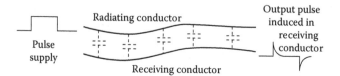

FIGURE 8.12
Pulse interferences through capacitive coupling between conductors.

Often protection engineers say that one-side grounding is sufficient for control cable shielding. Apparently, this resulted from such known measures as one-side grounding of current paths and one-side grounding of high-voltage cable shielding. Sometimes these two measures are used for grounding of control cable shielding regardless of the fact that grounding in these examples ensures electrical safety, but not protection from interference.

In fact, one-side grounding of the control cable shielding is effective only for capacitive interference (Figure 8.12; the so-called electrostatic protection), but not for inductive interferences (attenuation factor k = 1), since the shielding does not provide interference current closure circuits.

Two-side shield grounding provides an additional circuit (shield) with much lower signal impedance for high frequency than the ground. As a result, the operating signal is divided into two parts, one of which (low frequency) returns through the ground while the other (high frequency) goes through the cable shielding. Thus, for the high-frequency component, the shielding current is equal to the central cable's core current, going in the opposite direction, and is compensated by electromagnetic coupling between the shielding and the central cable's core. This also provides protection against high-frequency emissions from the central cable's core to the external space (i.e., to the neighboring cables) with the interference attenuation factor k = 3 – 20. This system is also effective with the external electromagnetic effects to the shielding, when a high-frequency signal induced to the shielding is close to the ground. When connecting the shielding to the ground bus, you should avoid wrapping the connecting cable over both the shielding and its coiling in order to reduce its length between the shielding and the ground bus. Each additional coil of the cable increases high-frequency impedance of the grounding system and significantly reduces its effectiveness.

Sometimes the sources of high-power interference at substations are unexpected and not obvious. For example, one Russian substation experienced false trips of one of the high-voltage circuit breakers (CBs) upon commands to the trip coil of the other CB. Control cables going to the trip coils of both CBs were not shielded and were installed in the same tray for approximately 25 m. Voltage oscilloscope studies showed that the pulses with an amplitude of 500–728 V may be induced to the trip coil of the CB1 by supplying 220 V control voltage to the trip coil of CB2. Sometimes the duration of such an induced pulse leads to a false trip of the CB (Figure 8.13a). The appearance of

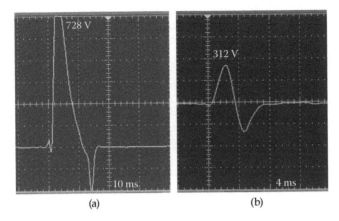

FIGURE 8.13
Oscillograms of an induced interference pulse from one control cable to another. (a) Non-shielded cables; and (b) one cable shielded and grounded on both sides.

such high-power pulse interference in the control circuits is confusing and even puzzling. Everything becomes clear if one recalls that the CB trip coil is equipped with a ferromagnetic core and has a relatively large inductance, while the CB is equipped with an auxiliary contact, breaking the current in the coil upon the CB trip. It is well known that the energy released upon the breaking of current circuit with inductance can be quite significant. After the control cable shielding at one of the CBs is grounded on both sides, the power of induced interference pulse at the second cable was significantly lower (Figure 8.13b), and false trips of the second CB were completely eliminated.

Two-side grounding may cause problems only if significant AC currents flow through the central conductor (typically, industrial frequency currents), generating high induced currents and heating the shielding. As a result, we should use larger wires (to reduce the heating of wire insulation) or ground one end of the shielding through the capacitor. The resistance of the capacitor is very high for the industrial frequencies, but it is very low for high-frequency interference.

In some cases, significant interference pulse currents can flow through a shield grounded on both sides, generating interferences into the central core. This can result, for example, from the high lightning strike current flowing through the grounding elements located near control cables, or occurring under the near short-circuit current (Figure 8.14). As shown in Reference 14, the peak voltage of interference at the central cable's core can reach up to 8.2 kV if the lightning's current is about 100 kA in the grounding conductor, even if the cable shielding is grounded on both sides. This is significantly higher than the DPR's resistance level. In these cases, we must either change the control cables' route (install them farther from the power-switching devices, lightning rods, and arrester) or reduce the potential difference between the

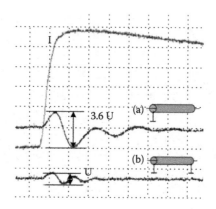

FIGURE 8.14

Voltage pickups at control cables with (a) one-side and (b) two-side grounding of the shielding during the flow of lightning's current pulse (I) through the grounding conductor.

grounded ends of the cable shielding during powerful pulse interference. In order to do this, copper bus should be installed along cables.

This so-called potential equalizing bus should be well grounded on both ends. The copper bus impedance at high frequencies is much lower than the ground impedance (and even the shielding impedance), and so the largest amount of high-frequency current of pulse interference will flow through this bus, rather than through the shielding. These measures, combined with the varistors mentioned above, will ensure the reliable protection of the DPR. Certainly these measures will be the most effective if adopted at the stage of design and construction of new substations, instead of when "patching up" old substations. The filter made of ferromagnetic (ferrite) ring or cylinder, which is mounted on the cable, has the opposite effect (Figure 8.15).

The impedance of the coil, which consists of one or two turns of control cable running through a ferrite ring, is very small for low-frequency operating signals and for the alternating current of industrial frequency, while it is very high for high-frequency (pulse) signals within a certain frequency range, which depends on the number of turns and the material of the ring. As a result, pulse and high-frequency interference, penetrating the cable, will be substantially reduced. Such filters can be widely used for relay protection

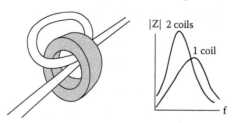

FIGURE 8.15

Ferrite ring filter and its frequency response.

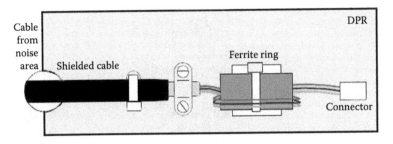

FIGURE 8.16
Installation of a ferrite ring filter on the control cable entering the DPR.

in supply circuits, logic signals transmission circuits, and the secondary circuits of current and voltage transformers (Figure 8.16). The effect of differential interference on the unshielded control cable (arising due to the difference between potentials induced on the direct and reverse wires) can be reduced by arranging conductors in such a way that the induced interferences will be similar, but opposite in voltage sign. This is achieved by twisting two AC conductors (direct and reverse). This method is effective, if frequencies do not exceed 5 kHz, while its effectiveness depends on the uniformity and density of the wire twisting.

8.5 Distortion of Signals in the Current Transformer Circuits

To discuss the distortions introduced by current transformers (CTs) in the signal controlled by protective relays, let's consider a typical design and characteristics of traditional CTs. High-voltage CTs, designed for installation in electrical networks (Figure 8.17), are equipped with a primary conductor (coil) 7, connected to the high-voltage line gap and several independent secondary windings with their own core (kern) 8. The basic insulation in such transformers is performed by multiple layers of special paper strips wrapped around the primary coil and interleaved with thin aluminum foil poured with liquid transformer oil (Figure 8.17).

Since all the secondary windings are completely independent of each other and have separate cores, they should have been called the *internal current transformers*. As a rule, these transformers have different electrical and magnetic characteristics, power, precision, and so on. Low-power high-precision windings are designed for connecting measuring instruments, while more powerful, but less accurate, windings are designed for connecting protective relays.

Like any other technical device, CTs show certain losses. Due to the loss the primary current is not completely transformed into the secondary circuit.

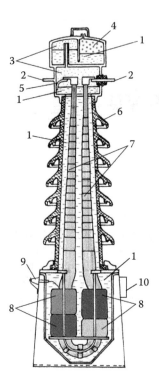

FIGURE 8.17
One of the common designs of a high-current transformer (160 kV) with paper-oil insulation.
1: oil; 2: copper plates to be included in the controlled current circuit; 3: containers with nitrogen; 4: air; 5: an internal jumper connecting one of the output terminals with the upper reservoir's metal casing; 6: a ceramic insulator; 7: an isolated primary conductor (primary coil); 8: four independent secondary windings with their magnetic cores; 9: an insulator on the outer wall of the container with the output terminal fixed to the container; and 10: a junction box with the secondary winding outputs.

These losses cause the CT *current error.* In addition, there is a phase shift in the secondary circuit in relation to the primary current, which causes the CT *angular error.* Losses in CTs heavily depend on the magnetic circuit condition.

A direct proportion between primary and secondary currents remains, until iron in the magnet core is saturated. An increase of primary current results in an increase in the magnet core iron saturation, and the corresponding characteristic begins to bend (Figure 8.18). Moreover, the increase in the CT secondary circuit load (i.e., an increase in load impedance) is followed by an increase in the characteristic's bending (since the effect of demagnetizing of the secondary current decreases).

It must be considered that these characteristics (Figure 8.19) are obtained during artificial CT testing and do not reflect the real relation between current and voltage under normal operating conditions of the CT. However, such artificial testing allows detecting various CT faults, and thus such CT

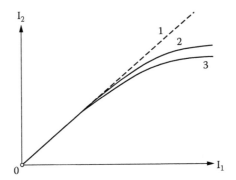

FIGURE 8.18
The relation of the secondary current (I_2) to the primary current (I_1) in the current transformers. 1: the ideal curve, 2: the real curve for the nominal load of the secondary circuit $Z_{2\,nom}$; and 3: the real curve for the heavy load of the secondary circuit $Z_2 > Z_{2\,nom}$.

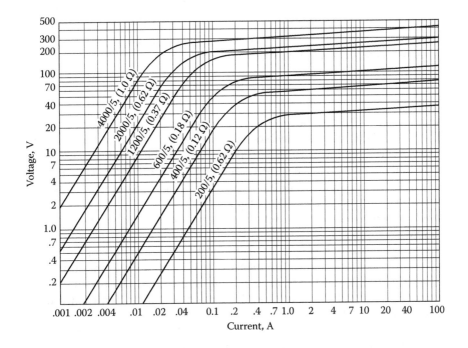

FIGURE 8.19
Actual current-voltage characteristics of the current transformers with different transformation coefficients represented in the manufacturer's documentation.

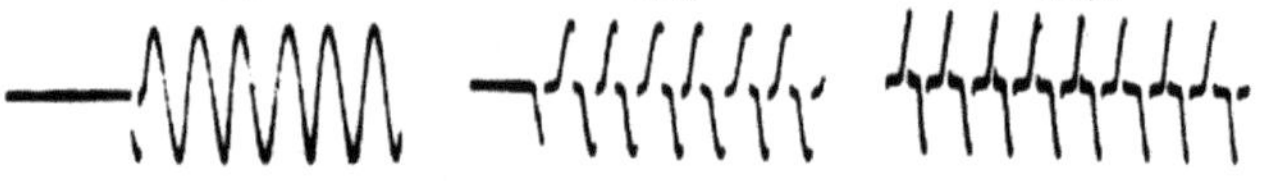

FIGURE 8.20
The form of the current transformer's secondary current under emergency mode (overcurrents). Error is shown in percentage.

characteristics are measured almost always upon the start-up of new equipment or during routine checks.

Measuring CTs operate within the rated current limits on the straight section of the characteristic, providing high precision of measurements. Measuring CTs are available with accuracy classes of 0.2, 0.5, 1, 3, and 5 (the class number corresponds to the error in %). Protection relay CTs operate in emergency mode, under currents far exceeding the rated level (i.e., on the magnetization curve arc), and their secondary current may be distorted (Figure 8.20). Therefore, the CT classes of relay protection include the primary current limit ratio related to its rated value at which the error still exists. For example, the *5P20* indication means that the error of the CT does not exceed 5% at primary currents exceeding the rated value up to 20 times.

Under otherwise equal conditions, the power of the load connected to the CT's secondary circuit should not exceed the rated CT power in order to maintain the defined error. For a given nominal current, for example 5 A, the load power is determined by its resistance:

$$P = Z_2 \times I_2^2,$$

where Z_2 is the load resistance, and I_2 is the secondary current.

Therefore, the lower the resistance of the external circuit connected to the CT (i.e., relays), the lower the CT load degree and the smaller its error. The nature of the load also has a significant effect on the CT error, as an increase in the inductive component of the load leads to the increase in current error and to the reduction of the phase displacement.

It would seem that the above-mentioned information about the CT operation and its main characteristics is basic and must be known by any protection engineer. Surprisingly, it was found that at some networks the relay protection was connected to the measuring CT and at the same time there were complaints about the failures of relay protection. The research report presented at the All-Russian Relay Protection Scientific Conference under the auspices of UES Federal Grid Company[15] seriously revealed that since measuring cores of CTs produces a highly distorted secondary current curve under the large primary current ratio, it was not recommended to connect relay protection to such cores. The report also recommended to "further investigate this issue." We hope that there are very few such pseudo-protection engineers in Russia.

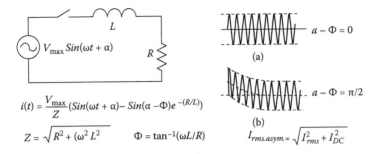

$$i(t) = \frac{V_{max}}{Z}(Sin(\omega t + \alpha) - Sin(\alpha - \Phi)e^{-(R/L)})$$

$$Z = \sqrt{R^2 + (\omega^2 L^2} \qquad \Phi = tan^{-1}(\omega L/R) \qquad I_{rms.asym.} = \sqrt{I_{rms}^2 + I_{DC}^2}$$

FIGURE 8.21
Transient process upon a live connection of power lines.

In fact, there were many CT distortion studies conducted. It is known that live connections of the power line lead to the occurrence of a transient process (Figure 8.21). It is known that the current of the transient process $i(t)$ upon circuit connection depends on periodic and nonperiodic components. The latter depends on the circuit parameters. Under the purely active load, the nonperiodic component equals zero, and the load current is absolutely sinusoidal (Figure 8.21a). The real power line has a certain inductance L, and therefore the nonperiodic component doesn't equal zero.

This component shifts the current sine wave in relation to zero (Figure 8.21b), at the first moment after the line switches on (or upon the sharp increase in short-circuit current) and until the component damping moment. This shifted sine wave can be mathematically represented as the sum of two components: the usual sine wave I_{rms} and DC component (I_{DC}), which represents the DC current.

Transformers with nonlinear magnetization characteristics (i.e., common CTs) are maladjusted for operating under the high nonperiodic component of the transient process. The direct current flowing through the CT's primary winding leads to a rapid saturation of the core and to a significant distortion of the periodic component's shape, and, subsequently, that of the secondary current (Figure 8.22). The CT's secondary current curve has the same shape when subjected to a magnetizing inrush current when the powerful supply transformer is turned on.

There are also significant distortions of the CT's secondary current, even if there is no DC component. At that, the saturation of the core occurs at a very high ratio of the primary current (Figure 8.23). At powerful electric energy systems with a large short-circuit currents ratio and significant time constant $L/R = 0.1$ s (Figure 8.21), the actual CT error can be very high (Figure 8.24), and the secondary current takes the shape of peaked pulses. Reduction of the CT's load and other measures have little influence on this error. The only measure that might reduce this error is a significant increase in wire section

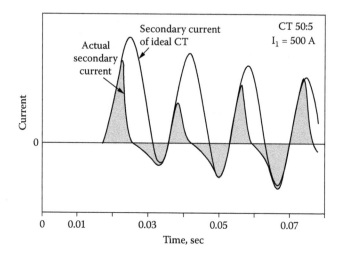

FIGURE 8.22
Distortion of the CT secondary winding's current under the constant component of the short-circuit current.

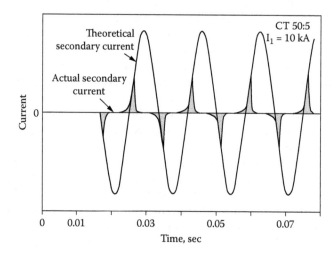

FIGURE 8.23
Distortion of current in the CT's secondary winding under the saturation at a high primary current ratio.

and mass of the CT magnet core iron. However, the calculations show that in this case, the size of the CT can reach unacceptable values.[16] The easiest way to reduce the CT error in transient mode is to use CTs with a primary rated current greater than the value obtained from the routine calculations. Thus, the DPR setup must include the real transformation ratio of the selected CT.

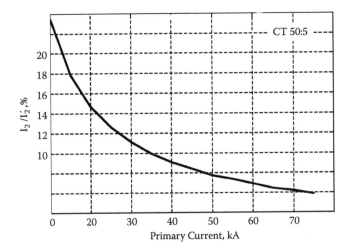

FIGURE 8.24
The amplitude of the real CT secondary current (I_2) in percentage of the ideal theoretical (I'_2) value, depending on the primary current.

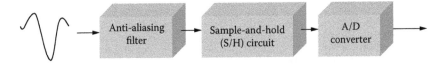

FIGURE 8.25
DPR analog signal input circuit (current and voltage).

It is known[17] that the input circuits of a DPR contain filter and analog-to-digital converters (ADCs) (Figure 8.25). The filter suppresses high-frequency signals, and since the distorted secondary current curve of the CT contains significant high-frequency components, the filter reduces the amplitude of the signal. Then, the ADC discretely measures the periodic signal at specific time intervals t (Figure 8.26). Because ADCs operate by sampling the input values at fixed intervals, it is clear that there is no way to determine the value of the input signal at intervals between such sampling (sectors 1–6 in Figure 8.26a).

If the secondary current curve is highly distorted, the error increases significantly (Figure 8.27). As a result, the signal generated at the ADC output has almost nothing to do with the real current. Time-current characteristics of DPRs are not maintained, sectors of remote protection are determined incorrectly, and so on. There are also specific DPR designs that provide fast computation of the second derivative of the current and use the obtained values for the threshold adjustment under the high nonperiodic component of the current.[19]

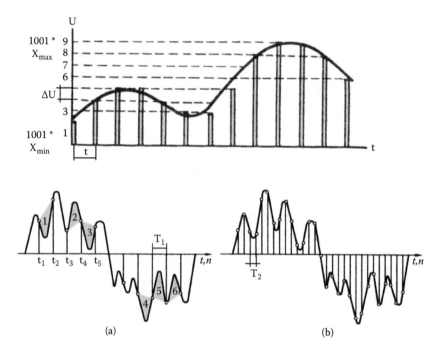

FIGURE 8.26
The operating principle of an analog-to-digital converter (ADC), and distortions that occur during transformation (sampling) of the signal.

8.6 Optical-Electronic Currents and Voltage Transformers

CTs and voltage transformers (VTs) of the optical-electronic type have been developed for a few decades in many countries (Figure 8.29). They are based on the application of Kerr and Pockels electrooptic effects (for voltage measurements) and the Faraday magneto-optic effect (for current measurements). The Faraday effect can be found in the rotation of the polarization plane of linearly polarized light in optically active material, caused by an external magnetic field (Figure 8.29c).

The angle for rotation of plane of polarization is given by

$$b = v\, Bd,$$

where β is the angle of rotation (in radians), B is the magnetic flux density in the direction of propagation (in Teslas), d is the length of the path (in meters) where the light and magnetic field interact, and v is the Verdet constant for the material (empirical proportionality constant varies with wavelength and temperature, and is tabulated for various materials).

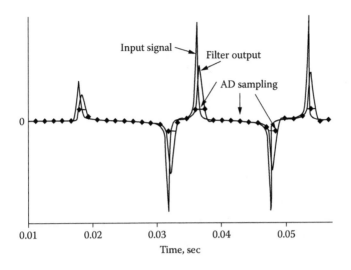

FIGURE 8.27
Transforming the secondary CT current at DPR input circuits. In order to compensate for this error, it is required to make the DPR input circuits significantly more complex, integrating new functional units (Figure 8.28[18]).

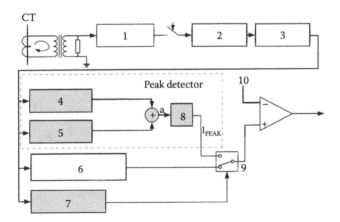

FIGURE 8.28
The structure of the DPR input circuits compensating the CT saturation error. 1: A filter; 2: an ADC; 3: storage of the last 16 samplings; 4: a detector of the maximum amplitude value; 5: a detector of the minimum amplitude value; 6: a peak meter for the basic frequency signal with a traditional filter; 7: a detector of the signal distorted by saturation; 9: an amplitude discriminator; and 10: a signal of the current cutoff element.

If the transducer is placed in the magnetic field of the measured current, one can determine the current strength by measuring the angle of rotation of the plane of light polarization. As a working substance in magneto-optical converters, glasses containing lead oxide (so-called flints or crowns) are usually used, and also fused quartz. Iron-garnet films are especially sensitive

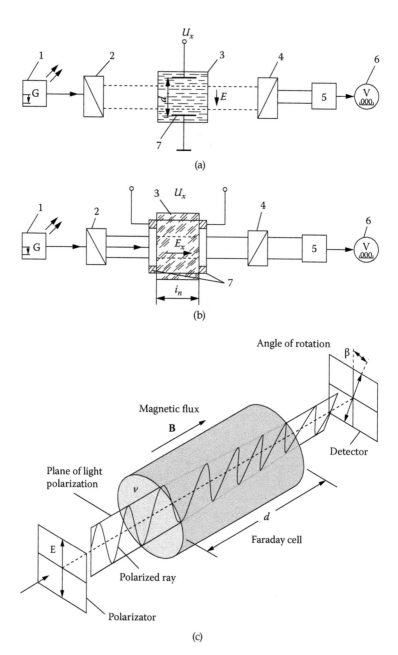

FIGURE 8.29
(a) Electrooptical transducers of Kerr. (b) Electrooptical transducers of Pockels. 1: A light source; 2: a light polarizer; 3: active material; 4: a polarization analyzer; 5: a photo detector; 6: an output element; and 7: electrodes to which the measured voltage is applied. (c) Electrooptical transducer of Faraday.

to magnetic fields. In this device (Figure 8.29c), the polarized ray from the grounded source comes through an optical fiber (or through any other type of light-guiding fiber) to the Faraday cell, placed on the high potential. In this optical cell the light flux changes its polarization vector, depending on the value of the magnetic flux affecting it. At that point the light ray, modulated in such a way, returns to the ground potential, where it is converted to electric current and is amplified.

In voltage transformers, Kerr or Pockels cells are used instead of Faraday cells (Figure 8.29a). The light flux in them is modulated not by the magnetic field, but by the electric field in the active material placed between the electrodes, to which the measured voltage is applied. The Kerr effect occurs in many isotropic substances (benzene, epoxy resin, etc.); very often nitrobenzene, which produces the biggest effect, is used.

The Pockels linear electrooptic effect can be observed in piezoelectric crystals placed into the electric field. This effect is seen most apparently in crystals of ammonium dihydro-phosphate ($NH_4H_2PO_4$) and potassium hydro phosphate (KH_2PO_4), in the longitudinal electric field created by ring electrodes (7) (Figure 8.29b). Pockels cells typically work with 5–10 times lower voltages than the equivalent Kerr cells.

Such devices have been designed for the last 30 to 40 years already, but only recently have optical transformers appeared on the scene. The NxtPhase Optical Current Transformer (Sensor) is intended for replacing conventional CTs in the range of 1 A (rms) to 63 kA (rms), from 115 kV to 500 kV (Figure 8.30). The sensor is based on the Honeywell Fiber Optic Gyro system. The sensor can be column mounted on an advanced polymeric insulating column or bus mounted with a suspension insulator to bring the optical fiber to ground. The NxtPhase Optical Current Sensor consists of a specialized opto-electronic converter of signal (item 1 in Figure 8.30) from a light-emitting diode into two linearly polarized signals, both sent through a polarization maintaining optical fiber to the sensing head. At the top of the column, there is a circular polarizer (2) that converts the two linear-polarized light signals into right and left circular polarizations. The light signals (3) travel around the conductor many times. The magnetic field created by the current flowing in the conductor slows one signal and accelerates the other (the Faraday effect). As the circularly polarized signals complete their path around the conductor, they are reflected by a mirror (4) and travel back through the fiber, the direction of their polarizations having been reversed. Along this reverse path, the effect is doubled. Both signals make their way back to the circular polarizer, which converts them back into linearly polarized light beams. Light 6 travels back down the column to the optical electronics (1). The difference in the speed of propagation has been translated into a phase shift between the linear signals. Since both signals have traveled an identical path, vibration and temperature changes have affected them equally—the highly accurate current measurement remains unaffected. Are the optical-electronic current

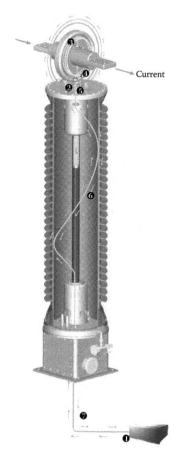

FIGURE 8.30
A NxtPhase T&D Inc. (Phoenix, AZ) combined optical current and voltage sensor.

transformers (OCTs) the panacea and solution for all of the above problems, or only a partial solution for some of the problems?

Traditional CTs have faithfully served the electric power industry already for more than 100 years. They comprise one of the simplest, most reliable, and most stable elements in the electric power industry. Therefore, when speaking of basic new devices and calling for them to replace the traditional ones, several arguments in favor of the new devices are put forward. We would ask if the arguments for the prompt transition to OCTs are really compelling. As a rule, the following reasons in favor of OCTs are brought out:

> *A wide pass band of signals (not less than 6 kHz), allowing making a complete analysis not only of the quantities, but also of the qualities of the electric power regarding harmonics (up to 100 harmonics) and transients for relay protection.*

Our question: What is the reason for such a pass band of CTs in the electric power industry, and why are 100 harmonics necessary?

> *High noise immunity to the electromagnetic interferences, allowing using OCTs in hard electromagnetic circumstances without preliminary analysis and correction.*

Our question: High noise immunity in comparison with what? With the traditional CTs? And is it true that traditional transformers had problems with noise immunity to the electromagnetic interferences? On the other hand, the OCTs contain rather complex microprocessor-based units for converting electric signals to optical and back and for realizing the Faraday effect, which, as well as all such devices, possess rather restricted noise immunity.

> *Longevity, long-term stability, and high repeatability of metrological parameters.*

Our question: Doesn't the traditional CT provide appropriate longevity, stability, and repeatability of results, sufficient for the needs of current measurement and relaying? Who has performed comparative tests for the reliability and parameters stability of the simplest traditional CTs and complex optical-electronic CTs with their microprocessor-based units?

> *A low susceptibility to vibrations and changes of temperature.*

Our question: Are the traditional CTs all that susceptible to vibrations and changes of temperature? Are the optical fiber and complex electronic systems more resistant against vibrations and changes of temperature than the copper windings with cast iron?

> *Simplicity and reliability of OCT design, high reliability, and self-diagnostics of electronic units minimize demands for maintenance servicing and testing of OCTs.*

Our question: From when were the complex optical systems with complex electronic units simpler than a copper winding on cast iron? On what theory is the basis that the most complicated electronic system is more reliable than a copper winding on the iron core? The necessity of self-diagnostics of electronic systems is more likely a negative index than a positive one because without such constant self-diagnostics, the system is not capable of providing the required reliability.

Now, it is true that expensive OCTs really have certain advantages, for example in the field of ultrahigh voltages when the cost of internal insulation of traditional CTs is very high, or at very high rates of primary current when the usual CT can be saturated. But in many other applications, the advantages of expensive and complex OCTs are very uncertain.

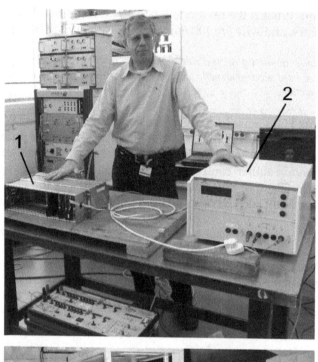

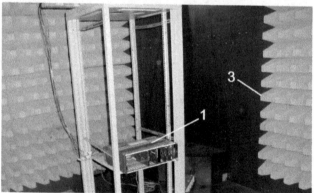

FIGURE 8.31

A fiber-optic communication for utility systems (FOCUS) unit during electromagnetic compatibility (EMC) tests in the specialized laboratory. 1: A FOCUS unit; 2: a generator of standard high-voltage pulses; and 3: a special covering in an anechoic chamber.

Recently, after the unpleasant incident (an inadvertent disconnection of a HV circuit breaker) with the fiber-optic communication for utility systems (FOCUS), used for converting electric relaying signals to optical, their transfer through optical cable and recovery of electric signals from optical on the end unit, we have tested it (Figure 8.31) for conformity to the EMC standards IEC60255 group: 60255-22-1, 60255-22-3, 60255-22-4, 60255-5, 60255-6, and

60255-11. During the tests, we were surprised at detecting that the system does not meet the requirements of most of the specifications of the standards and, hence, cannot provide reliable functioning of relaying under the effect of electromagnetic interferences. In particular, system functioning was broken, and communication between two FOCUS units completely stopped at applying high-voltage common mode pulses to the main power port and to the logical inputs of protective relays interface (PRI) ports. Moreover, in applying to PRI ports the standard high-voltage pulses (1.2/50 μs), according to IEC 60255-5, the internal breakdowns between electronic components on the printed circuit board occurred, and during the burst immunity test, according to IEC 60255-22-1, the internal communication setting was lost, requiring manual resetting to recover. All this confirms that the optical-electronic communication systems and optical-electronic transformers are not the panacea for power engineering, and have a narrow scope in some certain applications only.

8.7 The Harmonics Impact on the Measured Current and Voltage on DPRs

The harmonic distortion of the voltage and current entering the DPR input represent the components with higher frequencies, compared to the main frequency. Therefore, their effect on DPRs is similar to that described above, and is also associated with filtering and sampling errors in the ADC.[20] Studies by several authors demonstrated that the third harmonic has the greatest influence on DPRs. Among others this harmonic usually has the greatest value of voltage and current, which only exacerbates the situation. The study of microprocessor voltage relay with definite time delays[21] showed that this delay heavily depends on this harmonic distortion (Figure 8.32). Voltage

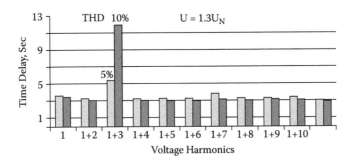

FIGURE 8.32
Effect of voltage harmonics distortion (5% and 10%) at the DPR analog input on the response time.

control elements are used in many DPR types, including such critical ones as distant line protections, which make this dependence on the harmonic distortion existence very dangerous.

As for the supply voltage harmonic distortion, their influence on DPRs is much lower, as virtually all DPR types are supplied with switched power supply with a filter and rectifier installed at the input. Rectified and filtered voltage is converted into high-frequency voltage, and changes in magnitude are stabilized, and then rectified and passed through the filter again.[22] All these conversions in switched power supplies reduce the effect of supply mains harmonic distortion on DPRs to zero.

8.8 The Quality of Voltage in the Supply Mains

As shown above, supply mains voltage harmonic distortions do not have a significant effect on DPRs. Also, they are less sensitive to voltage drops than electromechanical contactors and relays. Unfortunately, some authors provide incorrect data on DPR sensitivity to voltage drops,[23] making far-reaching conclusions and forecasts based on these data, which they didn't check.

We have made some research to examine the real stability of DPRs to voltage drops. For this purpose, we have selected several types of microprocessor protection relays from the world's leading manufacturers. Test results are shown in Table 8.1. In the last two cases it was not possible to determine the voltage drop time due to a high inrush current of the relay exceeding the power of the test equipment (Power System Simulator DOBLE 2253).

TABLE 8.1

Impact of Voltage Disturbances on DPR Operations

Type of Microprocessor Protection Relay and Manufacturer	Maximum Blackout Time Supply Voltage Withstands without Faults (s)	Maximum Supply Voltage Reduction to Ensure Safe Operation of the Relay with a Rated Voltage of 230 V (V)
SIPROTEC 7UT6135 (Siemens)	1.6	78
SIPROTEC 7UT6125 (Siemens)	1.6	36
SIPROTEC 7SJ8032 (Siemens)	3.8	44
SEL-311L and SEL421 (Schweitzer Electrical Laboratories)	0.6	45
T60 (General Electric)	—	80
P132 (AREVA)	—	45

As can be seen from Table 8.1, the microprocessor-based protection relay is capable of maintaining safe operations long enough both under full (i.e., down to zero) supply voltage drops and during prolonged deep supply voltage reduction down to 20–40% of the rated value. This can be explained by the integration of the internal switched power supplies equipped with inertia-less DC voltage regulators ensuring deep pulse-phase adjustment, a high capacity of electrolytic capacitors at the output, and very low consumption of modern DPRs.

With regard to the impact of pulse-switching surges of mains, they are less dangerous for DPRs as they arrive at an internal switched power supply, which includes input filters, high-voltage electronic components, and varistors. The only thing to ensure is that parameters of varistors and input components are correct. Conventional supply voltage harmonic distortion (harmonics up to 25) also has little effect on the switched power supply, as it is equipped with a full-wave rectifier and smoothing filter, installed directly at the input which converts AC harmonic voltage (of not very high frequency) into DC voltage.

8.9 Intentional Destructive Electromagnetic Impacts

As the title implies, this section is devoted to the intentional distant impact of electronic equipment to electromagnetic radiation aimed at its destruction or faulty operation. The difference between such effects and the above-discussed switching noise or pickups resulting from the lightning current is explained by the fact that when power is comparable to that of lightning discharge, these effects may be as close to sensitive equipment as sources of relatively low switching noise.

8.9.1 Background

In 1928, American physicist Arthur Compton theoretically predicted that the adverse factor of a nuclear explosion is represented by electromagnetic pulse. The test explosion of the first hydrogen bomb over the Pacific Ocean proved Compton's effect in 1958. The explosion was followed by such unexpected complications, observed hundreds miles from the explosion point, like street lights in Hawaii going out, the failure of radio navigation systems in Australia, and faults in radio communication systems in many other regions. This was where Compton was recalled. That caused panic: it turns out that a powerful stream of electrons generates such a high electromagnetic pulse that it can cause electric and electronic devices located at great distances to stop operating. Therefore, it can be used as a separate kind of weapon! Obviously the history of electromagnetic weapons began at that moment.[24]

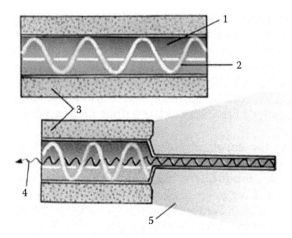

FIGURE 8.33
1: an electromagnetic resonator; 2: a standing wave; 3: an explosive substance; 4: a directed electromagnetic emission; and 5: scattered explosion products.

The first theoretical ideas about the nonnuclear shockwave super-powerful pulse generators (SPGs) were described by Andrei Sakharov, the member of the Soviet Union Academy of Sciences, during his research of nuclear fusion reactions in the early 1950s.[25] In the 1960s not only scientists but also politicians of the USSR understood that this source of super-power electromagnetic pulses could be the basis for creating a new type of weapon. This was revealed in addresses of N. S. Khrushchev in the 1960s, where he referred to some "fantastic weapons." Certainly, it took some time to create new weapons based on purely theoretical studies. The first report about SPGs, as independent weapons generating super-powerful electromagnetic pulses, was officially announced by the Soviet scientist A. B. Prischepenko after successful tests were made on March 2, 1984, at the weapon range of the Krasnoarmeisky Research Institute "Geodesiya" in the Moscow region (now the Federal Research Center "Geodesiya"). Later, Prischepenko formulated general principles for combat use of electromagnetic weapons. The SPG shown in Figure 8.33 generates a powerful pulse of electromagnetic emission at the moment of sudden contraction of resonator cavity.

A specific external source, even a low-power one, initiates a standing wave in the resonator cavity. The wave can either be maintained during a certain period of time, or be created a few moments before the explosion. An explosion of conventional explosives results in very quick contraction of the resonator cavity (either uniformly along the whole lateral surface, or at the end).

The electromagnetic field limited by the resonator cavity is quickly compressed, increasing the frequency of its oscillations. This way the explosion energy transforms into the energy of electromagnetic oscillations. Compared to the initial power, the power of electromagnetic oscillations increases a thousandfold. At this point, the pyro-cartridge destroys one of the resonator ends, and the standing wave transforms into the traveling wave with huge pulse power. According to published data, this pulse lasts tens or hundreds of microseconds, while the amplitudes of the emerging current reach tens of millions of amperes. Just to compare: the strength of lightning current during a storm discharge usually does not exceed 20,000–30,000 A, and very rarely it can reach 100,000 A.

According to Reference 26, in the 1980s the Soviet Union had repeatedly conducted experiments with electromagnetic weapons in space, which resulted in numerous faults of power systems in various U.S. states. About 20 years ago, I attended an advance defense of a doctoral thesis on the theoretical aspects of the transfer of energy from a space super-high-frequency source to the Earth. In those years, the USSR simultaneously conducted experiments on generating super-power electric discharges (which were the powerful sources of electromagnetic emission).

Many American newspapers and magazines of that time reported unusually powerful electric discharges over the territory of the USSR, which had never been seen before, and at zero storm activity. Some 25 years ago, I personally saw a picture of a super-long horizontal discharge between two towers above the houses of one village. Today, the information about the possibilities of generating such pulses can be found on the website of Istra Branch of the All-Russian Electrotechnical Institute (REI; Figure 8.34).

In the years of perestroika, Russian scientists A. Prischepenko, V. Kiselev, and S. Kadimov informed the world about a new type of weapon developed in the USSR in commemoration of the new era of relations with Western countries. Their report presented at the International Conference in France[27] was entitled "Radio Frequency Weapons at Future Battlefield." That report created a furor and became public domain. Thereafter, other reports on the achievements of Russian scientists in this area appeared in print.[28–30] Today, the issues on electromagnetic war and terrorism are freely discussed in the press, and at foreign and Russian scientific conferences.[31,32]

The first experiments on nuclear explosions in the atmosphere were not forgotten as well. Recent studies have shown that a nuclear explosion set off in near space (at the altitude of 200–300 km) would hardly be noticed by the population of the territory over which it was set off. However, all life-support systems (the power system, water supply, telecommunications, communication, etc.) would be put out of operation within a few moments. For this purpose, there are special IEC standards (see, e.g., Reference 33), which detail the methodology for testing the steadiness of electric network equipment to high-altitude electromagnetic pulse (HEMP). Special mobile simulators generating pulses similar to those that are induced in power line wires under HEMP were designed for such testing.

According to data cited in this document, the overvoltage in dead power lines under HEMP becomes so high that it causes a breakdown of even the linear insulators of 35 kV class and, naturally, of all lower-class insulators.

FIGURE 8.34

Experimental equipment of the Istra Branch of the All-Russian Electrotechnical Institute (REI). Left: A KTZ-MV device: the AC voltage in continuous mode is 3 million volts, and the switching pulse amplitude is 4 million volts; tests at AC voltage of 3 MV showed a 50-m-long electric arc, and at switching pulse it was 80 m. Right: GIN-9MV; this pulse generator with an output voltage of 9 MV generated anomalous sparks 150 m long.

If the same pulse penetrates the live power line, even 110 kV insulators are broken down (Figure 8.35).

Thus, there is no need to say what happens with all the rest of equipment that has direct, inductive, or capacitive couplings with power line wires. Today, the work on electromagnetic weapons is concentrated in Russia, mainly in three major research centers: the Joint Institute for High Temperatures (OIVTRAN, Moscow), led by V. E. Fortov, member of the Academy of Sciences; the Institute of High Current Electronics (IHCE SB RAS, Tomsk), led by G. A. Mesyats, member of the Academy of Sciences; and the Troitsk Institute for Innovation and Fusion Research (TRINITY), led by Professor V. E. Cherkovets.

Moscow OIVTRAN is creating magnetic explosion generators of super-power electromagnetic pulses (Figure 8.36).[34] Tomsk IHCE SB RAS develops super-power ultra-broadband oscillators of directed nonexplosive electromagnetic emission (Figure 8.37).[35] And TRINITY generates super-power plants and studies the physical effects of the super-power electromagnetic pulses (Figure 8.38). There is no need to have any special imagination to imagine how one can put mobile generators weighing 300–400 kg (Figure 8.37) in a light truck or a van with a plastic body in order to affect the electronic

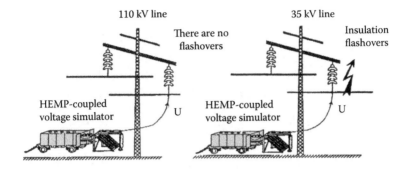

FIGURE 8.35
Illustration from the international standard IEC/TR 61000-1-3 representing tests of 35 kV and 110 kV power lines with a high-altitude electromagnetic pulse (HEMP) simulator.

FIGURE 8.36
Equipment of OIVTRAN (Moscow). Left: Magnetic explosion generator with an output current of 1 million amperes and a voltage of 1 million volts (power is 800 kJ). Right: Apparatus 13YA3 "SPHERE" designed for an explosion equivalent to 1 ton of TNT.

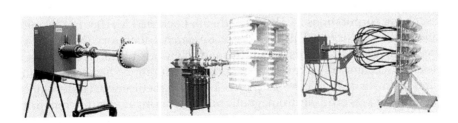

FIGURE 8.37
Super-power ultra-broadband pulse oscillators of a directed super-high-frequency electromagnetic emission of IHCE SB RAS with output power up to 1 billion watts, which is consistent with the capacity of nuclear power plant unit.

FIGURE 8.38
ANGARA-5-1 Unit designed for generating pulses of super-high electric power. The unit is based on an eight-module generator of electric pulses. The system is designed for generating electrical pulses of up to 12 billion watts.

equipment of substations and power plants, computer centers, mission control centers, and so on from a certain distance.

Several pulses of such emissive power will be enough to burn out all electronic devices, including the DPRs, of course. This was, obviously, the suggested application of the latest device developed by OIVTRAN, which has recently been proudly announced by the director of the institute, Academy of Sciences member V. E. Fortov, to ITAR-TASS. This device is the explosive electromagnetic generator with the pulse power of 1 billion watts, which can be put in a small suitcase while it is capable of burning out all the electronics within a radius of many hundred meters. And according to some reports, it damages even electronics that are switched off at the moment of action. Surely, many countries work on the same issues, including Israel, China, India, and Iran. In the United States, for example, this field is intensively researched by such big corporations as TWR, Raytheon, Lockheed Martin, Los Alamos National Laboratory, and the Air Force Research Laboratory (Kirtland Air Force Base, New Mexico), as well as by many civic organizations and universities. In Germany, Rheinmetall Weapons and Munitions has been leading in this field for many years. In particular, the Americans have developed generators of powerful electromagnetic pulses working under different principles (Figure 8.39).

It turns out that old written-off radars are similarly dangerous and can also be used for directed impact on electronic equipment. Why fool around with radars, if you only need to rework your microwave oven to destroy the

FIGURE 8.39
Left: FEBETRON-2020, a super-power mobile high-voltage pulse generator. The output voltage is 2.3 million volts, and the output current is 6000 amps. In the center: a powerful microwave generator mounted on the car. Right: a transportable powerful microwave generator.

TV of your neighbor? Such, I may say, "recipes" with drawings and detailed descriptions can also be found on the Internet today.

According to *Popular Mechanics* magazine, today even an amateur can make quite a battle-ready electromagnetic weapon with a radius of impact of several meters for about $400. The "spread" of such technologies is heavily based on the fact that they are widely used in the army and by police. Such "toys" as a hand emitter (Figure 8.39) can not only stop a car of criminals by burning out all of its on-board electronics and ignition system, but also completely destroy the electronic control systems of energy facilities, security systems, communications, and so on that are in the hands of criminals or terrorists.

The main channels of destructive force impact on the following electronic equipment: the power lines of all voltage classes, control cables and wire communication lines, and air. Since DPRs are connected to external power lines, branched networks of control cables and wires, and power line antenna-cables (by VTs and CTs) and computer networks, the destructive effect on them can be both very high and at the same time hidden.

The reticence of electromagnetic attack is increased by the fact that the analysis of damage in the destroyed equipment does not allow identifying the cause of damage, since the same damage can be a result of either intentional (attack) or unintentional (e.g., lightning induction) destructive forces. This circumstance allows an attacker to use this technique repeatedly with success.

High-frequency sources of high-power emission, operating in the centimeter and millimeter range, have an additional mechanism for the penetration of energy into equipment through so-called back doors, that is, even through small holes, openings, windows, and cracks in metal housings and through poorly shielded interfaces. Any hole that leads inside the equipment acts as a crack in the microwave cavity, allowing microwave emission to form an extensional standing wave inside the equipment. The components located at opposite assemblies of the standing wave will be exposed to strong electromagnetic fields and overvoltage. Memory elements and modern highly integrated microprocessors are especially sensitive to this kind of impact.

Thus, it is clear that it is not that easy to protect ourselves from all these "troubles." And even such well-known noise-resistant technology, like optical fiber, is prone (as strange as it may seem) to the impact of powerful electromagnetic pulses. First, optical fiber lines are equipped with micro-electronic-based and even microprocessor-based terminals designed to convert electric signals into light and vice versa. Second, it is known that the light polarization vector in optical fibers can be changed under an external magnetic field (strictly speaking, this is a basic principle for magneto-optical CTs available in the market today). Due to this, the signals of relay protection systems and communications transmitted by optical fiber, built in a power line wire (a very common technology today), will be subjected to distortions under high pulse currents flowing through the same wires and creating pulsed magnetic fields.

This is far from being a theoretic survey only. Today there are faults of such systems registered during the lightning current in power lines.

Certainly there is no way to protect high-sensitive electronic equipment of modern power stations and substations from natural and, particularly, intentional electromagnetic effects completely. However, modern technologies (special cabinets, conductive inserts and greasing, filters, etc.; see Figure 8.40) can significantly decrease the effect of external electromagnetic emission

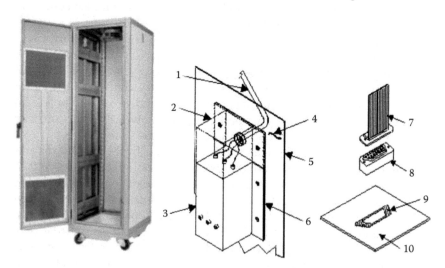

FIGURE 8.40
Special metal cabinet with ventilation windows protected against penetration of radio waves, and elements of filters ensuring advanced protection against external electromagnetic fields of a wide range of frequencies. 1: shielded power cable; 2: compartment shielding power cable's input and splicing; 3: powerful mains filter with overvoltage protection elements; 4 and 5: internal and external surface of the cabinet; 6 and 9: a special conductive rubber seal; 7: a signal cable; 8: a special high-frequency filter; and 10: a cabinet wall.

within the wide frequency range on high-sensitive apparatus, such as DPRs. Today such cabinets are produced by following companies: R.F. Installations, Inc.; Universal Shielding Corp.; Eldon; Equipto Electronics Corp.; European EMC Products Ltd; Amco Engineering; and many others.

It is obvious that the use of special technologies to protect DPRs will lead to additional price increases of relay protection. But this is the price that one has to pay for machinery progress. But if we do not do it today, we can find ourselves in a situation when it will be late to do it, since the dependence of our life on electronics, computers, and microprocessors has become so high that carelessness in the field of protection of these systems from the intentional influence of direct electromagnetic emission is comparable with crime and can lead to unpredictable consequences. It is noteworthy that several years ago, the mass media were reluctant to publish articles on this topic, being afraid to attract attention of terrorists and criminals. However, after the last extensive damage in the U.S. power system, terrorists have paid attention to the dependence of modern Western civilization on the power industry in a range of their statements and threats. Afterwards, there was a considerable number of articles devoted to the issue of exposure of society's life-support systems to electromagnetic terrorism in *The New York Times* and other mass media. For example, in Reference 36 it is obviously shown that power systems are the main target of terrorist attacks.

Hence, the carelessness of power systems management and workers, who have shut their eyes to this problem for many years, is somewhat appalling. It is nothing short of criminal carelessness to neglect the huge amount of publications on this topic in special technical journals, the mass media, the Internet, and even books (Figure 8.41).[37–41] I was shocked that not only ordinary power energy specialists but also managers do not have even minimal knowledge on this topic. Moreover, my attempts to raise this topic in articles and on relay protection forums lead to mockery and "pooh-poohing."

The power industry's actual tendency to broaden the implementation of microprocessor-based devices of relay protection which control power equipment, on the one hand, and the tendency to increase the number of elements in microchips (which is accompanied by the decrease of their steadiness to external electromagnetic impacts), on the other hand, form a rather dangerous vector in the background of progress in the field of creation of distant destructive impact devices. And the "ostrich" policy of unwillingness to know and to understand future dangers has never had any good consequences.

It is fair enough to say that in Russia, which is actively developing devices for electronic equipment defeats, there are specialists in the areas of power industry and relay protection who understand the high level of danger and have taken necessary measures. For example, one of the core centers of advanced computer (intellectual) technologies implementation in the power industry of Russia, which was created on the base of Velikiy Ustyug electric networks Vologdaenergy and which comprises 35 substations, accepted a model, which initially presupposed not to take electromechanical protection

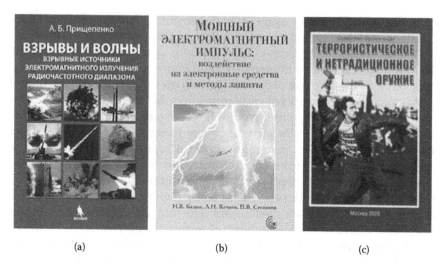

(a)   (b)   (c)

FIGURE 8.41

Some of the books published in Russia on the topic of intentional destructive electromagnetic impacts. (a) *Explosions and Waves: Explosive Sources of Electromagnetic Radiation in the Radiofrequency Range* by A. B. Prishchepenko[37]; (b) *The Powerful Electromagnetic Pulse: Effects on Electronic Equipment and Methods of Protection* by N. B. Baluk, L. V. Kechiev, and P. V. Stepanov; (c) *Terrorist and Nonconventional Weapons* by J. P. Sullivan.[43]

devices to the dump, but instead to develop and create new panels of relay protection on the base of new electromechanical protection relays, which are deemed to activate in critical situations, when a whole set of computer equipment can be knocked out of action.

Besides, the intellectual system of automatic control itself is especially developed for this proving area of the Russian power industry by the defense industry enterprises based on technologies applied in spaceships manufacturing. According to employees of this center, the reliability of their system far exceeds that of microprocessor-based systems manufactured by the leading relay-manufacturing concerns of the world. We think that such a visionary policy and such approaches should be taken into consideration all over the world. This way, we'll manage to avoid new disasters and huge financial losses.

References

1. Shalin, A. I. The Effectiveness of the New Relay Protection Device. *Energy and Industry of Russia*, vol. 1, no. 65, 2006.
2. Prokhorov, A. Intelligence Is Our Main Competitive Advantage (interview with General Director of OJSC ChEAZ M.A. Shurdov). The Equipment, Market, Supply, Price, no. 4, 2003.

3. Kuznetsov, M., Kungurov, D., Matveev, M., and Tarasov, V. Input Circuits of Relay Protection Devices. Issues on the High Surge Overvoltage Protection. *News of Electrical Engineering*, vol. 6, no. 42, 2006.

4. Borisov, R. Negligence to the EMC May Be Disastrous. *News of Electrical Engineering*, vol. 6, no. 12, 2001.

5. Pravosudov, P. Trabtech: Technology for the Surge Overvoltage Protection. Components and Technologies, no. 6, 2003.

6. Moore, G. E. Cramming More Components onto Integrated Circuits. *Electronics*, vol. 38, no. 8, 1965.

7. Nailen, R. L. How to Combat Power Line Pollution. *Electrical Apparatus*, December 1984.

8. Matsumoto, T., Kurosawa, Y., Usui, M., Yamashita, K., and Tanaka, T. Experience of Numerical Protective Relays Operating in an Environment With High-Frequency Switching Surge in Japan. *IEEE Transactions On Power Delivery*, vol. 21, no. 1, 2006.

9. Matsuda, T., Kobayashi, J., Itoh, H., Tanigushi, T., Seo, K., Hatata, M., and Andow, F. Experience with Maintenance and Improvement in Reliability of Microprocessor-Based Digital Protection Equipment for Power Transmission Systems. Paper 34-104, SIGRE Session, 30 August–5 September 1992, Paris.

10. Sowa, A. W., and Wiater, J. Overvoltages in Protective and Control Circuits due to Switching Transient in High Voltage Substation. Bialystok, Poland: Electrical Department, Bialystok Technical University, 2007.

11. Wiggins, C. M., Thomas, D. E., Nickel, F. S., Wright, S. E., and Salas, T. M. Transient Electromagnetic Interference in Substations. *IEEE Transactions on Power Delivery*, vol. 9, no. 4, 1994.

12. Carsimanovic, S., Bajramovic, Z., Ljevak, M., Veledar, M., and Halilhodzik, N. Current Switching with High Voltage Air Disconnector. Paper no. 229, International Conference on Power Systems Transients (IPST'05). 19–23 June 2005, Montreal, Canada.

13. Mohana Rao, M., Joy Thomas, M., and Singh, B. P. Transients Induced on Control Cables and Secondary Circuit of Instrument Transformers in a GIS During Switching Operations. *IEEE Transactions on Power Delivery*, vol. 22, no. 3, 2007, pp. 1504–1513.

14. Kuznetsov, M., Kungurov, D., Matveev, M., and Tarasov, V. The Problems with Protection Input Circuits in Relay Protection Equipment against High Power Impulse Overvoltages: Relay Protection and Substation Automation of Modern EHV Power Systems, 10–12 September. Moscow: Chebocsary, 2007.

15. Stolnikov, M. I. Malfunctions of Relay Protection at Using Current Transformers with Low Magnetization Curve. Report on XVII Scientific Conference, Relay Protection and Substation Automation, 16–19 May 2006, Moscow.

16. Afanasiev, V. V., Adoniev, N. M., and Dgalalis, L. V. Current Transformers. S.-P. *Energia*, 1980.

17. Gurevich, V. I. Microprocessor-Based Protective Relays: How Are They Constructed? *Electrotechnical Market*, nos. 4–6, , 2009, and, nos. 1–2, 2010.

18. Benmouyal, G., and Zocholl, S. E. *The Impact of High Fault Current and CT Rating Limits on Overcurrent Protection*. Pullman, WA: SEL Publications.

19. Zisman, L., and Gurevich, V. Fast Over Current Microprocessor Protective Relay: Theory and Practice. Thesis presented at the Electricity 2007 International Scientific-Technical Conference, SEEEI, Eilat, Israel, November 2007.

20. Zamora, I., Mazón, A. J., Valverde, V., Torres, E., and Dyśko, A. Power Quality and Digital Protection Relays. International Conference on Renewable Energies and Power Quality (ICREPQ'04), March–April 2004, Barcelona.
21. Gencer, Ö. Ö., Basa Arsoy, A., Özturk, S., and Karaarslan, K. Influence of Voltage Harmonics on Over/Under Voltage Relay. Izmit, Turkey: Department of Electrical Engineering, Kocaeli University.
22. Gurevich, V. I. Secondary Power Supplies: Anatomy and Application. *Electrotechnical Market*, no. 1, 2009.
23. Kartashov, I. Voltage Sags: Reality of Forecasts and Schematic Solutions for Protection. *News of Electrical Engineering*, vol. 5, no. 29. 2004.
24. Gurevich, V. I. Electromagnetic Terrorism: The New Reality of the 21st Century. *World of Technics and Technologies*, no. 12, 2005, pp. 14–15.
25. Sakharov, A. D. Magnetic Explosion Generators. *Advances in Physical Sciences*, vols. 83–84, no. 4, 1966.
26. The Shocking History of Soviet Russia's Electromagnetic (EM) War Attacks on the United States. http://www.bayside.org/news8/sovietelectromagnetic-attacksonunitedstates.htm
27. Prischepenko, A. B., Kiseljov, V. V., and Kudimov, I. S. Radio Frequency Weapon at the Future Battlefield: Electromagnetic Environment and Consequences. In Proceedings of the EUROEM94, Bordeaux, France, 30 May–3 June 1994, part 1, pp. 266–271.
28. Kadukov, A. E., and Razumov, A. V. The Basics of Technical and Prestrategic Application of Electromagnetic Weapon. *Journal of Electronics of St. Petersburg*, no. 2, 2000.
29. Russia Markets: The Weapon of the Future. *Independent Military Review*, vol. 39, no. 261, 19–25 October 2001.
30. Prischepenko, A. B. New Challenge of Terrorists: Electromagnetic. *Independent Military Review*, November 5, 2004.
31. Bogdanov, V. N., Zhukovskiy, M. I., and Safronov, N. B. The Electromagnetic Terrorism: The State of the Problem: Presented by the Scientific and Technical Center "Atlas" of FSS of Russia. In Proceedings of the Informational Safety of Regions in Russia: 2005 conference, St. Petersburg, 14–16 June 2005.
32. Daamen, D. Avant-Garde Terrorism: Intentional Electro Magnetic Interference: On Methods and Their Possible Impact. Report. Spring 2002.
33. International Electrotechnical Commission (IEC). IEC/TR 61000-1-3: Electromagnetic compatibility (EMC): Part 1-3: General: The Effects of High-Altitude EPM (HEMP) on Civil Equipment and Systems. Geneva: IEC.
34. Fortov, V. E., ed. *Explosion Generators of Powerful Electric Current Impulses.* Moscow: Science, 2002.
35. Mesyats, G. A Generation of Powerful Nanosecond Impulses. Moscow: Soviet Radio, 1973.
36. Commission to Assess the Threat to the United States from Electromagnetic Pulse (EMP) Attack (EMP Commission). Report of the Commission to Assess the Threat to the United States from Electromagnetic Pulse (EMP) Attack. http://www.empcommission.org/docs/empc_exec_rpt.pdf
37. Prischepenko, A. B. The Explosions and Waves: Explosion Sources of Electromagnetic Radiation of Radio Frequency Range. Course Book Means of Defeat and Ammunition, 170103 Department. Moscow: BKL Publishers, 2008.

38. Prischepenko, A. B. The Weapon of Unique Possibilities. *Independent Military Review*, no. 26, 17–23 July 1998.
39. Gannota, A. The Object of Defeat: Electronics. *Independent Military Review*, no. 13, 2001.
40. Gazizova, T. R., ed. *Electromagnetic Terrorism at the Edge of Centuries*. Tomsk, Russia: Tomsk State University, 2002.
41. Dobrykin, V. D., Kupriyanov, A. I., Ponomaryov, V. G., and Shustov, L. N. *Radio-Electronic Struggle: Forced Defeat of Radio-Electronic Systems*. Moscow: Vuzovskaya Kniga, 2007.
42. Baluk, N. B., Kechiev, L. V., Stepanov, P. V. *The Poweful Electromagnetic Pulse: Effects on Electronic Equipment and Methods of Protection*, Moscow, 2008.
43. Sullivan, J. P. *Terrorist and Nonconventional Weapons*, 2009, Moscow.

9

Alternative (Nondigital) Protective Relays

9.1 Universal Overcurrent Protective Relays

Overcurrent protection relays are the basic components in a majority of types of powerful electrical and electronic devices and also have uses in power engineering. Analysis of the trends in relay technology development shows that the major relay developers do not share a solid and consistent direction for simple current relay design improvement. For instance, some experts reason that certain families of electric-mechanical relays currently in mass production need to be replaced by static integral microcircuit (IC) relays. At the same time, they claim that regular electric-mechanical relays (including the simple overcurrent relays) are the most reliable and affordable for many electric utility companies, and will, therefore, be used in the majority of control and protection systems. It is important to note that the modern electric-mechanical relay is a high-speed device, which is insensitive to pulse and high-frequency interference and surge voltage. It exhibits a very robust behavior in overload modes and has a satisfactory reset ratio. One has to agree, though, that electric-mechanical relays usually consist of many expensive, high-precision components, the production of which becomes inefficient for the relay manufacturer. A dynamic measuring system with exposed contacts reduces the required relay reliability in a dust- and gas-intensive environment under a constant vibration factor. Besides, the need to clean and adjust the contacts contributes to increased labor-intensive maintenance.

Static relays, however, have a lower complexity and a better assembly factor, since they consist of standard electronic components mounted on printed circuit boards (PCBs). They require zero maintenance and are decently robust when subjected to environmental and mechanical impacts. At the same time, the threshold components, such as IC triggers and comparators as well as the transformer by means of which ICs are connected to the high-current circuits, cause an entirely new range of problems related to the interface immunity issue. The threshold components happen to be extremely sensitive to high-frequency signal interference, pulse peak interference coming through the feeding circuitry, and the like. Therefore, it is difficult to filter out the useful signal in the wide spectrum noise background for these components.

In compliance with the recommendations of the *International Electrotechnical Commission* (IEC), static relays are to be subjected to obligatory special noise immunity tests. At the same time, manufacturers of such relays do not recommend performing tests that include the application of electronic relay inputs having high-voltage (HV) pulses and powerful high-frequency signals. Moreover, it is not recommended even to use a megohmeter for insulation tests of such relays. Similar tests are resolved only for the short term, when all relay inputs are connected together. In this case, there is no sense in such tests!

The input relay transformer—an interface between the highly sensitive electronic module and the high-current circuit—transforms the useful signal as well as the noise. Besides, in many cases, the transformer itself becomes a source of the interference. While this issue is very critical for static relays, the electric-mechanical relays are well compatible with transformers. In this way, the transformed-based interfaces have found a wide range of applications in relay protection engineering.

We have used the above considerations as a basis for the new methodology in current relay design to combine the advantages of the two relay concepts (electric-mechanical and static).[1] The basic guidelines of our methodology are as follows:

- The threshold element, which in principle is a measuring organ, has to have an electric-mechanical structure to ensure interference immunity.

- It is expedient to use a reed switch equipped with a special module to move the former relative to the control coil.

- To ensure the compatibility of the reed switch specifications with the output commutation component, an interface unit should be implemented with the discrete electronic components (not ICs!) having wide current and voltage margins; the number of these components has to be minimal, and their schematics should not result in a threshold circuitry (such as a trigger, comparator, monostable vibrator, etc.).

9.2 Reed Switches Are New Perspective Elements for Protective Relays

Many engineers have come across original contact elements contained in a glass shell (Figure 9.1). However, not everyone knows that reed relays differ from ordinary ones not because of the hermetic shell (sealed relays are not necessarily reed ones), but because of the fact that in a reed relay a thin plate

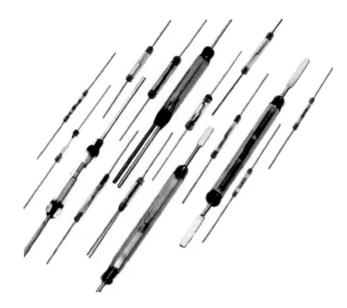

FIGURE 9.1
Different types of modern sealed reed switches.

made of magnetic material functions as contacts, a magnetic system, and springs at one time.[2] One end of this plate is fixed, while the other end is covered with some electroconductive material and can move freely under the effect of an external magnetic field. The free ends of these two plates directed toward each other are overlapped from 0.5 to 2.0 mm and form a basis for a new type of switching device—a *hermetic magnetically controlled contact* (in Russian) or *reed switch* (in English). Such a contact is called a *magnetically controlled contact* because it closes under the influence of an external magnetic field, unlike contacts of ordinary relays, which are switched with the help of mechanical force applied directly to them. The original idea of such a function mix, which was in fact the invention of the reed switch, was proposed in 1922 by a professor from Leningrad Electrotechnical University named V. Kovalenkov, who lectured there on magnetic circuits from 1920 until 1930. Kovalenkov received a USSR inventor's certificate registered under no. 466 (Figure 9.2).

In 1936 the American company Bell Telephone Laboratories launched research work on reed switches. Already in 1938, an experimental model of a reed switch was used to switch the central coaxial cable conductor in a high-frequency telecommunication system, and by 1940 the first production lot of these devices, called *reed switches*, was released (Figure 9.3).

Reed switch relays (i.e., a reed switch supplied with a coil setting up a magnetic field; see Figure 9.4.), compared with electromagnetic armature relays which are similar in size, have higher operation speeds and durability, a higher stability of transient resistance, and a higher capability to withstand

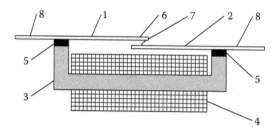

FIGURE 9.2
Kovalenkov's relay. 1 and 2: contact elements (springs) made of magnetic material; 3: the external magnetic core (the core of the relay); 4: the control winding (an external magnetic field source); 5: dielectric spacers; 6: the ends of contact elements; 7: a working gap in the magnetic system and between contacts; and 8: contact outlets for connection of the external circuit.

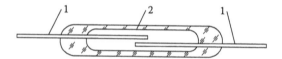

FIGURE 9.3
Construction of a modern reed switch. 1: contact elements (springs) from Permalloy; and 2: a glass hermetic shell.

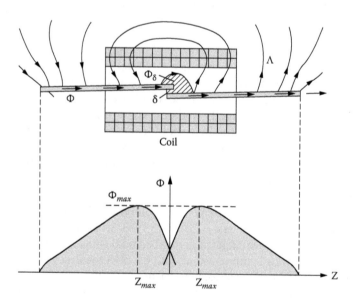

FIGURE 9.4
Magnetic field in a sealed reed relay. δ: The working (magnetic and contact) gap.

impacts of destabilizing factors (mechanical, climatic, and specific), in spite of their relatively low switching power.

At the end of the 1950s, some Western countries launched construction of quasi-electronic exchanges with a speech channel (which occupied over 50% of the entire equipment of an exchange) based on reed switches and control circuits—on semiconductors. In 1963 the Bell Company created the first quasi-electronic exchange of ESS-1 type designed for an intercity exchange. In a speech channel of such an exchange, more than 690,000 reed switches were used.

In the ensuing years, the Western Electric Company (New York, NY) arranged a lot production of telephone exchanges based on reed switches with a capacity from ten up to 65,000 numbers. By 1977 about 1000 electronic exchanges of this type had been put into operation in the United States. In Japan, the first exchange of ESS type was put into service in 1971. By 1977, the number of such exchanges in Japan was estimated in the hundreds.

In 1956, Hamlin (Lake Mills, WI) launched lot production of reed switches and soon became the major producer and provider of reed switches for many relay firms. Within a few years this company built plants producing reed switches and relays based on them in France, Hong Kong, Taiwan, and South Korea. Under its licensed plants in Great Britain and Germany, it also started to produce reed switches in those countries. By 1977 Hamlin produced about 25 million reed switches, which was more than a half of all its production in the United States. Reed switches produced by this firm were widely used in space-qualified hardware, including man's first flight to the Moon (the Apollo program). The cost of each reed switch thoroughly selected and checked for this purpose reached $200 a piece.

In the former Soviet Union, lot production of reed switches was launched in 1966 by the Ryazan Ceramic-Metal Plant (RCMP; Ryazan, Russia). Plants of the former Ministry of Telecommunication Industry (its Ninth Central Directorate in particular) were also involved in the production of weak-current relays based on reed switches. At the end of the 1980s, there were 60 types of reed relays produced in the USSR. The total amount of such relays reached 60–70 million a year. Economic crises in Russia led to a steep decline in production both of reed switches and of reed relays. In 2001 plants producing relays (those that were still working in Russia) ordered only about 0.4 million reed switches for relay production. Depending on the size of a reed switch, the working gap between contact elements may vary within 0.05 and 0.80 mm (and more for HV types), and the overlap of ends of contact elements may vary within 0.2 and 2.0 mm. Due to the small gaps between contacts and a small total weight of movable parts, reed switches can be considered to be the most high-speed type of electromagnetic switching equipment with a delay of 0.5–2.0 ms capable of switching electric circuits with frequencies of up to 200 Hz.

Hermetic Switch, Inc. (now HSI Sensing, Chickasha, OK), produces the smallest reed switches in the world (Figure 9.5). The tiny oval-shaped glass

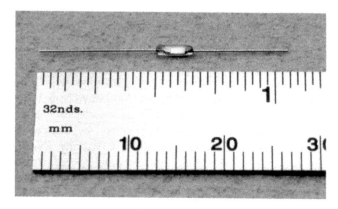

FIGURE 9.5
Smallest reed switch in the world, HSR-0025 type (Hermetic Switch, Inc.).

balloon of the HSR-0025 reed switch measures a mere 4.06 mm long, 1.22 mm wide, and 0.89 mm high. The maximum switching rating of the HSR-0025 is 30 V, 0.01 A, and 0.25 W. Sensitivity ranges from 2 to 15 ampere turns. Bigger reed switches can switch higher current, as the contacting area of the contact elements, their section, their contact pressure, and their thermal conductivity increase.

Most reed switches have round-shaped shells (balloons), because they are cut from a tube (usually a glass one), the ends of which are sealed after the installation of contact points. Glass for tubes should be fusible with softening temperatures and coefficients of linear expansion similar to the ones for the material of contact elements.

Contact elements of reed switches are made from ferromagnetic materials with coefficients of linear expansion similar to glass. Most often it is Permalloy, an iron-nickel alloy (usually 25% nickel in alloy). Sometimes Kovar, a more high-temperature alloy, is used. It allows the application of more refractory glass for tubes (560–600°C), and as a result, more heat-resistant reed switches are obtained. To provide a better joint with the glass, contact points are sometimes covered with materials providing better joints with glass than Permalloy. Sometimes contact elements have more of a complex cover consisting of sections with different properties. Contact elements may also contain two parts, one of which joins well with the tube glass and has the required flexibility, and the other one having the necessary magnetic properties. The contacting surfaces of contact elements of average-power reed switches are usually covered with rhodium or ruthenium, low-power reed switches designed for switching of dry circuits are covered with gold, and high-power and HV reed switches are covered with tungsten or molybdenum. Covering is usually made by galvanization with further heat treatment to provide diffusion of atoms of a cover to the material. It can also be carried out by vacuum evaporation or other modern methods. Contact elements of high-frequency reed switches are fully covered with copper or silver to avoid loss of or attenuation of high-frequency signals, and after that the contacting surfaces are also covered with gold.

The tube of a medium- or low-power reed switch is usually filled with dry air or a mixture of 97% nitrogen and 3% hydrogen. A 50% helium-nitrogen mixture, carbonic acid, and other mixtures of carbon dioxide and carbonic acid can also be used.

Carefully selected gas environments effectively protect contact elements from oxidation and provide better quenching of spark as low powers are being switched. Reed switches designed for switching voltages from 600 V up to 1000 V have a higher gas pressure in the tube, which may reach several atmospheres. HV reed switches (more than 1000 V) are usually vacuumized.

The facts that there are no rubbing elements, that there is full protection of contact elements from environmental impact, and that there is the possibility to create a favorable environment in the contact area provide switching and mechanical wear resistance of reed switches estimated in the millions and even billions.

Reed switches which are in mass production and are widely used in practice can be classified by the following characteristics:

1. By size
   - Normal or standard reed switches with a tube about 50 mm in length and about 5 mm in diameter
   - Subminiature reed switches with a tube 25–35 mm in length and about 4 mm in diameter
   - Miniature reed switches with a tube 13–20 mm in length and 2–3 mm in diameter
   - Microminiature reed switches with a tube 4–9 mm in length and 1.5–2.0 mm in diameter
2. By type of magnetic system
   - Neutral
   - Polarized
3. By type of switching of electric circuit
   - Closing or normally open: A type
   - Opening or normally closed: B type
   - Changeover: C type
4. By switched voltage level
   - Low voltage (LV) (up to 1000 V)
   - HV (more than 1000 V)

5. By switched power
   - Low power (up to 60 W)
   - Medium power (100–1000 W)
   - High power (more than 1000 W)
6. By types of electric contacts
   - Dry (the tube is filled with dry air or gas mixture, or is vacuumized)
   - Wetted (in the tube, there is mercury wetting the surface of contact elements)
7. By construction of contact elements
   - Console type (symmetrical or asymmetrical) with equal hardness of the movable unit (Figure 9.3: main type of reed switches)
   - With a stiff movable unit
   - Ball type
   - Powder type
   - Membrane type
   - Others

9.3 Polarized and Memory Reed Switches

Polarized reed switches are those reed switches that are sensitive to the polarity of the control signal applied to the control coil, in other words to the vector direction of the magnetic field F (Figure 9.6). Such sensitivity is caused by an additional static magnetic field produced by a permanent magnet placed nearby (or by an additional polarized winding, which is rarer). The external magnetic field of the control signal may have the same direction as the magnetic field of the permanent magnet. In this case, as their fluxes are summed up, causing operation of the reed switch, the sensitivity of the reed switch to the control signal increases considerably. If vectors of the magnetic fluxes have opposite directions, the resultant magnetic flux is so small that the reed switch cannot be energized.

One of the most important applications of such polarized reed switches is obtaining an opening (normally closed) contact out of a standard normally open one. In this case, the magnet is selected in such a way that its magnetic field is enough to energize and hold a standard normally open reed switch in such a state. If the direction of the control magnetic field of the coil is opposite to the direction of the magnetic field of the permanent field, the total value of the magnetizing force affecting contact elements will be less than their elastic force, and they will open, affected by these forces.

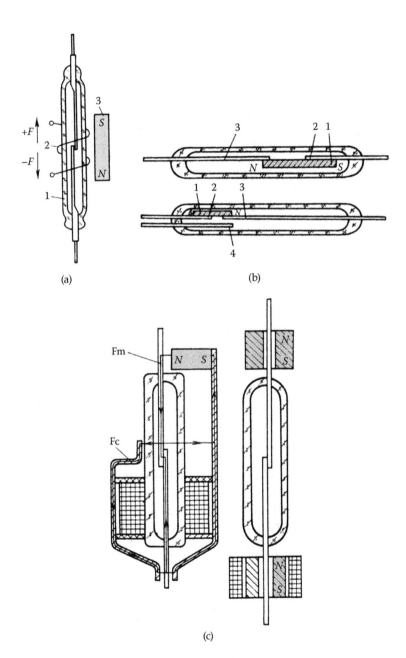

FIGURE 9.6
(a) polarized reed switch. 1: a neutral reed switch; 2: a control coil; and 3: a polarized permanent magnet. (b) polarized reed switches with an internal arrangement of magnets. 1: a permanent magnet with electroconductive covering; 2: a stationary contact element to which a magnet is welded; 3: a movable contact element; and 4: a second stationary contact element. (c) polarized reed switches with an external arrangement of magnets. Fm: the magnetic flux of the permanent magnet; and Fc: the control magnetic flux.

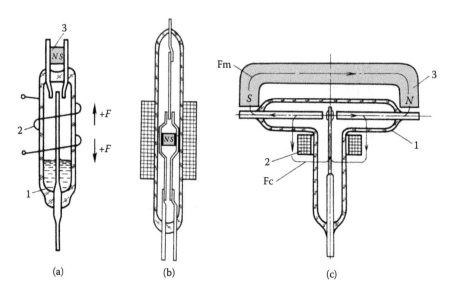

FIGURE 9.7

Three-position polarized reed switches. (a) a mercury reed switch with an external magnet; (b) a dry reed switch with an internal magnet; and (c) a high-frequency reed switch. 1: a glass tube; 2: a control coil; and 3: a permanent magnet with external insulating covering that can also be made of ferrite.

As far as construction is concerned, a magnet can be placed not only along the tube but also outside it, as is shown in Figure 9.6. There are a number of different constructions with original combinations of control coils and permanent magnets, some of which are shown in Figure 9.6. With the help of permanent magnets, it is possible to produce a three-position reed switch with a neutral midposition, which would switch this or that way under the effect of the magnetic field of the control coil of this or that polarity (Figure 9.7).

Using several control coils placed in different parts of a reed switch, instead of one, it is possible to produce reed switches capable of carrying out the standard logical operations AND, OR, NOT, EXCLUSION, NOR (OR-NOT), NAND (AND-NOT), XOR (EXCLUSIVE-OR), and others (Figure 9.8). If such multiwound reed switches are combined with permanent magnets (Figure 9.9), it is possible to obtain quite complex functional elements with adjustable operation thresholds, and remote switching of certain options. The number of such combinations is practically endless. This allows designers to implement almost fantastic projects.

Taking into account that the reset ratio of reed switches is less than 1 (i.e., for operation a stronger magnetizing force is needed than for release), one may try to choose a magnet of such a strength which is sufficient for operation of a reed switch, and at the same time capable of holding closed the contact elements which have already been closed by the control coil field. In this case, the reed switch is switched ON by a short current pulse in the

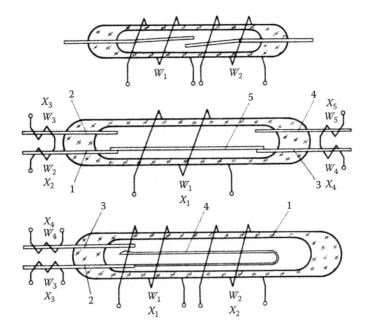

FIGURE 9.8
Multiwound reed switches designed to carry out standard logical operations.

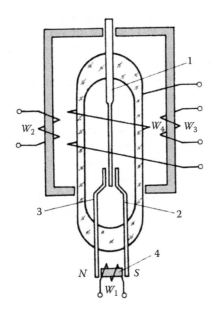

FIGURE 9.9
Combining a switching logical device on a reed switch. 1–3: Contact elements; and 4: a permanent magnet.

control coil and remains in this state even after the control impulse stops affecting it (i.e., it "memorizes" its state). The reed switch can be switched OFF by applying a control pulse of the opposite polarity to the coil.

Such a switching device, though in fact capable of operating, is not used in practice. There are several reasons for this. First, such a device requires very accurate adjusting because the slightest excess of magnetizing force of the permanent magnet will cause spontaneous closing of a reed switch. If the magnetizing force is not strong enough, the reed switch will not remain closed after the control impulse stops affecting it. Taking into account great technologic differences between parameters of reed switches, magnets, and control coils, each device will require individual adjusting, which is impossible for mass production. That's why even a device adjusted beforehand to a certain temperature may malfunction at other temperatures.

In 1960 A. Feiner and his colleagues from Bell Laboratories published in the *Bell System Technical Journal* an article entitled "Ferreed: A New Switching Device," in which they suggested that the way to overcome these difficulties was by creating memory reed switches. The idea was that the permanent magnet should become a magnet only when it and the reed switch are affected by the control pulse of the coil. Other details were technical. They chose magnetic material with a medium coercive force, which could be magnetized during the time of affecting of the control pulse, and remain magnetized for a long time, until the pulse of the magnetic field of the opposite polarity affects it (such material is called *remanent*). This device, consisting of a reed switch and a ferrite element, was called *ferreed* (by the first letters of the words *ferrite* and *reed switch*). Later, for advertising purposes, some firms began to name devices operating on the same principle in a different way, such as *remreed* and *memoreed*.

It turns out that ferrite can be remagnetized for 10–50 µs, while closing of contact elements requires a time of 500–800 µs. This allows us to use very short pulses for ferreed control (in practice, pulses with a reserve of up to 100–200 µs for closing are used). This means that contact elements are not only held after magnetization of the ferrite but also used for the contacts' closing process where the remanent magnetic flux of the ferrite after the short control impulse stops affecting it.

It is obvious that ferreeds with one control winding will be critical to the amplitude of the control pulse. When the amplitude of the switching current pulse (for switching OFF) is not high enough, the core in the control coil is not fully magnetized and the contacts will remain in a closed position.

If the control signal is quite strong, the core can be reversely remagnetized and obtain the opposite polarity. In that case, the contact elements will remain in a closed position also. To avoid this, two control windings are applied (Figure 9.10). The magnetomotive force of each winding is not enough to magnetize the core to the degree necessary for closing of the contact

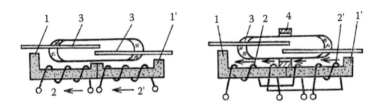

FIGURE 9.10
Ferreeds with two control windings. 1: a core from remanent material; 2 and 2′: control windings; 3: contact elements; and 4: an additional magnetic shunt.

elements. Only when switching-ON current pulses of opposite polarity are applied to both windings is the total magnetizing force enough to magnetize the core, closing the contact elements.

To open the contact elements, switching-ON current pulses of the same polarity are simultaneously applied to both windings. The polarity of magnetization of the halves of the ferrite core will be opposite, and both contact elements are likely to be magnetized; therefore, a repulsive force causing the contacts to open will arise between them. An additional shunt (4) made from soft magnetic material enhances configuration of the magnetic field in the overlap of the contact points, and reliability of operation of the device.

In a ferreed with a so-called orthogonal control (Figure 9.11), in order to change the state of the magnetization the vector turns by an angle of 90°, instead of 180° as in the previous case. The first such solution was patented by A. Feiner from Bell Laboratories (U.S. Patent 2,992,306). In his construction the magnetic flux of the winding (item 2 in Figure 9.11) for switching ON passes through a magnetic gap between the contact elements, and the magnetic flux of the winding (3) for switching OFF does not pass through the gap between the contact elements, providing reliable switching OFF of the reed switch.

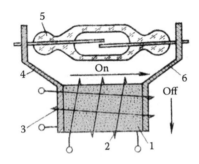

FIGURE 9.11
Ferreed with orthogonal control. 1: a core from remanent material; 2: a switching-ON winding; 3: a switching-OFF winding; 5: a reed switch; and 6: the magnetic core.

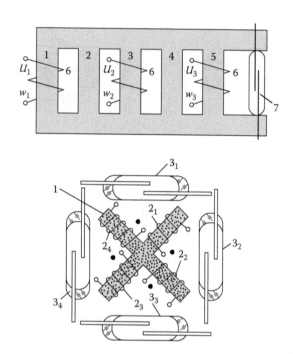

FIGURE 9.12
Automation logic elements on ferreeds.

As in multiwound reed switches, on ferreeds it is quite easy to implement single-circuit or multicircuit automation logic elements (Figure 9.12). For example, in multicircuit relays with a cross-shaped core (Figure 9.12), there are 16 possible combinations of closed and opened reed switches, depending on what windings are switched ON. In some constructions, one can enable or disable memorization options with the help of additional control signals (Figure 9.13). In the construction described above, a memory element of the external type is used.

Since the 1970s one can observe the rapid development of ferreeds with internal memory, which were produced by Hamlin, FR Electronics (Italy), and Fujitsu (Kawasaki, Japan). Their external design was practically identical to that of dry reed switches, but their contact elements were made of special alloys, providing sealing of the reed switch after it was affected by the pulsed magnetic field. Thus, no external elements are needed for such ferreeds.

Originally, contact elements of such ferreeds consisted of two parts, an elastic one and a hard magnetic one (from remanent material), but there were also excess joints with increased magnetic resistance. Later on, hard magnetic alloys were invented and used, so that the contact elements could be made more flexible and elastic enough. Such an alloy consists of 49% cobalt, 3% vanadium, and 48% iron, or of 30% cobalt, 15% chromium, 0.03% carbon, and the rest iron. There are also bimetallic contact elements

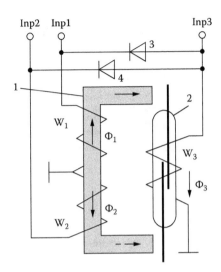

FIGURE 9.13
Device in which the memorization option can be disabled.

(U.S. Patent 3,828,828), the internal rod of which consists of an alloy of 81.7% iron, 14.5% nickel, 2.4% aluminum, 1% titanium, and 0.4% manganese, with the shell of the section made of an alloy containing 42% iron, 49% cobalt, and 9% vanadium.

9.4 Power Reed Switches

The American branch office of the Yaskawa Company (Waukegan, IL; main office, Tokyo, Japan) advertises its R14U and R15U type power reed switches, produced under the brand name Bestact™ (Figure 9.14), which belong to the same class of reed switches with partially detached magnetic and contact systems. This reed switch has a current carrying capacity up to 30 A. It can break similar current (as an emergency one) 25 times in an AC circuit with a power factor of 0.7. Switched power in the AC circuit is 360 VA (inductive load), maximum switched current is 5 A, and maximum switched voltage is 240 V. The electric strength of the gap between contacts is 800 V AC, and the mechanical life is 100,000 operations for R15U and 50,000 for R14U. Operating and release time is 3 ms.

On the basis of such reed switches, the Yaskawa Company produces a great number of different types of switching devices: relays, starters, push buttons, and so on. Relay designers are well familiar with the technical characteristics of these reed switches: a high level of protection from environmental

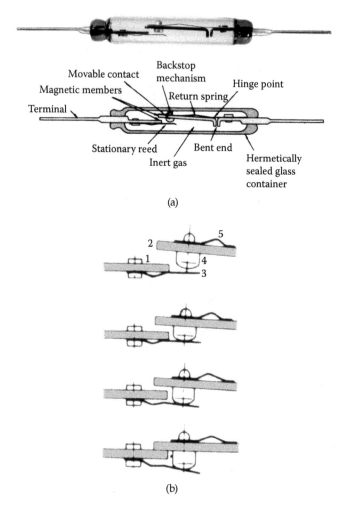

FIGURE 9.14
(a) the power reed switch Bestact™ produced by Yaskawa Electric America. Dimensions of the glass tube: diameter 6 mm, and length 37 mm. (b) the closing and opening process of the power reed switch Bestact™. 1 and 2: the main silver-coated contacts; 3 and 4: auxiliary contacts from tungsten; and 5: the spring.

impact, extreme reliability, a large communication resource, and zero maintenance requirements. It is less known, however, that their reset ratio in the AC magnetic field is about 0.8...0.9, the pickup ratio (not to be confused with its statistical variance) is relatively stable, and its adjustment function has been technologically resolved. The problem of the statistical variance of the magnetomotive force becomes irrelevant with the introduction of the reed switch adjustment module. The initial value of the relay can be defined at the manufacturing stage.

9.5 Overcurrent Protective Relays with Reed Switches

The mentioned principles have been implemented in a whole new family of relays, including the universal overcurrent relay, arc protection relay, short circuit indicator, and current relay with a nontransformer HV interface. Some of these developments are described in this section.

The Quasitron is a multipurpose protection relay, based on a hybrid (reed electronic) technology, with very high noise immunity (Figures 9.15 and 9.16). One relay unit may be used simultaneously with different current sensors, LV (Figure 9.17) and HV (Figure 9.34), each of which has a different current trip value. A current sensor may be mounted into the relay unit (as shown in Figure 9.15) or mounted outside the relay unit on an additional plate (Figure 9.17b). The relay unit has three time and current characteristics (T1–T3; Figure 9.18), one of which can be selected by a customer by means of jumpers on resistors R1 and/or R2 (see Figure 9.16). For this purpose, one (or two) jumper(s) can be cutting.

FIGURE 9.15
Hybrid overcurrent protection relay, Quasitron series (without a protection lid).

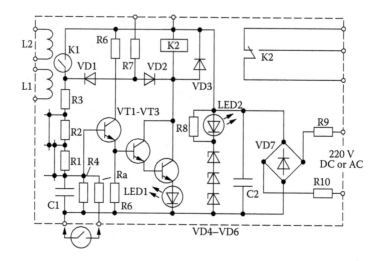

FIGURE 9.16
Circuit diagram of the Quasitron relay. K1: a reed switch; L1, L2: input current coils; and K2: an output auxiliary relay.

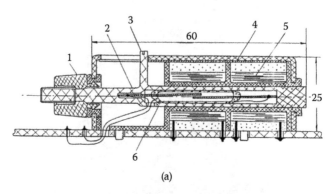

(a)

FIGURE 9.17
(a) the Quasitron current sensor with an adjustable current trip level. 1: a limb; 2: a movable dielectric capsule; 3: a level indicator of the current trip; 4: a ferromagnetic screen; 5: a coil; and 6: a reed switch. (b) external LV current sensors to Quasitron relays. 1: a cutting-circuit sensor type 1 for current trip 0.01 to 100 A; and 2: a sensor type 2 for busbar and cable installation (30 to 10,000 A). (c) outside dimensions of external LV current sensors. 1: external wires of a current circuit; 2: a plate; 3: a fixative element; and 4: a limb. "Output" is connected to the Quasitron relay input. (d) outside dimensions of sensor type 2 for busbar and cable installation. (e) circuit diagrams of type 2 sensors. a: for current level 100 A and more; and b: for low current levels.

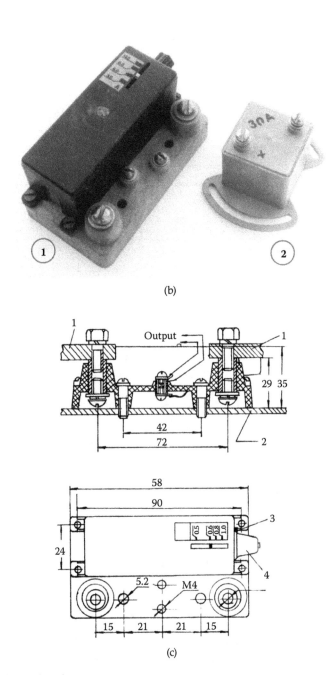

(b)

(c)

FIGURE 9.17 (continued).

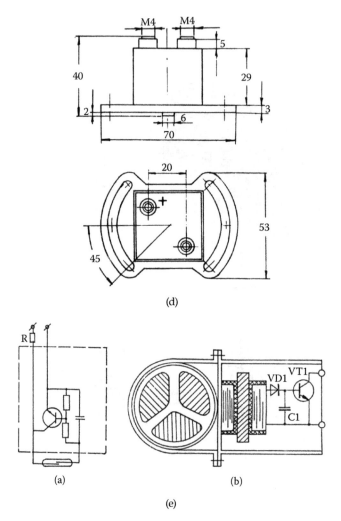

FIGURE 9.17 (continued).

All sensor outputs are connected to the relay unit via an LV wire. As the circuit configuration (Figure 9.16) implies, it does not contain ICs; its active solid-state components (transistors) do not constitute a threshold element and are merely used as an amplifier. An interface between the electronic circuit and the outside network bus is implemented via an insulated interface, based on reed switch K1, which also plays the role of threshold elements and starts vibrating with the double network frequency when the relay trips. The contact erosion-free capacity of the reed switch (about 10^6–10^8 operations) along with the short period of the maximum current relay's ON-state ensure the required commutation resource of the relay.

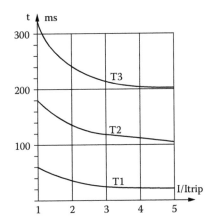

FIGURE 9.18
Time and current characteristics of the Quasitron relay.

The amplifying module of the base circuit (Figure 9.16) is nothing else but a compatibility link between the integrating couple L1-L2C1, and the output auxiliary relay K2 provides for the stability of the ON-state of the relay under the K1 vibration conditions. The feeding voltage of transistors (2N6517, 2N5657, or a similar series) does not exceed 50...70 V, the nominal operating range being 350 V. The control coil (K2) current under the tripping relay condition is as low as 15 mA, while the maximum collector current limit for these transistors is 500 mA. This magnitude of the current and voltage margins ensure a high level of the relay's operational reliability.

The high frequency and short pulse interference at the relay input cannot migrate to the electronic module, since K1, being the interface link, does not react to the high-frequency control signals due to the inherent inertia. Neither does it respond to the transient interference from the power circuit commutation. Therefore, the whole relay becomes very robust to the power circuit pulse interference.

The effect of the magnetic component of the dissipation fields can be neutralized by introducing the ferromagnetic screen into the relay design (see Figure 9.17). The 1.5-mm screen shields the reed switch in the fields with an intensity much higher than that of the dissipation fields under the actual operation conditions. For different applications of the Quasitron device, several types of output modules are available (see Figure 9.19).

Perspective improvement of this device is a current protection relay with a built-in timer (Figure 9.20). The time dial is adjustable in the range of 0.1 to 25 (±2.5%) seconds with a grade of 0.1 s. Dimensions of all modifications of relay units Quasitron and Quasitron-T device are 110 × 65 × 150 mm. All types of relays were tested. Some oscillograms are shown in Figure 9.21.

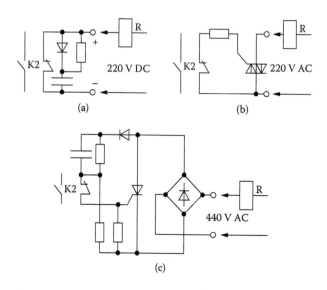

FIGURE 9.19
Output modules for Quasitron relays. K2: Contact of an output auxiliary relay, mounted on a PCB in the relay unit; R: a load of RL type; (a) with spark protection, for a DC load with large inductance; (b) with a power amplifier, for a power AC load (up to 500 VA); and (c) for an AC load, connected to a power supply with voltage more than the switching voltage of an output auxiliary relay.

9.6 Simple, Very-High-Speed Overcurrent Protection Relays

Overcurrent and overload protection functions for both LV and HV consumers of electric power (and also electric networks) are usually realized on current relays with dependent or independent time delay characteristics or on high-speed differential relays or impedance (distance) relays (for power line protection).

In some situations, however (at close short circuits and high-power sources), the multiplex overcurrent passing through the protected object is capable of causing destruction of the object, even when it is protected with one of the above-mentioned protection relays.

For such cases, special very-high-speed relays are stipulated. Usually the time delay of such high-speed relays, both electromechanical (e.g., KO-1, produced by ABB, Oakland, CA) and microprocessor based (e.g., SEL-551C from Schweitzer Engineering Laboratories, Pullman, WA; BE1-50 from Basler Electric, Highland, IL; and RCS-931A/B from Nari-Relays, Nanjing, China), is within 20 to 40 ms (as stated by manufacturers). In addition, electromechanical protection relays with instantaneous pickup characteristics frequently provide even higher speeds (18–25 MC) than microprocessor-based relays.

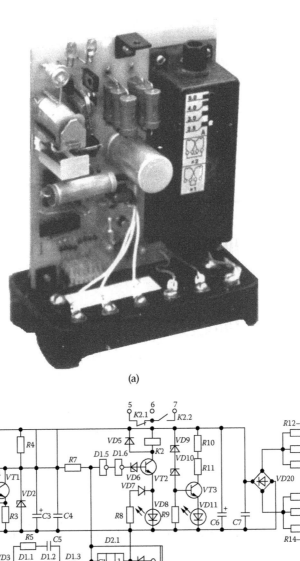

(a)

(b)

FIGURE 9.20
Overload protection relay Quasitron-T type with the built-in timer.

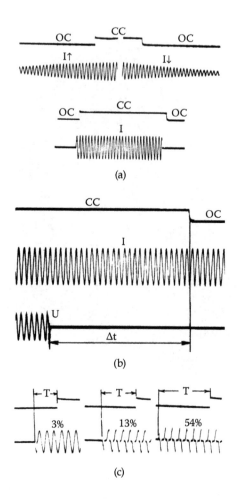

FIGURE 9.21

(a) some oscillograms, received during the test process. Sudden increasing and decreasing of control current (I). OC: an opened contact; CC: a closed contact; I↑: slow increasing of control current; and I↓: slow decreasing of control current. (b) some oscillograms, received during the test process. Unset of relay after tripping in case of sudden disconnecting from the power supply with voltage U (U = $U_{NOMINAL}$; frequency: 50 Hz). (c) some oscillograms, received during the test process. Influence of distortion of the measured current in relay input (simulation of overload of current transformer) on a response time (T) of the relay.

Promotional materials may sometimes be found which claim that a specially constructed high-speed microprocessor relay is capable of operating with a time delay of less than one period (less than 20 millisecond) (Figure 9.22). Such small operating time delays really can be realized sometimes for microprocessor relays with the injection of a high current with an artificially fixed phase for the first half-cycle (as on the oscillogram; Figure 9.22). Unfortunately in practice such extreme artificially

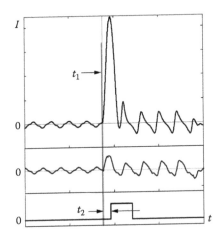

FIGURE 9.22
Oscillogram of an operation of a high-speed microprocessor relay of the SEL-487B type. According to promotional materials, its operation time is less than 20 ms.

created conditions are rarely achieved; therefore, such unique operation times look more like an advertising gimmick than a parameter provided under real operating conditions.

Many companies are engaged in the development and production of actual high-speed relays, for example the Israeli company SATEC.[3] The analysis of real transients of short circuits with high DC components and strong current transformer (CT) saturation (Figure 9.23) has brought some researchers to the conclusion that it is impossible to provide relay protection for operating times of about one half-cycle (10 ms). These researchers offer a new algorithm based on the measurement of first (di/dt) and even second ($di2/dt$) current derivatives.

The FOR2007 (Figure 9.24) operation is based on the fast two-stage overcurrent protection for each phase with an operation time of less than 7.5 ms. The first low-set protection stage has a pickup current equal to the protection pickup current and 3-ms operation delay. It responds to the peak current magnitude. The second high-set stage has a pickup current equal to 1.5 ms of the protection pickup current and a 0.5-ms operation delay. It responds to the instantaneous current. In the event that the input current exceeds a stage's pickup current, the stage operation timer is launched. If a stage operation time elapses before the current fault is cleared, the tripping command is sent to the output trip relay. The trip contacts are closed all the time while the fault is present and the programmable postfault time after the fault is cleared. The fault event is recorded in the device event log, and the fault current waveforms are stored in the waveform log file.

The protection of each phase consists of the following blocks and units (Figure 9.24). A common unit is the input hardware low-pass filter 1. The block 1CR is basic and consists of units 2 (calculation of the first derivative

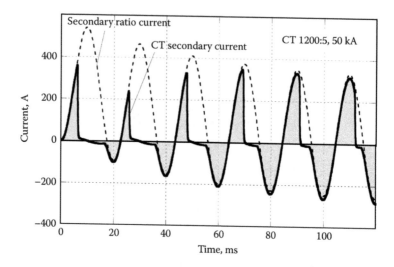

FIGURE 9.23
Shows the relation between the current transformer (CT) secondary current, applied to input of the relay, and the secondary ratio current at close short-circuit mode.

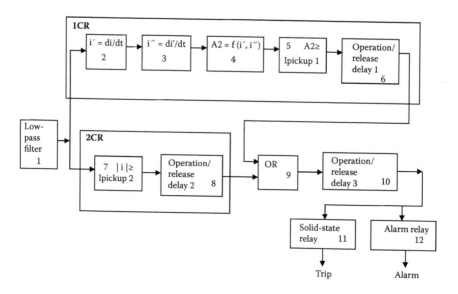

FIGURE 9.24
Block diagram of the fast digital protective relay FOR2007 (for single phase).

of a current), 3 (calculation of the second derivative of a current), 4 (calculation of a phase current amplitude A), 5 (a comparator, which compares this amplitude to a pickup current), and 6 (operation and release time delays). If a output signal on 1CR block is designated as T1, the operation conditions can be presented as follows:

- If T1 = 0 and A ≥ $I_{pickup1}$, T1 = 1,
- If T1 = 0 and A < $I_{pickup1}$, T1 = 0 (4),
- If T1 = 1 and A ≥ $k_{r1} \cdot I_{pickup1}$, T1 = 1, and
- If T1 = 1 and A < $k_{r1} \cdot I_{pickup1}$, T1 = 0,

where $I_{pickup1}$ is a pickup current of a protection, and $k_{r1} < 1$ is the release coefficient of the measuring block 1CR. The unit 1CR can operate at an instantaneous current more than 1.5 IN at least during 0.5 ms. An alternative (nonmicroprocessor) fast overcurrent relay on a reed switch is shown on Figure 9.25.

In reality, experimental oscillograms of transients (Figure 9.23) confirm the stability of such parameters as a current derivative (speed of change of current, or in other words an inclination angle of the front of the first pulse of a current at short circuit) even with high DC components contained in the current.

On the basis of this research, SATEC has developed a microprocessor relay with this algorithm. Thus the relay has turned out to be relatively complex because measurement of only the second current derivative is insufficient for realizing necessary relay stability. Inserting special elements for blocking of excessive relay operations is required because of the excessive sensitivity of the relay to some operating modes, as revealed. In addition, as the current derivative depends on a relation between an initial current before failure and a pickup current at failure, it appears that the relay does not always work

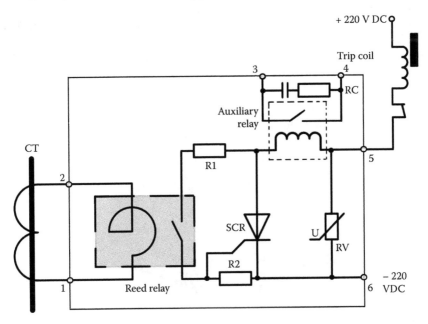

FIGURE 9.25
Basic circuit diagram of a simple very-high-speed overcurrent relay on reed switch.

properly if a relatively high load current is preceded to failure mode, and vice versa; excessive relay operations sometimes take place for great current changes (from zero value up to high values, but less than the pickup value).

Despite some technical problems, preliminary tests of the relay prototype have confirmed its high speed. For the most difficult cases, the time delay displayed was 8.4 ms, which is much less than any microprocessor relays existing today in the market. The electromagnetic interference compatibility and some other important relay parameters have not been investigated yet; nevertheless, the possibility of creation of the overcurrent microprocessor relay with an operation time of about a half-period has been confirmed.

I offer an alternative variant of a very simple and low-priced high-speed overcurrent relay with an algorithm based on the measurement of *instantaneous* value of a current. The relay is so simple that it can be produced by the staff of power systems. The offered overcurrent relay is based on a reed switch with an HV thyristor as an electronic amplifier (Figures 9.25 and 9.26).

The basic sensitive element in this device is the reed switch, which begins to vibrate at a pickup with a frequency of 100 Hz. Its first pulse opens a powerful thyristor, or silicon-controlled rectifier (SCR), which energizes a circuit breaker trip coil. The thyristor remains in the conductive condition, despite reed switch vibration, so long as the circuit is not turned-OFF by auxiliary contact of the circuit breaker. An addition of an auxiliary relay with a low impedance current coil and a spark-protected power reed switch may be used for energizing external electromechanical relays of automatic or signal systems.

Subminiature HV vacuum reed switches of the MARR-5 (Hamlin) or MIN-21 (Binsack Reedtechnik GmbH, Mühlheim, Germany) types, with withstanding voltage of 1.5–2.0 kV and turn-ON times of not more than 0.6–0.7 ms, are used as metering elements that provide high reliability of the

FIGURE 9.26
Construction of very-high-speed overcurrent relay. 1: a reed relay module; 2: a capacitor; 3: varistors; 4: a ferromagnetic screen of auxiliary reed relay; and 5: a thyristor.

relay. A thyristor of the 30TPS16 type was also chosen with a large reserve for current (30 A) and for voltage (1600 V), making it possible to choose the varistor (RV) for protection from overvoltages with a large reserve (clamping voltage of about 800 V DC) regarding rated voltage (220 V DC), providing both higher reliability and longer durability.

The reed switch module can be provided with different methods of pickup adjustment: by means of a moving reed switch inside the coil or by using different modules with different fixed values of pickup current. The last variant is quite acceptable, as this module is very simple and low priced. After adjustment of the reed switch position in the coil, it must be fixed by means of silicon glue.

Output auxiliary relays are also made as reed relays (without adjustment) because their winding is not standard relay winding but is designed as current winding (80–100 turnings) for current values suitable to the trip coil of circuit breakers. For such purposes, power reed switches can be used, for example R14U and R15U (Yaskawa Electric America); MKA-52202 (Russia); GC 1513 (Comus Group, Clifton, NJ); and DRT-DTH (Hamlin), provided with spark protection (RC circuit).

As usual, an HV circuit breaker (CB) has separate drivers and separate trip coils for each phase. All these coils may be connected in parallel. In an actual emergency mode, the single protective relay must disconnect not just one but also several CBs simultaneously. Otherwise the current of all trip coils may exceed the maximal permissible current for a single thyristor. In this case, a demultiplexor with single input is used, connected with a trip relay contact, and several separate outputs for circuit breakers, as shown in Figure 9.27. The prototype model of a 10 A pickup reed switch module (Figure 9.28) without a thyristor amplifier (the thyristor switch-ON time is less than 10 μs, which does not affect in any way the general time delay of the device) and without an auxiliary relay has been submitted to tests.

Tests were performed by artificial simulation of various modes on a current by means of a Power System Simulator F2253 (Doble Engineering, Watertown, MA) and also by injection in the module, by means of the same simulator, of real secondary currents of short-circuit transients restored from COMTRADE files of the real failures in a 160 kV power network, extracted from microprocessor-based transient recorders.

In the first series of experiences, the operation time of the module was measured at instant change of current on an input of the module in a range from 0.2–0.8 I_{PICKUP} to 1.2–5.0 I_{PICKUP}, with various random phases of current transition and also with a zero phase of current sinusoid (Figure 9.29). The tests verified that the lower-limit predetermined current value preceded to pickup current does not affect the operating time (Figure 9.30), as against microprocessor-based relay reactions to current derivatives. Research also affects harmonics (contained in a current) on operating time at different phase transitions of a current (Figure 9.31) and verified that even the high harmonics content does not affect operating time. The main factors are still

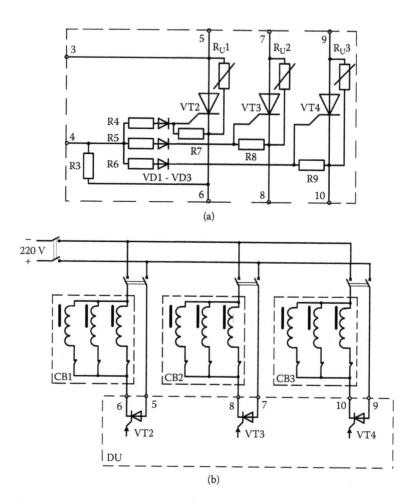

FIGURE 9.27
(a) a demultiplexing unit; and (b) a connection diagram.

FIGURE 9.28
Unit of a reed switch, submitted on tests with a rated pickup current of 10 A.

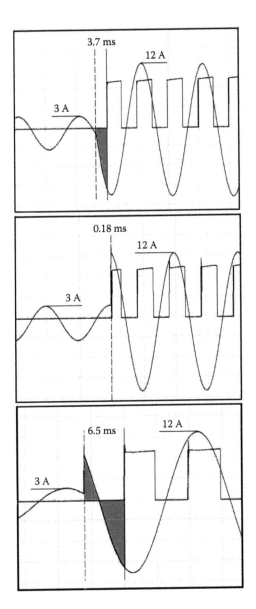

FIGURE 9.29
Some oscillograms of the operation of a reed switch unit at instantaneous changes of current with various phases of current transition. Nonoperating zones of the relay are marked.

the phase and magnitude of a current transient. For the most difficult case, that is, at small current $I = 1.2\ I_{PICKUP}$ and with a switching current phase at close to 45°, maximal operation time can reach 7–8 ms.

Heavier testing appealed for real secondary currents of short-circuit transients containing a high DC component, causing displacement of a sinusoid

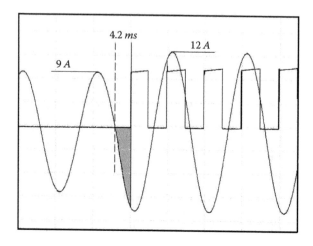

FIGURE 9.30
An oscillogram of operation of a relay with a previous nonoperating current (9 A) near a pickup (10 A).

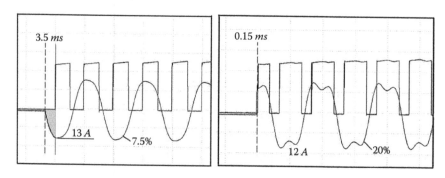

FIGURE 9.31
Oscillograms of operation of a relay at high harmonic content in current (for contents of the third harmonic of 7.5% and 20%).

of current concerning an axis (Figure 9.32). The maximal operation time fixed at these tests reached 9.4 ms. In addition, in some experiences with high DC components, pickup current decreased by as much as 0.7 of the rated pickup current. This occurred when the relay pickup phase occurred at the moment corresponding to the maximal displacement of the first half-wave of a current sinusoid.

For such conditions, relay picked up at much smaller current than at a normal sinusoid in the continuous mode. In our opinion, this phenomenon is not so essential, as the basic purpose of such high-speed relays is not exact current measure, but only detection of the presence of a dangerous short circuit for acceleration of action of basic relay protection. In other words, at adjustment of the relay for a primary current, for example 20 kA, it is possible

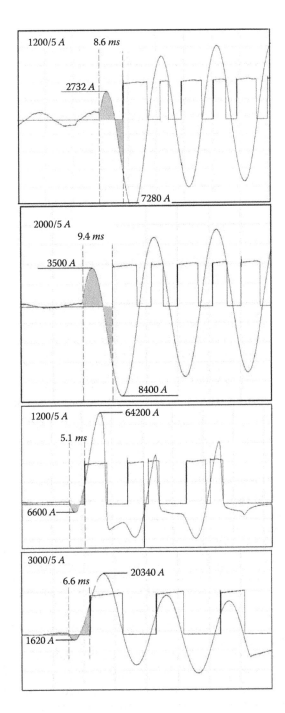

FIGURE 9.32
Oscillogram of operation of the relay for actual short-circuit transients containing a high DC component.

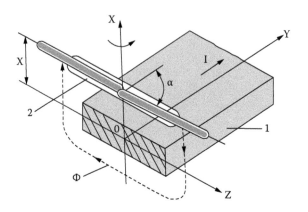

FIGURE 9.33
Principle of a winding-free overcurrent reed relay. 1: A current-carrying busbar; and 2: a reed switch with pickup depends on distance X from the busbar and on angle α for the longitudinal axis Y.

to achieve pickup in some cases at a current of 14 kA that also specifies a dangerous short circuit, as well as pickup of a 20 kA current. Nevertheless, in some cases this phenomenon can limit the application of reed relays.

For such cases, we have developed reed relays which do not demand use of the CT (the CT is the reason for pickup relay decrease). Such relays can be applied for both LV and HV (up to 24 kV) electric networks. By a principle of action it is a winding free relay, sensitive to a magnetic component of the current carrying the busbar (Figure 9.33). The reed switch has been placed in an internal cavity of the polymer isolator in the form of a glass (Figure 9.34) and filled with epoxide compound for direct installation on the HV busbar (Figure 9.35).

Due to use of the reed switch as a sensitive threshold element, the high-speed overcurrent relay developed is not only very simple, low priced, and accessible to manufacturing even by technicians, but also highly steady to external electromagnetic influences: to distortions of a current, to voltage spikes, to powerful high-frequency radiations, and so on. Such sensitive elements on a reed switch, adjusted on operation at the high rate of a current, can also be built in various microprocessor protection relays (or can be connected to them outside, through a separate input) as the bypassed element of the microprocessor for accelerator tripping of the circuit breaker.

The winding-free reed relay can be used also as self-sufficient over the current relay. For this purpose, the device must be equipped with a simple electronic converter. The magnetic field of the current-carrying bus excites the reed switch, whose pulses are then converted into a standard binary signal compatible with the relay protection devices. Figure 9.36 shows the solid-state converter circuitry with a reed switch as a triggering component.

The sensitivity of this module is directly proportional to the sine of the angle α between the longitudinal axes of the reed switch and the HV bus,

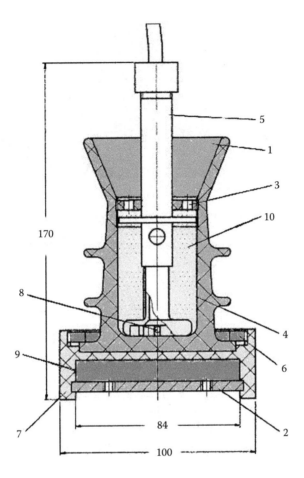

FIGURE 9.34
The design of an HV overcurrent reed relay not requiring CT. 1: the main insulator; 2: a fixative plate; 3: an inside nut; 4: a semiconductive cover; 5: bushing; 6: a fixative nut; 7: a fastener; 8: a reed switch; 9: an HV busbar; and 10: an epoxide compound.

and is inversely proportional to the distance h between these axes. Keeping in mind that for a reed switch operating in the magnetic field of the current-carrying bus, the operative (F_o) and the release (F_r) magnetomotive forces are respectively adequate to the operative and the release currents in the bus. One can say the following:

$$I_o = \frac{F_o}{K_h \sin \alpha};$$

(9.1)

$$I_r = \frac{F_r}{K_h \sin \alpha}$$

(9.2)

FIGURE 9.35
External view of an HV (24 kV) overcurrent reed relay not requiring CT.

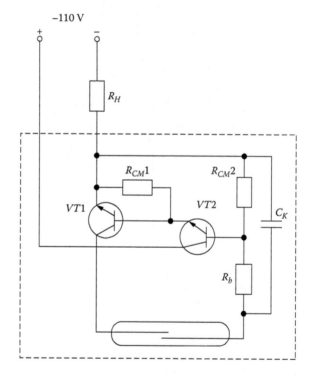

FIGURE 9.36
A circuit diagram and external view of the built-in converter.

where I_o, I_r are the values of the current in the bus causing the triggering and the release of the reed switch, respectively; and K_h is the remoteness factor accounting for the distance between the longitudinal axes of the reed switch and the bus. Therefore, by rotating the current transducer with respect to its longitudinal axis, one can set up the operative current at differential

FIGURE 9.36 (continued).

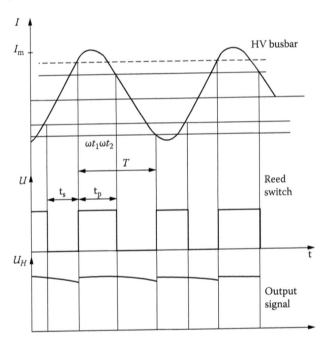

FIGURE 9.37
An AC operational oscillogram.

values. The solid-state converter is based on a transistor filter with a peculiar parameter vector, which is explained in this section.

Excited by AC magnetomotive force, the reed switch generates rectangular pulses (Figure 9.37); the duration of this is t_p, and the space is t_s. Capacitor C_k is supposed to gain the full charge during period t_p, that is, the full charge time must satisfy the following condition:

$$\pi R_H C_k < t_p$$

which defines the following capacity:

$$C_k \leq \frac{t_p}{\pi R_H}. \tag{9.3}$$

During the discharge period of C_k, the transient voltage free component attenuation on the R_b resistor is not supposed to exceed a given load pulsation factor K_p:

$$K_p = U(t_s)/U$$

where U, $U(t_s)$ are the nominal voltage and voltage drop on the load at the end of the pulse space, respectively. (If the load R_H is represented by a control coil of an auxiliary relay, then this pulsation factor can be expressed through the relay reset ratio.)

The above requirement can formally be expressed as

$$K_p \geq \exp\left(-\frac{t_s}{R_b C_k}\right) \tag{9.4}$$

where $\tau = R_b C_k$ is the discharge time constant.

The resistance of R_b should, therefore, satisfy the following condition:

$$R_b \geq \frac{t_s}{-\ln K_p C_k}. \tag{9.5}$$

The oscillogram in Figure 9.37 shows that the duration of pulses generated by the reed switch is

$$t_p = \omega^{-1}\,(\omega t_2 - \omega t_1) \tag{9.6}$$

where the current phases ωt_2 and ωt_1 are, respectively, given by

$$\omega t_1 = \arcsin\,(I_o/I_m) \tag{9.7}$$

$$\omega t_2 = \pi - \arcsin\,(I_r/I_m) \tag{9.8}$$

where I_m is the current amplitude.

Based on (9.1), (9.2), (9.7), and (9.8), the reed switch pulse duration takes the form of

$$t_p = \omega^{-1}\,[\pi - \arcsin\,(I_o/I_m) - \arcsin\,(I_r/I_m)]. \tag{9.9}$$

The pulse duration and space are related to each other by

$$t_p = T - t_s$$

where T is the pulse period, which under the sinusoidal form of the current in the busbar is equal to π.

Taking into account equation (9.9),

$$t_s = \omega^{-1} [\arcsin(I_o/I_m) + \arcsin(I_r/I_m)]. \tag{9.10}$$

By incorporating equations (9.9) and (9.10) into (9.3) and (9.5), we finally get the following:

$$C_k \leq \frac{\pi - \arcsin\left(I_o/I_m\right) - \arcsin\left(I_r/I_m\right)}{\pi \omega R_H}$$

and

$$R_b \geq \frac{\arcsin\left(I_o/I_m\right) - \arcsin\left(I_r/I_m\right)}{C_k \omega \left(-\ln K_p\right)}. \tag{9.11}$$

Thus, equation (9.11) defines the parameter vector for the above solid-state converter.

9.7 Reed-Based Devices for Overcurrent Protection of HV DC Equipment

HV equipment (10–100 kV) has become very popular over the last few years. It is utilized in military and civil radar stations, powerful signal transmitters for communication, broadcasting and TV systems, technological lasers, X-ray devices, powerful electronic and ion devices, devices for inductive heating and melting of metals, technological electron accelerators for material irradiation, electrophysical and medical equipment, and industrial microwave ovens, among other uses.

Technical difficulties caused by the presence of functional components isolated from each other, not permitting direct connection owing to a high difference of potentials, are encountered when designing systems for control and protection against emergency conditions (such as overcurrent and sparks) in modern power HV equipment. To guarantee information and electrical compatibility, as well as to implement the required algorithms for

interaction of functional components of equipment, special control instruments are required that have been called *interface relays* or *insulating interfaces*
(in the technical literature).

Apart from problems connected with the transmission of commands
between parts with opposite potentials of HV equipment, there are also
problems of current overload protection (level current trip) of such devices,
caused by HV circuit insulation breakdowns or breakdowns in the HV
devices. These problems still remain acute. The first is related to unfavorable
conditions that cause moisture and dust to penetrate the equipment, and
the second is related to unpredictable internal breakdowns in HV vacuum
electronic elements (klystrons, tetrodes, etc.) or semiconductor elements
(HV rectifier).

Current overload protection in such devices is usually resolved by the
inclusion of current sensors and electronic relays into the LV or grounded
circuits. However, such protection is not necessarily efficient and in itself can
cause many problems, which is why high-performance systems of protection
of HV equipment from current overloading are based on interface relays.

The general principle of design of interface relays is the presence of a special galvanic decoupling unit between the receiving and final controlling
systems of the relay. High-voltage reed relays (HVRRs) are a new type of
HV interface relays that I designed[4] for automation systems (e.g., overload
protection, fault indicating, and interlocking of HV equipment) as well as for
the transfer of control signals from ground potential to HV potential (reverse
connection).

The RG (Relays of Gurevich) series consists of the following devices: RG-15,
RG-25, RG-50, and RG-75, which are designed to operate under voltages of
15, 25, 50, and 75 kV DC, respectively (Figure 9.38). The operation of these
devices is based on the separation between the electric and magnetic electromagnetic field components. Each device is based on a magnetic field source
(coil), connected in a high potential current circuit, a reed switch, and a layer
of HV insulation, transparent for the magnetic component of the field and
completely insulating the reed switch from the electric field component
(Figures 9.39 and 9.40). The current trip levels can be adjusted up to 50% (for
each subtype). Material for all construction elements is Ultem-1000 (General
Electric, Fairfield, CT), and leads are HV Teflon cable (etched) 178-8195 type.

The RG-75 (and RG-50) relay (Figure 9.41) is composed of the main insulator (item 1 in Figure 9.42) formed as a dielectric glass, whose cylindrical part
is extended beyond the flange (2). The flat external surface of the bottom (3)
of this glass smoothly mates with the extended cylindrical part (4) having
threaded internal (5) and external (6) surfaces. The relay also includes a control coil (7) with a Π-shaped ferromagnetic core (8) located inside the main
insulator and reed switch (9) located in an element for reed switch rotation
through 90° (10). This element (10) is formed as an additional thin-walled
dielectric glass with walls grading into the bottom and mating with the

FIGURE 9.38
HV reed relays of the RG-series: RG-15, RG-25, RG-50, and RG-75.

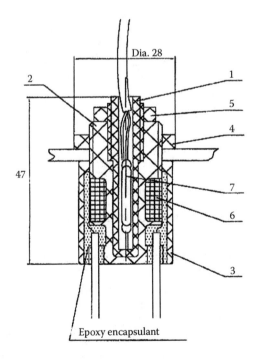

FIGURE 9.39
An RG-15 series design for compact HV power supplies. 1: a moving insert; 2: the main insulator; 3: external bushing; 4, 5: nuts; 6: a coil; and 7: a reed switch.

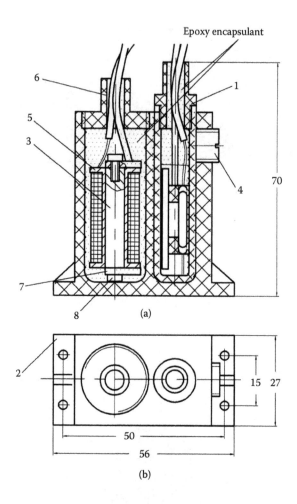

FIGURE 9.40

(a) The RG-25 series design for power lasers, industrial microwave ovens, and medium-power radar. 1 and 6: bushing; 2: the main insulator; 3: a ferromagnetic core; 4: plastic screw; 5: a coil; and 7: a pole. (b) The revolving assembly part of RG-25. 1: a reed switch; 2: an insulator; 3: bushing; 4: support; and 5: a ferromagnetic plate.

inner surface of the cylindrical part (4). These mated surfaces are coated with conducting material (11).

Reed switch outputs (9) are conveyed through an additional insulator (12) formed as a tube extending beyond the reed rotation element body (10). The lower end of this tube is graded into an oval plate (13) covering the reed formed with the conducting external coating. Control coil (7) outputs are also conveyed through a tube-shaped insulator (14) extending beyond the main insulator. The reed switch position fixation element is formed as a disk (15) with a threaded side surface and a central hole with an insulator (12) conveyed through it. External attachment of the device is effected with a

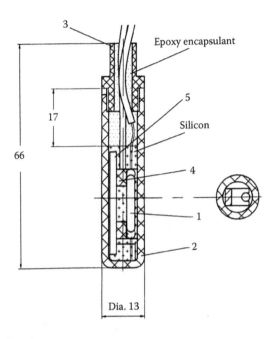

FIGURE 9.40 (continued).

dielectric nut (16). The lower layer (17) of epoxy compound filling the main insulator to the control winding performs conduction by the addition of copper powder (60–70% of the volume). The rest of the filling compound (18) has been made dielectric. Element space (10) is filled with the same dielectric epoxy compound. The shape of the main insulator and the reed switch rotation element are chosen so that their mating surfaces, which contact with the conducting coating, do not form sharp edges emerging on the main insulator surface and, at the same time, provide for safe shunting of the air layer between them and remove the thin conducting sharp-edged layer from the design. Significantly reducing the field intensity generated by the sharp outputs of the reed switch is achieved by adding one more tube-shaped insulator extending beyond the main insulator used to convey the reed switch outputs and executing the inner end of this tube as a plate with conducting coatings covering the reed switch.

Applying the lower layer of epoxy compound, which fills the main insulator conducting space (holding the control coil with a ferromagnetic core), thus reduces the intensity of the field generated by the winding outputs and neutralizes the action of the air bubbles remaining between the coil windings. Implementing the reed fixation element as a simple threaded disk, which is threaded into the respective part of the main insulator, forces the reed rotation element.

Use is made of an additional dielectric nut threaded on the appropriate part of the main insulator as an element of the relay external attachment

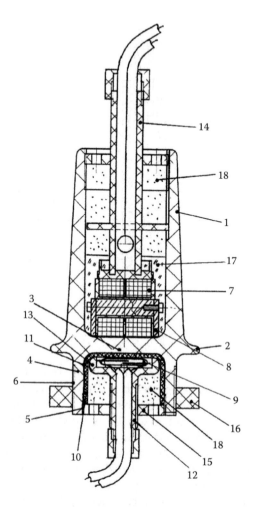

FIGURE 9.41
The RG-75 and RG-50 series relay design.

assembly, and the main insulator flange is used as a stop for this attachment assembly. Device operation is based on the action of the magnetic field of the control coil (penetrating through the bottom [3] of the HV insulator [1]) to the reed switch (9). When the reed switch threshold magnetic flux value is attained, it becomes engaged and appropriately switches the external circuits of the installation.

The reed switch engagement threshold value is adjusted by changing its position relative to the magnetic field source. This change is effected by the rotation of element (10) with a reed switch (9) by an angle of 90° relative to the poles of a Π-shaped ferromagnetic core (8). The position of element (10) with the reed is fixed by the forcing element (10) as a disk (15) is screwed in.

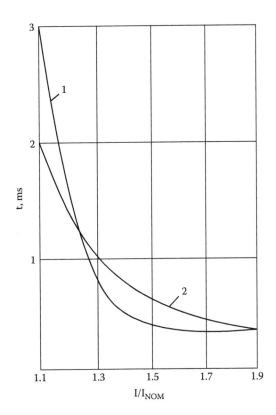

FIGURE 9.42
Dependence of operating time of the RG from the current in the control coil. I: The operating current; I_{NOM}: the nominal operating current; 1: RG for I_{NOM} = 0.25 A; and 2: RG for I_{NOM} = 3A.

Each device from this series (Table 9.1) functions as four separate devices:

- A current-level meter in an HV circuit
- A trip-level adjustment unit
- A galvanic isolation assembly between the HV and LV circuits
- A fast-response output relay in the LV circuit

The devices withstand the action of external environmental factors according to MIL-STD-202 requirements:

- Operation temperatures range from −55°C to +85°C
- Cyclical temperature change in range from −55°C to +85°C
- Air humidity 87% at a temperature of 40°C

TABLE 9.1

Main Parameters of the RG Devices

RG Relay Type	RG-15	RG-25	RG-50	RG-75
Nominal voltage (kV DC)	15	25	50	75
Test DC voltage 1 min (kV)	25	35	70	90
Control signal power (W)	0.2...0.4	0.2...0.5	0.5	0.9
Maximal switching voltage in the output circuit (V):				
DC		600		
AC		400		
Maximal switching output circuit current (A)		0.5		
Maximal response frequency (Hz)		100		
Maximal response time (ms)		0.5...0.8		
Maximal dimensions (mm)	$\varnothing26 \times 47$	$56 \times 27 \times 70$	$\varnothing75 \times 150$	$\varnothing75 \times 190$
Weight (g)	45	130	370	620

- Low air pressure with HV applied −87 mmHg
- Vibrations resistance 10 g at an oscillation amplitude frequency in the range of 10–500 Hz
- Repeated shock: 55 g, duration 2 ms, and 10,000 impacts
- Single mechanical impacts: 30 g/s, S sinusoid, with duration 11 ms

In current overload protection systems, the RG relays are usually connected as series circuits between the rectifier bridge and filter capacitor in the HV power supply, when the acting current does not exceed 10 A (pulsating current amplitude up to 30 A). However, when the current is above 10 A, they are connected by means of the shunt. The LV RG output is usually connected to the trip circuit of the LV solid-state contactor on the LV side of the power supply. The RG relay is triggered when the current in the HV circuit exceeds the trip level (Figure 9.42). The RG response time (Figure 9.43) depends, to a large degree, on the overload-to-nominal current ratio (I/I_{NOM}).

As an example, we shall observe application of the RG for current overload protection of the electronic vacuum tube (ET) in a powerful high-frequency transmitter with HV power supply. The decision about the construction principle of efficient ET protection should be mainly based on the ET's power. For example, as a rule, low-power ETs (such as X-ray tubes) do not need any special current protection, since the discharge current limitation in the ET is provided by high internal resistance of the power supply.

For some higher-power ETs, serial connection of a nonlinear current-limiting element proved to be sufficient, while it has been possible to restrict the

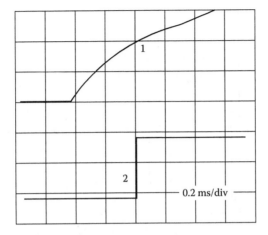

FIGURE 9.43
Time-current characteristics of an RG relay. 1: a current in a control circuit; and 2: a current in an output circuit.

short-circuit current ratio at a level of three- to fourfold of the nominal value. As the order of magnitude of the operating current in an ET reaches several amperes, relay-type protection means should be used.

The simplest and also sufficiently efficient devices of this type have an LV switching element in the LV power supply circuit, which is controlled by the commands originating from a sensor connected to the HV circuit. When the ET and consequently the power supply capacity are increased, the latter incorporates reactive filter elements accumulating energy sufficient for damaging the ET in case of breakdown, even after the LV supply circuit is disconnected. In this case, a more sophisticated protection system is to be used having a quite expensive HV fast-response switching element (shorting device) with an autonomous control system, bypassing the ET in case of breakdown. In this case, in order to prevent overload and power supply system elements failure, such a shorting device can be used only in addition to rather than in place of them.

ET protection systems related to disconnecting the power supply should have an automatically repeated connection for better "survivability" of the equipment. In autonomous systems, all the mentioned problems become even more complicated because of the severe weight and dimension requirements, which necessitate special system solutions, alternative elementary bases, and so on. The use of the developed elementary basis offers higher efficiency of the protection system's operation.

The simplest single device is a current relay in the LV circuit of the power supply and a contactor, which disconnects the step-up transformer from the mains upon overload. Such a protection system is simple and inexpensive; however, it does not offer adequate ET protection (e.g., low speed of

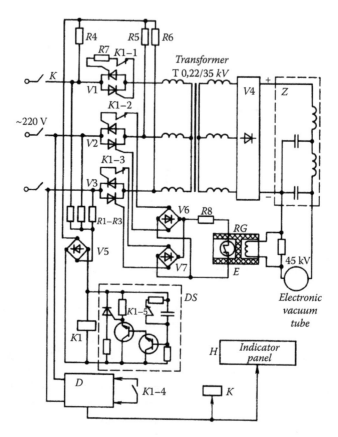

FIGURE 9.44
Circuit diagram of a protection system for medium-power supply.

operation, inability to distinguish between the internal ET breakdowns and other damages in the power supply system, the operation current depends on the resistance of the grounding buses' circuits, etc.).

Apart from these, in HV power circuits, current restriction elements (such as special throttles) are used. Elements described in this book are designated for substantial improvement of ET protection. Medium-power ET protection device operation is as follows (and see Figure 9.44).

In normal operation with closed contacts of the LV contactor (item K in Figure 9.44), the ET operating current in the winding (RG E) generates a weak magnetic field, which is too low for the reed switch operation. Short spikes of operating current, which depend on the ET modulation regime, do not put reed switches into operation since their duration is two to three orders of magnitude lower than the reed switch–operating time. The ET noncontact switch thyristor control circuits V1–V3 are closed, the thyristors are open, and the voltage across the K1 relay winding is nearly zero. Internal breakdowns in ET operate the RG E (the contacts become open), thus disconnecting the diode bridges V6 and V7 of the thyristor switches V2 and V3's control circuits. Once the sinusoidal alternating current first passes the zero point, the thyristors of the switches V2 and V3 are cut off, causing deenergizing of the primary winding of transformer T. Hence, even under the most adverse conditions, the maximal time for complete power disconnection does not exceed a single half-period duration. Once thyristor switches V2 and V3 are cut off at the zero point of the star-connected resistors R4–R6, phase voltage is generated, while the voltage at the zero point of the second "star" formed by resistors R1–R3 remains zero. The voltage difference causes K1 relay operation. Its contacts produce additional disconnection of the thyristor-switched control circuits. Thus, renewed power supply after the discharge of filter Z elements is prevented, and the HV reed switch interface E returns to its initial state. At the same time, the K1–4 contact of this relay is engaged, sending an input signal to the operations counter D; the contact R1–5 is broken, which disconnects the condenser of the time relay timing circuit DS. After an elapsed time of $T = RC$ sufficient for deionizing the vacuum gap generated in the ET and perfect reconstruction of its dielectric properties (about 0.5 c), the time relay thyristor DS is opened, thus shunting the K1 relay winding. Once the K1 relay is disconnected, all its contacts are returned to their initial state, and the thyristors of switches VI–V3 are reconnected in order to power the transformer again. The circuit returns to its initial state, except for counter D.

If the first cutoff does not result in self-elimination of the ET failure, the protection mechanism is operated again in a similar way. Once a preset number of pulses are accumulated in the operation counter (say, three), a signal for electromagnetic contact K disconnection and the *ET Failure* display are generated at the counter output.

9.8 The New-Generation Universal-Purpose Hybrid Reed: Solid-State Protective Relays

In the past few years, small-sized standard case TO-247 and TO-220 thyristors and transistors intended for soldering on the PCB for the switching of current of tens of amperes at voltages of 1200–1600 V have appeared. Various companies manufacture miniature high-speed (fractions of milliseconds) vacuum reed switches that can withstand voltages of 1000–2500 V, which can serve as precision threshold (pickup) elements in the protective relays. The Japanese company Yaskawa and its branches manufacture a series of middle-sized powerful reed switches for switching currents of up to 5 A at a voltage of 250 V (Figure 9.14).

When using reed switches, it should be kept in mind that their high reliability will be guaranteed only when observing the restrictions imposed by the switching ability determined in the technical specifications. As in semi-conductor switches, the reed switches quickly fail when the allowed switch-ing parameters are exceeded even for a short time. At the same time, even though modern reed switches are electromechanical elements, their reli-ability and number of switching cycles are closer to that of semiconductor elements, and so are many of their parameters, such as withstanding electro-magnetic interferences, surge capability, and so on. It should be pointed out that they considerably surpass semiconductors in withstanding surges.

Because of the extraordinary features of the reed switch relays, not pos-sessed by usual electromechanical relays, such as high speed, precise and stable pickup value, high release factor on an alternating current, and the like, many devices for protection and automation systems in the indus-try, power engineering, and military techniques have been developed on their basis.

The combination of the reed switches with magnetic circuits and semicon-ductor elements opens new avenues in the development of interesting and promising devices distinguished by simplicity and low cost.[1] For example, a simple device such as a reed switch with two operating coils (Figure 9.45a) can be a basis for creation of the differential protection, logic elements, the threshold summing element, and so on. A reed switch with a special mag-netic circuit (Figure 9.45b) appears to be insensitive to the DC (aperiodical) component of the current in the coil. The reed switch, connected to a sim-ple circuit (Figure 9.45c), responds to the voltage asymmetry. In the circuit (Figure 9.45d), the reed switch picks up only at rapid changes of current (volt-age) in a control circuit which is distinctive for emergency modes and does

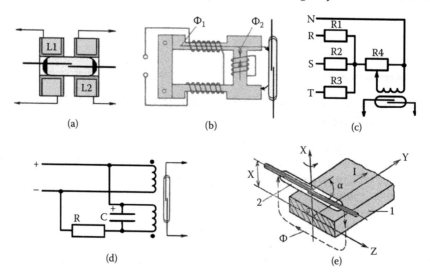

(a)   (b)   (c)

(d)   (e)

FIGURE 9.45
Examples of various applications of reed switches in the protection devices.

not respond at slow changes of the current, related to the changes in load, and as described in this section. The reed switch is also directly responsive to the magnetic field of the current passing in the busbar without additional windings (Figure 9.45e).

Let us consider concrete examples of the most widespread kinds of protective relays based on the suggested technology.

9.8.1 Instantaneous Current Relay

This relay is shown in Figure 9.46. The overcurrent relays without time delays are widely used for the protection of electric networks and electric equipment against overloads. This version of the relay is intended for directly energizing the trip coil of the HV circuit breaker.

The sensitive threshold (pickup) element of the device is the Rel1 relay made with a miniature high-speed vacuum reed switch. Its coil contains 2050 turns of 0.16-mm wire. At pickup, this reed switch starts to vibrate at double the current frequency. Upon the initial closing of the circuit by the reed switch, the thyristor VT will turn-ON and energize the CB trip coil.

The thyristor only switches this coil ON; it is switched OFF by its own auxiliary contact of the CB. Rel2 is an auxiliary relay, intended for signaling or blocking circuits, and it uses a medium-capacity reed switch such as GC1513. Its coil has very low resistance, and it is designed for the short-term carrying of a direct current in a range from 0.5 up to 15 A (typical currents of CB trip coils of various types) in which this reed switch operation is reliable.

Adjustment of pickups (coarsening the relay) is carried out with the help of potentiometer R1. In the relay, a thyristor such as 30TPS12 (in the case of TO-247AC) is used with a rated current of 30 A, a maximal withstanding voltage of 1200 V, and a miniature vacuum reed switch such as MARR-5. The input CT is made on a low-frequency ferrite ring with an external diameter of 32 mm.

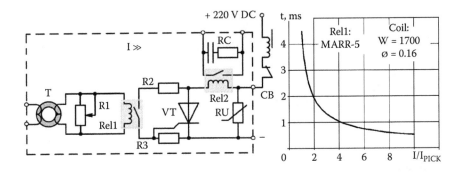

FIGURE 9.46
The simplest hybrid protective relay: instantaneous current relay. The basic circuit diagram and experimental time-current characteristic curve. I/I_{PICK}: multiples of pickup setting.

The RC circuit serves as protection of the auxiliary contact (reed switch) from spark erosion at the switching of inductive loads. A varistor RU such as SIOV-Q20K275 protects the device from spikes in the DC circuit. Its clamping voltage does not exceed 350–420 V DC. This voltage level should be higher than the rated voltage of a DC network, but lower than the maximal withstanding voltages of the thyristor and the reed switch. As shown in the experimental time-current characteristic curve, the relay speed is higher than that of the electromechanical, static, or microprocessor-based devices; it does not need power supply; it is insensitive to high-frequency interferences and spikes in a current circuit; and it remains reliable at strong distortions of current.

9.8.2 Instantaneous Current Relay with a High Release Ratio

This is shown in Figure 9.47. In this relay design, a powerful reed switch, such as R15U (Yaskawa), is used as a contact of tripping output relay Rel2 instead of power thyristor VT. The second CT (T2) serves as an energy source necessary for the operation of the power reed switch.

Closing and opening the reed switch (Rel1) with double the current frequency at energizing do not suit the operation of relay Rel2. Therefore, a special active filter can be used between pickup element Rel1 and output relay Rel2. The filter is formed with capacitor C2 (22 µF), resistors R2 and R3, and transistor VT, which can be any low-power transistor for voltages not less than 100 V with current gain (h_{FE}) not less than 100, for example ZTX753 or ZTX953.

With a low-power Darlington transistor (for example, ZTX605), as shown in Figure 9.47, the capacity of the capacitor C2 can be considerably reduced. By means of this filter, the current pulsation in the reed switch circuits of Rel1 will be transformed to a stable current in the coil circuit of relay Rel2.

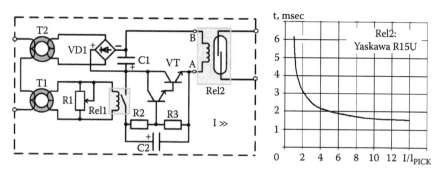

FIGURE 9.47
An instantaneous current relay with a high release ratio: the basic circuit diagram and experimental time-current characteristic curve. I/I_{PICK}: Multiples of the pickup setting.

The release ratio of reed switch Rel1 is close to 0.99 at alternating current. For a lower release ratio of the relay (0.7–0.6), it is sufficient to connect the Rel1 coil through a rectifier bridge and to transfer capacitor C2 to a different location, in parallel to this bridge. Since the capacity needed to feed the powerful reed switch is much greater than for a miniature reed switch, CT (T2) is formed with two identical transformers, similar to transformer T1 in which the secondary windings are connected in parallel, and the primary windings are communicating, covering both ferrite rings. The total power consumed by the relay from the current circuit (at a current of 5 A) does not exceed 4 W. The winding of relay Rel2 consists of two coils placed on the reed switch and connected between them in series.

Each of them contains 7600 turns of a 0.08-mm wire. Experimental time-current characteristic curves (Figure 9.47) were obtained for a series of consecutive pickups of the relay, during the time intervals between which the charge of capacitor C1 was kept unchanged. At the initial pickup of the relay with an uncharged capacitor, the time delay is approximately twice as long. Such an acceleration of operation in the case of repeated pickups at short circuit is a positive property of the protective relay. Even in view of increasing the operation time at the initial pickup, the relay speed still remains very high. Modern insulated gate bipolar transistors (IGBTs) and complete modules used for their so-called drivers enable the realization of a very simple switching output unit of the relay on a contactless basis (Figure 9.48).

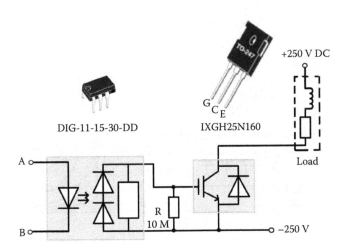

FIGURE 9.48
An embodiment of an output switching unit of the relay based on the modern IGBT (IXGH25N160) and specialized driver with the dynamic discharging (DIG-11-15-30-DD) for this transistor control.

9.8.3 Current Relays with Independent and Dependent Time Delays

This is shown in Figure 9.49. Similar to the above design, the relay contains two independent current transformers: the first one, T1, is used as a source of control value for the pickup module on the reed switch, Rel1; and the second, T2, for feeding the time delay unit.

When the microswitch S is switched on, the Zener diode VD3 is connected to the output of rectifier bridge VD1 and provides a constant level of a voltage (on an input of the time delay unit time) which is independent of the input current in the current pickup range. In this case, the relay works with the constant time delays, which are determined by capacitance C2 and resistance R2. As this capacitance is charged to a certain voltage value, the thyristor VT1 is turned on and capacitance C2 is completely discharged through low resistance (81 Ohm) winding (2050 turns by a 0.16-mm wire) of relay Rel2, activating the reed switch. In order to turn this device into a relay with time delay depending on the current, it is necessary to turn the microswitch S to OFF.

In this way, the voltage charge of capacitor C2 will depend on the input current level: the higher the current, the higher the voltage applied to capacitor C2, and the shorter the time of their charging up to a voltage level at which thyristor VT1 turns ON, forming a typical time-current characteristic curve (Figure 9.49) of a relay of this kind. If a second reed switch is removed from the center of the coil and mounted in the coil of relay Rel1 (so that its pickup will be 10–15 times higher than that of the first reed switch) and is

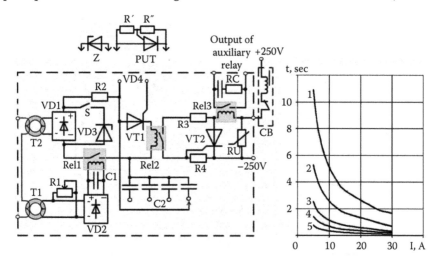

FIGURE 9.49
Universal protective current relay with the time delay: the basic circuit diagram and set of experimental time-current characteristic curves. For the relay with the dependent time-current characteristic curves, the various values of capacity C2 (in μF) are as follows: 1: 4400; 2: 3200; 3: 2200; 4: 1000; and 5: 300.

connected in parallel to a reed switch of the relay Rel2, the device pickups will be instantaneous at high rates of the input current and energize the trip coil of the CB within 3–4 ms. A turned-ON thyristor VT1 was used as the threshold element VD4, and a standard Zener diode was used in the relay prototype; however, the best results can be obtained with a so-called pro-grammable unijunction transistor (PUT), for example the 2N6027 or 2N6028 types. This element of the structure and its characteristics are similar to those of a thyristor with a very low leakage current (microamperes) through a gate junction that allows more efficient use of capacitance C2. Its turn-ON voltage can be adjusted (i.e., "programmed") by means of resistors R′ and R″.

9.8.4 Relay of a Power Direction

This is shown in Figure 9.50. Even such a complex function as detection of power direction can be realized very simply by means of hybrid technology. As is known, the power direction is determined by the angle of phase dis-placement between the current and the voltage; therefore, actually, the power direction relay responds to the change of angle between the current and the voltage. It turns out that application of two equivalent phase-shifted volt-ages to two primary windings of the intermediate transformer T3 causes the output voltage on the third winding to depend very strongly on the phase displacement between these voltages (Figure 9.50).

This is necessary only in order to prevent the effect of a change of the input voltage supplied from the current transformer T1 and voltage transformer T2 at a level of the output voltage of transformer T3. The simplest solution of this problem is provided by means of two back-to-back connected Zeners, as shown in Figure 9.50. A pickup relay can be adjusted by means of the potentiometer R.

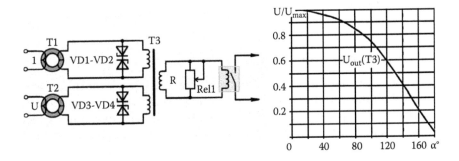

FIGURE 9.50
The relay of power direction: a basic circuit diagram of measuring the threshold module and experimental dependence of an output voltage of transformer T3 on an angle shift between two voltages on its primary windings.

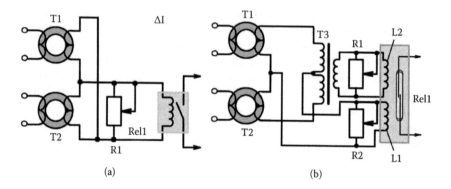

FIGURE 9.51
Measuring modules for the relay of differential protection. (a) the simplest embodiment. (b) an embodiment with restraint.

9.8.5 Relay of Differential Protection

This is shown in Figure 9.51. The use of two current transformers (T1 and T2) connected to the input of the pickup module of any of the devices described above enables realizing a two-input relay of differential protection (Figure 9.51a). Interesting opportunities are provided with the use of two separate auxiliary transformers with the secondary windings connected in series in this device.

To allow more complex functions, such as decreasing the relay sensitivity with increasing the current carried directly through protected object (so-called restraint), an auxiliary transformer T3 is included in the relay. In addition, the output reed relay Rel1 consists of two windings (L1 is differential and L2 is restraint), which shift the working point of the relay proportionally to the current carried directly through the protected object (Figure 9.51b).

9.8.6 Current Relay with Restraint

As is known, during the first moment after switching a power transformer on, almost all the current is expended in the magnetization of the iron core. This current is unipolar (Figure 9.52) and leads to a large sinusoidal displacement of the alternating current (consumed by the transformer) relative to the zero value during the first moment after being switched on.

A similar sinusoidal displacement is caused in the aperiodical component of a transient in the electrical power network when it is short-circuited (Figure 9.53). Because of the high rate of the short-circuit current, the line CT causes an additional deformation of this already displaced sinusoid because of its core saturation (Figure 9.54). It is obvious that the protective relay

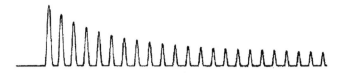

FIGURE 9.52
DC component of the magnetization current.

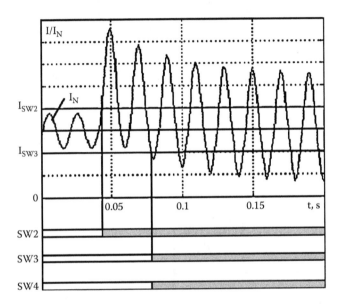

FIGURE 9.53
Displacement sinusoid of the short-circuit current at an aperiodical component and time diagram of the operating reed switches. I_N: The maximal working current in a power line; and I_{SW2} and I_{SW3}: pickup currents of reed switches SW2 and SW3, respectively.

(current, differential) should not pick up an increasing current value caused by the displacement of a sinusoid.

In practice, various methods preventing false operations of the relay are used: from simple relay interlocking for some fixed time (time of transient), up to distinguishing a second harmonic of a current by means of filters with electronic amplifiers and the formation of a restraint signal from it. In microprocessor relays, even more complex algorithms (which are carried out by the microprocessor) are used.

The hybrid relay (Figure 9.55) employs another principle of the relay interlocking for the aperiodical component and inrush current. The relay consists of the three independent, high-speed current relays on reed switches SW1, SW2, and SW3; one two-half-period rectifier VD1 and two one-half-period rectifiers (separately for positive VD4 and negative VD5 half-waves);

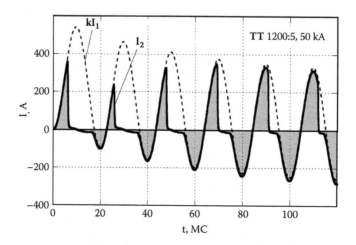

FIGURE 9.54

Deformation current sinusoid in conventional CTs caused by core saturation. I_1 and I_2: Theoretical and actual curves of the secondary CT currents.

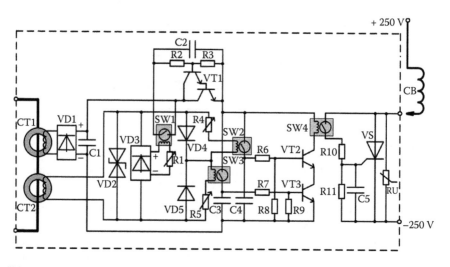

FIGURE 9.55

Circuit diagram of the current relay with restraint.

two current transformers, CT2 (as a source of an input signal) and CT1 (as a power supply of transistors), based on low-frequency ferrite rings; the filter on the capacitor C2 and the Darlington transistor VT1; two-input logic elements, transistors VT2 and VT3; and also an output-switching unit on the basis of an HV reed switch SW4 and the HV thyristor VS with a working voltage of 1200 V for directly switching the circuit breaker trip coil on.

This relay works as follows. As the voltage increases on a secondary winding of the current transformer CT2 up to some threshold value, corresponding to the maximal allowable value of the rated current I_{IN} (Figure 9.53), the reed switch, SW1, starts vibrating at a frequency of 100 Hz. The voltage on the transistor VT1 base, acted upon by SW1, remains stable (due to capacitor C2), which leads to the unlocking of this transistor. VT1 switches on the power supply of the logic elements of transistors VT2 and VT3, putting them into a ready-state mode. Each of the reed switches, SW2 and SW3, controls only one half-wave of the current, which diodes VD4 and VD5 extract for each reed switch.

The pickup values of both reed switches are identical and correspond to the amplitude of the established value of a short-circuit current (or close to it). At the initial moment of a short circuit (when positive and negative half-waves of a current have different amplitudes because of the displacement of the current sinusoid relative to the zero value), only one reed switch is picked up, namely, the one whose polarity corresponds to the half-wave having greater amplitude relative to zero. It is, in fact, the reed switch SW2 (in Figure 9.53) corresponding to a positive half-wave of the sinusoid displaced upward. Therefore, the transistor VT2 switches on and keeps conducting at the vibration frequency of the reed switch SW2 due to the capacitor S4. The reed switch SW1 at this time does not work, as the amplitude of a negative half-wave of the sinusoid (displaced upward) is not yet sufficient for the picking up of this reed switch. The output reed switch SW4 remains open. In the process of attenuation of the aperiodical component, the sinusoid of a current gradually becomes symmetric, and thus the amplitude of the negative (bottom) half-waves increases (Figure 9.53). When it reaches a pickup threshold of the reed switch SW3, it starts to vibrate at a frequency of 50 Hz and unlocks the transistor VT3. Now both transistors, VT2 and VT3, enter a conducting condition that leads to the switching-ON of the output stage with relay SW4 and thyristor VS and energizes the trip coil of the circuit breaker. The device is supplied with a standard spike protection element, VD2 and RU, and some other auxiliary elements. Requirements for solid-state elements and reed switches are the same, as for other hybrid relays described in this chapter. Adjustment of the relay is carried out by potentiometers R1, R4, and R5.

Descriptions of quite interesting and promising devices based on the suggested elements could be continued. However, the purpose of this volume is not to present the advantages of reed switches, but to prove that on the basis of a combination of modern reed switches and modern power semiconductor elements, a new generation of hybrid protective relays, not including complex mechanisms, can be easily created and can replace the out-of-date electromechanical relays at the upper level while retaining their high noise and surge stability, maintainability, and other positive features. The use of the new generation of the relays would allow sparing considerable financial expenses connected with the necessity of purchasing the expensive microprocessor-based protective devices. Thus, further perfection of automatic control systems in electric networks through equipping them with microprocessor recorders of emergency modes, optical communication systems, and other modern systems can be gradually accomplished, during the accumulation of financial resources independently of the relay protection. In my opinion, the examples of the protective relays discussed in this chapter that I developed and tested can support this conclusion.

References

1. Gurevich, V. *Electronic Devices on Discrete Components for Industrial and Power Engineering*. Boca Raton, FL: CRC Press, 2008.
2. Gurevich, V. Electric Relays: *Principles and Applications*. Boca Raton, FL: CRC Press, 2006.
3. Zisman, L., and V. Gurevich. Microprocessor Fast Overcurrent Protective Relay. In Report on SEEE Electricity 2007 Conference, Eilat, Israel, 14–16 November 2007.
4. Gurevich, V. *Protection Devices and Systems for High-Voltage Applications*. New York: Marcel Dekker, 2003.

10

Problems with International Standards

10.1 Introduction

Standards of the International Electrotechnical Commission (IEC) are the major documents affecting international technical policy and also the technical policies of the various countries whose national standards are based on IEC standards. Therefore, any inaccuracy, half-words, or unintelligible formulations in IEC standards can lead to very serious aftereffects. The question arises if the standards in effect today are ideal. We shall try to examine this question on the basis of the critical analysis of some basic standards in the fields related to digital protective relays (DPRs), in particular on electromechanical, solid-state, and reed switch–based relays widely used in existing DPRs and described in this book for perspective models.

10.2 The Basic International Standard on Electromechanical Relays (IEC 61810-1)

Electromechanical relays are the major elements of each DPR and are also used in non-microprocessor-based constructions. Therefore, great demands should be made of the electromechanical relays' basic standards. But what do we see actually?

10.2.1 Terms and Definitions

In the section of IEC 61810-1 entitled "Terms and Definitions" (items 3.7.1 and 3.7.2), two terms, *functional insulation* and *basic insulation*, are defined and are further used throughout the standard. According to IEC 61810-1, functional insulation is the insulation necessary only for the proper functioning of the relay, and the insulation protected from the electric shock is defined as basic insulation. As an explanation of the difference between these two kinds of insulation, in remarks to Tables 10(h) and 11(e) of IEC 61810-1, an instance

347

of "functional" insulation is given as the insulation between contacts of the relay, necessary (as affirmed in the standard) only for the proper functioning of the relay. It is impossible to agree with this assertion, however, for it is abundantly clear that the same insulation can be "basic" or "functional" depending on the application of the relay. For example, if contacts of the relay make switching in the electric circuits inaccessible to a contact by the person, the insulation between contacts of the relay really is clearly functional, but if contacts of the relay disconnect a voltage source of a part of the electrical installation to which there is an access of a person (direct or mediated, through other electric circuits), this is already "basic" insulation. On the other hand, the relay is often used for galvanic decoupling circuits with the different potential in the equipment; thus, insulation between the coil and contacts of the relay has no relation to the safety of the person and is clearly functional, whereas in other cases of relay application it is "basic." Thus, it can be asked, "How can the insulation marking in the relay be generally defined, that is, without connecting it to the concrete application?" To establish various demands to electric strength of insulation of the relay only by these definitions determined in advance is impossible. So what is the necessity for defining these terms in general?

10.2.2 Rated Values of Currents and Voltages

Sections 5.1 and 5.7 of IEC 61810-1 provide the ranges of rated values of a coil voltage:

   1.5; 3; 4.5; 5; 9; 12; 24; 28; 48; 60; 110; 125; 220; 250; 440; 500 Volt DC

and

   6; 12; 24; 48; 100/3; 110/3; 120/3; 100; 110; 115; 120; 127; 200; 230; 277; 400; 480; 500 Volt AC.

And, accordingly, for contacts of the relays working on a resistive load:

   4.5; 5; 12; 24; 36; 42; 48; 110; 125; 230; 250; 440; 500 Volt DC or AC.

   In Tables 16 and 17 of IEC 61810-1, different ranges of rated values of voltage are given:

   10; 12.5; 16; 20; 25; 32; 40; 50; 63; 80; 100; 125; 160; 200; 250; 320; 400; 500; 630

and

   12.5; 24; 25; 30; 32; 42; 48; 50; 60; 63; 100; 110; 120; 125; 127; 150; 160; 200; 208; 220; 230; 240; 250; 277; 300; 320; 380; 400; 440; 480; 500; 575; 600; 630.

First, for a correct designation of the value of an AC voltage, it needs to be specified about which value is being spoken (amplitude, average, or root mean square [RMS]), as the standard does not say. Second, there is bewilderment due to the essential differences in the ranges of rating values of voltage. In our opinion, these are absolutely unjustified and illogical; as a rule, both contacts and coils of the relay join in electric circuits of the same equipment having a certain range of rating values of voltage. Why these ranges should be different for circuits of contacts, circuits of coils, and internal voltage sources in the same equipment is not clear. Third, ranges of currents and voltages for contacts of the relay in the given standard mismatch classes of the contacts loading, as determined in the standard IEC 61810-7, "Electromechanical Elementary Relays: Part 7: Test and Measurement Procedures."

In Section 5.7 of IEC 61810-1, a minimal value of a load for contacts of the relay is specified as 4.5 V and 0.1 A. At the same time, as is well known, electronic circuit voltages much below 4.5 V (0.5–1 V) are used, as are currents much less than 0.1 A (0.005–0.010 A). The microelectromechanical relays with the bifurcated gilt contacts are widely used for switching such loads. What should be done when such relays, which really are present in the market, are widely used but mismatch IEC 61810-1? On the other hand, the current range 100 A, specified as the maximal value of the rated contacts current, is more characteristic for powerful contactors than for the relay.

The maximal values of rated voltages 400–440 V are not correct, in our opinion, as they do not reflect an existing reality. On the one hand, there are standard voltages 660 V and 1140 V, widely used in the industry; on the other hand, many companies make the small-sized open electromechanical relays for voltages 4–5 kV (e.g., Hehgsler, Aldingen, Germany; Italiana Rele, Cormano, Italy; SPS Electronic GmbH, Schwäbisch Hall, Germany; and Magnecraft, Des Plaines, IL), and also gas-filled and vacuum relays for voltages 70 kV and above (Kilovac/Tyco, Berwyn, PA; Gigavac, Santa Barbara, CA; and Jennings Technologies, San Jose, CA). Many companies also manufacture high-voltage reed switch relays for voltages 10–20 kV.[1] Low-current relays with insulation between the coil and contacts up to 120 kV[2] were already developed many years ago. Such relays are widely used in powerful electrical-physical, radio-electronic, and medical equipment, test systems, and so forth. In view of this, we can see that in the market there is really a large group of electromechanical relays that have not been embraced by the existing overall standard in spite of the fact that applicability of this standard is all-embracing, as is affirmed in the first section of the standard, and it does not divide the electromechanical relays into low and high voltage. A logical solution to this situation would be to change of the name of this standard (e.g., "Low-Voltage Electromechanical Elementary Relays") and restrict the area of its application only to the low-voltage relays with rated voltages up to 1000 V.

In Section 5.7b of IEC 61810-1, it is noted that rated values of currents and voltages for inductive loads should match those given in Annex B. However, in Annex B there are no rated values of currents and voltages. In this annex, provisions for the testing of relays with respect to making and breaking capacity and electrical endurance for inductive contact loads are specified.

Actually, the load kinds AC-15 and DC-13 are so-called utilization categories specified in Table B of IEC 61810-1, which characterize loadings as the coils of the switching electromagnetic apparatus: the relays, contactors, and starters. The rated current level for these kinds of loads is given in shares (fraction of amperes), while IEC 61810-1 concerns itself with current levels in tens of amperes (up to 100). Actually, IEC 61810-1 does not determine the commutating ability of the relay for an inductive load, and only confuses the situation as the classes AC-15 and DC-13 are not suitable for currents in tens of amperes.

10.2.3 The Documentation and Marking

According to IEC 61810-1, the major parameters of the circuit-breaking type provided by contacts of the relay should be reflected in the catalog or the service manual. According to N 3g of item 7.1 of Table 4, one of the following types of circuit breaking should be specified: "micro-interruption" (Section 3.5.16), "micro-disconnection" (Section 3.5.17), or "full disconnection" (Section 3.5.18). As follows from the IEC 61810-1 section "Terms and Definitions," differences between these types of circuit breakers consist in the size of the contact gap, that is, in the final reckoning, in a dielectric strength of a contact gap. Why was it required to invent special terms, poorly clear to users of the relays, and to incorporate these into the engineering specifications on the relays rather than specify the dielectric strength for a contact gap?

Another obligatory parameter that should be reflected in the technical specifications, according to Table 4 of IEC 61810-1, is the type of insulation, including functional or basic. There and then in remarks that follow, it is noted that it depends on the concrete relay application (as we already have shown in this chapter). But if it depends on concrete application of the relay and in advance cannot be ascertained, then should it be specified in the specification?

Along with the extremely dubiousness of the informativeness and definiteness in parameters which standard IEC 61810-1 demands to be specified in technical specifications, there are major parameters of the relay not mentioned at all, parameters such as operating time and release time, contact bouncing time, contact resistance, time constant of the operating coil (R/L), and the minimal values of switching voltage and current. Other problematic areas include the way of designating the contact loads (in item 7.1, Table 4, N 3a, of IEC 61810-1) by means of the instructions for contact type, current, and voltage without any elaboration of what is offered: maximum or rated.

In item 7.4's (Table 6 of IEC 61810-1) instances of marking the switching ability of relay contacts, a designation only rated values of switching current and voltage that are brought in the form of 16 A 230 V (or 16/230), even without the instructions of type of loading (power factor on AC, or time constant L/R on DC). It is necessary to note that such a designation does not give users information about true switching capability of the relay contacts and is capable of only confusion.

First, without an obligatory marking designation of loads, it is simply impossible to size up the switching ability of the relay because it strongly depends on the kind of load. For example, for the power relay G7Z type (Omron), the admissible switching current is 40 A at cleanly active (resistance) loads of up to 22 A at the mixed load with PF = 0.3.

Second, in item 7.1 of IEC 61810-1, a switching current and a voltage are spoken of, and item 7.4 discusses *rated values* of a switching current and a voltage, but it can be all the same; according to an explanation of item 3.3.16 of the standard, the rated value is determined as the value matching specially stipulated conditions. That is, the "rated switching current" is a current under certain, stipulated conditions. Such conditions can be a voltage across the contacts, frequency, a load kind, and so on. However, in IEC 61810-1, there are no explanations for the terms *rated switching current* or *rated switching voltage*, which makes impossible for all practical purposes to determine the correct use of these terms and the values connected with them. For example, what is "a rated switching current 16 A"? Is it a current at a voltage across contacts 250 V or no more than 125 V? Is it a current only for pure resistive loads, or also for mixed loads? The possibilities go on.

Third, as the concept "rated" (or nominal) in the standard is not stipulated, a designation in the case of the relay switching ability of contacts in the form of "16 A 230 V" in all cases designates that the relay contacts can switch a current 16 A at a voltage 230 V.

In many cases, there is mention of values of a current and the voltage, which relay producers characterize in the technical specifications as the "maximum values" (for purposes of advertisement). Thus, the maximum switching voltage and the maximum current of the switches are often marked. As a rule, the maximum switching power is not equal to the product of the maximum current and the maximum voltage (see Table 10.1). This is because the value

TABLE 10.1

Switching Parameters of Contacts of Some Types of Electromechanical Relays of Wide Application

Relay Type and Manufacturer	Maximal Switching Current	Maximal Switching Voltage	Maximal Switching Power	Production of the Current and Voltage
750-523 (Wago)	16 A AC	440 V AC	5000 VA	7040 VA
J114FL (CIT Relays)	16 A	440 V AC	4000 VA	7040 VA
		125 V DC	480 W	2000 W
CT (NAiS)	8 A AC	380 V AC	2000 VA	3040 VA
G2RL (Omron)	12 A AC	440 V AC	3000 VA	5280 VA

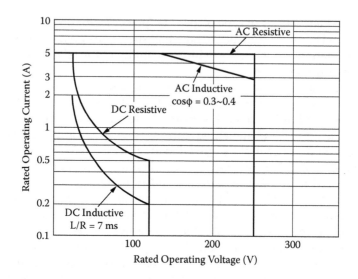

FIGURE 10.1
Typical curve loading of contacts of the relay.

of the admissible switching current, as much as possible, to a large extent depends on the value of a voltage on the contacts, especially in connection with direct current, and from a kind of load (see Figure 10.1).

Unfortunately, in IEC 61810-1 such "subtleties" are not mentioned at all, and this essentially complicates the standard's practical use.

10.2.4 Test Procedure

In item 8.2 (e and f) of the standard, it is emphasized that test of the relay for heating is made at the powered coil (coils) of the relay and loading by a current of *all contacts*. In practice, it is impossible to realize this requirement because if in the relay there are available both normally open and normally closed contacts, it is impossible to load a current simultaneously on both contacts (and how is this reconciled with the standard?).

In p.10.3 of IEC 61810-1, a procedure for dielectric strength is presented. Thus, as a 1-minute test of voltage, it is recommended to apply a variable sine wave voltage with a frequency of either 50 or 60 Hz or a DC voltage having a value selected from Table 10 or 11 of IEC 61810-1. Comparing the two tables, it can be seen that the values of voltage presented in them are absolutely identical for the same kinds of connections in the circuit diagram. But, in fact, one table refers to RMS values of the AC voltage, and the other DC voltages! As is known, 1000 V RMS voltage affects insulation quite differently from 1000 V DC. From the point of view of the effect on insulation, even in the most elementary case, even neglecting the known physical effects connected with the effect of frequency of an alternating voltage on insulation, it is necessary to inject, at least, crest factor (1.41) as a relationship between the r.m.s. and peak voltages. IEC 61810-1 makes no mention of this.

For tests of electrical endurance (IEC 61810-1, p. 11) as a criterion of an estimation of the condition of the relay, the standard uses such concepts as *make malfunctions* or *break malfunction* in contacts. And *malfunction* is defined (in IEC 61810-1, 3.5.21) as a single event where an item does not perform a required function. The certain quantity and sequence of malfunctions during tests characterize serviceability or failure of the relay. Alongside this criterion for an estimation of serviceability of the relay, its duplicate test of dielectric strength is applied. However, as is well known, many cycles of wear under the maximum current can essentially change not only the dielectric strength of insulation inside the relay, but also the contact resistance (owing to erosion of contact surfaces). It is also known that in the use of contacts of relays in low-current circuits of the electronic equipment, the essential increment of resistance of contacts is one of the frequent reasons of failure of the electronic equipment. In this case, it is possible to ascertain that the relay is not in any condition to carry out its functions (that, is to connect circuits), and to it we shall apply the term *failure*. Hence, contact resistance is a major criterion for estimating the serviceability of the relay and should be applied as one more criterion during tests of the relay for electrical endurance.

10.3 The International Standard on Solid-State Relays (IEC 62314-1)

10.3.1 Subject and Scope

In this section, I focus on the scope of the standard. According to the standard, the scope is limited to solid-state relays with rated voltages of up to 750 V and AC currents of up to 160 A. In the remarks, it is noted that requirements for solid-state relays with DC output circuits are under consideration. These remarks are a little bewildering: what is the necessity to release a standard on the solid-state relays that covers only a small fraction of the solid-state relays that are available in the market today? In fact, many types of solid-state relays with the output element in the form of two powerful metal-oxide semiconductor field effect transistors (MOSFETs) can switch both an AC and a DC current (Figure 10.2). The absurdity of the situation becomes apparent when one considers that only half of such a relay (AC) is covered by IEC 62314, while the other half of the same relay (DC) is not covered. Even the standard itself (in note to Section 3.1.8) recognizes that "the same solid-state relay may be ... rated for both a.c. and d.c."

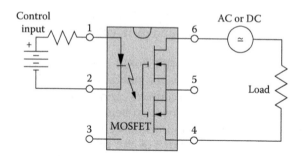

FIGURE 10.2
Solid-state relay rated for both AC and DC loads.

Also, it is not clear to which current and voltage values it refers; usually for semiconductor devices the peak values are specified, while for electromechanical relays the RMS values are specified. Such ambiguities in the international standard are, in my opinion, inadmissible.

Again, in the "Scope" section of IEC 62314-1, it is noted that the "solid-state relays are components (not stand-alone devices) and, as such, do not perform a direct function. Therefore, no EMC requirements are included in this standard." In my opinion, this is rather debatable. First, very often solid-state relays are used for direct switching of electric motors, solenoids, heating elements, lighting lamps, and so on, in which case all control equipment consists of this relay only, without any additional electronic control circuits, and as such it is a stand-alone electronic device incorporated into power equipment which does not contain any other low-signal electronic elements. The electromagnetic compatibility (EMC) of such power equipment is determined completely by the solid-state relay only. The consumer has to know about the EMC of the equipment (solid-state relays in this example). Second, in itself even the simplest solid-state relay contains many complex built-in electronic components—light-emitting diodes (LEDs), photo diodes or phototransistors, electronic amplifiers and triggers, synch circuits with a network voltage, elements of overvoltage protection, and so on—that characterize it more as a complex electronic *device* and not a simple *element*. Third, today in the market there are so-called intelligent solid-state relays with expanded functions, supplied by complex electronics and sometimes even a built-in microprocessor.

For the above reasons, the EMC requirements should of necessity be included in the standard of such rather complex electronic devices as solid-state relays. It is not only our opinion. In Reference 3, the authors introduce research results for influence of voltage harmonics and EMC on solid-state relays in industrial applications. From the test results, it can be generally concluded that the effect of harmonics and random high-frequency distortion would lead to a delayed opening and closing of the solid-state relays. Any discontinuity or line disturbance that interferes with the triggering and latching process can cause the solid-state relays to mistrigger by dropping out half-cycles.

According to IEC 62314-1, "solid-state switching device[s] with monolithic structures fall within the scope … and are not covered in this standard"; however, it does not explain what is exactly the "monolithic structure." Besides, the definition for the "solid-state relay" given in Section 3.1.1 (the analysis of this definition is given later in this chapter) does not expel the presence of the monolithic structure in the solid-state relay. In fact, the terms *solid-state* and *monolithic* are almost synonymous!

Such absolutely nonvalid restriction of a scope is, in our opinion, inadmissible. For example, it is possible to buy solid-state relays from a widely known company, Crydom (South San Francisco, CA), which for many years has specialized in the production of such relays (Figure 10.3), which represent monolithic structures, embedded in plastic or filled with epoxy resin. In some cases, as a uniform monolith it is integrated as well with a heatsink. According to standard IEC 62314, it is claimed that these relays fall within the scope only because they are filled with epoxy resin. Whatever logic is in this is not clear!

The standard states that semiconductor contactors fall within the scope, but does not give any definition for the term *contactor*. Without precise definitions of the differences between the *relay* and the *contactor*, it is impossible to understand precisely what devices fall within each scope. Analysis that we have performed has shown that today there are no precise boundary lines between the contactor and the relay, and no precise definitions allowing one to refer to this or that concrete switching device as a *contactor* or as a *relay* univocally.

The opposite is more true; in the technical literature today, there is complete arbitrariness in the assignment of the names *relay* and *contactor*. By what

FIGURE 10.3
Solid-state relays from Crydom with monolithic structures embedded in epoxy resin and integrated with a heatsink.

R100 and R300 series

Semiconductor   Conductors
20 A, 30 A, 45 A

R111 and R31 series

Solid-state relays
90 A (single phase), 75 A (three phase)

FIGURE 10.4
Solid-state contactors for low currents and solid-state relays for high currents (produced by ABB).

criterion, for example, has the world leader in the field of electrical equipment, ABB, used when referring to some of its solid-state switching devices as belonging to a relay class, and others as belonging to a contactor class (see Figure 10.4)?

The single basic difference that we have detected by analyzing the technical parameters of these two product "classes" is that a contactor is equipped with the heatsink, and a relay is not. Of course, this doesn't prevent a relay's user from adding a heatsink to the relay at a switching current of more than 5–10 A. In this case, does it "magically" become a contactor? We think not; a solid-state contactor is the exactly the same as a solid-state relay, but installed on a heatsink?! We have even seen a website where this was acknowledged. However, the adoption of such a criterion as the absence or presence of a heatsink for differentiating between a *relay* and *contactor* for a switching device is patently absurd (Figure 10.5). Besides, we have to keep in view that at rated currents of ten and more amperes, any semiconductor switching device (whether labeled a contactor or relay) demands a heatsink. Without

FIGURE 10.5
The absence or presence of a heatsink for differentiating between a *relay* and *contactor* for the same switching device seems patently absurd.

a heatsink, these devices are capable of carrying only a small percentage of the rated current.

Thus, with the absence of a clear boundary between a *contactor* and *relay*, it remains absolutely unclear what product is covered by the standard, and what products fall within the standard as the upper limit of currents established in the standard (160 A) characterizes. It seems to be that it is more likely to be power contactors rather than relays. If such a boundary is not established, it is obviously impossible that the given section should restrict the scope of the standard, and this should be removed as absolutely incorrect.

10.3.2 Terms and Definitions

10.3.2.1 Solid-State Relays

In Section 3.1.1 of IEC 62314-1, the definition for the term *solid-state relay* is given. According to the standard, it is an "electrical relay in which the intended response is produced by [an] electronic, magnetic, optical or other component without mechanical motion." What is the "intended response" in this definition? We understand this as the response upon an input value.

Generally, any electric relay has an input-sensing element that responds upon an input value, and an output element that performs the switching in an external circuit. In a typical electromechanical relay, these elements are a coil with a magnetic system and contacts.

Apparently, from the definition given in IEC 62314-1, the principle of construction of an input control circuit stipulates only that it should not contain any moving components, while the output power-switching circuit is not characterized in any way. Therefore, under the definition given in the standard, any solid-state relay may be a suitable device that consists of a semiconductor input-sensing transducer (e.g., optocoupler), a semiconductor amplifier, and an output electromagnetic relay. On the other hand, a well-known relay design in which it is embedded in epoxy resin includes a semiconductor output-switching element (the thyristor, the transistor) controlled by a reed switch. The control coil of the reed switch forms a magnetic control circuit of the relay. The relays based on this well-known principle are widely used.[4] But the reed switch contains internal moving components, so such a device cannot be referred to as a solid-state relay in spite of the fact that it meets all requirements of IEC 62314 except the definition of the term *solid-state relay*. Another example is a unit containing two back-to-back connected power thyristors (the natural solid-state switch) controlled by an external push button (Figure 10.6); is this a solid-state relay, or not?

In our opinion, the definition of the term *solid-state relay* presented in IEC 62314 is not correct and requires a revision with an orientation to the "solid-state" character of the output element that is carrying out the switching of an external circuit (load) instead of to the design of a control circuit.

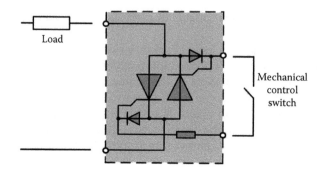

FIGURE 10.6
A solid-state relay controlled by a mechanical switch (e.g., push button).

10.3.2.2 Rated Insulating Voltage

In Section 3.1.4 of IEC 62314-1, *rated insulating voltage* is defined as the "value of voltage to which dielectric tests and creepage distances are referred."

In our opinion, it is rather strange and is not, per se, a defining formulation. The definition of *rated insulating voltage* should be absolutely precise, and the circuits to which this voltage refers (applies) must be clearly stipulated. For a solid-state relay, it is usually acceptable to use terms such as *withstand voltage* or *dielectric strength* to relate the connection between the internal input control element and the output switching element connected into the external load circuit, and also between a metal plate on the relay body (intended for contact with the heatsink) and all other current-carrying terminals of the relay. In our opinion, these would be a more correct definition for the term under consideration in that they are more precise and widely used in technology. In addition, the standard has to stipulate the parameters of the "voltage" as RMS or peak value, AC or DC, and time interval for voltage applying. Without all of these, it is impossible to understand and use the term *rated insulating voltage*.

10.3.2.3 Rated Impulse Withstand Voltage

In Section 3.1.5 of IEC 62314-1, the definition of the term *rated impulse withstand voltage* is given as the "peak value of an impulse voltage of prescribed form and polarity which the solid-state relay is capable of withstanding without failure under specified conditions of test and to which the values of the clearances are referred."

And again, as in the previous case, it is not certain to which circuits of the relay the term is to be applied. The question is that in the solid-state relay, there are some independent circuits isolated from each other to which the term *rated impulse withstand voltage* can be applied, and these circuits have essentially different levels of insulation and, hence, withstanding voltages, leading to uncertainty in the use of this term. For example, in the solid-state relay of type R111/25 (with a nominal operating voltage 280 V AC), the withstanding peak voltage across terminals of the switching element can reach 650 V, and between an input control circuit and an output circuit of the switching element it can reach 4000 V. In the low-power solid-state relay H11D1 type, the withstanding peak voltage across the terminals of the switching element can reach 300 V, and between an input control circuit and an output circuit of the switching element it can reach 7500 V. So, what is the rated impulse withstand voltage in these examples?

10.3.2.4 Rated Operational Current

In Section 3.1.6 of IEC 61810-1, *rated operational current* is defined as the "normal operating current when a solid-state relay is in the ON-state and takes into account the rated operating voltage, the rated frequency (4.3), the load category (4.4) and the overload characteristics at 40 °C ambient temperature unless otherwise specified."

In this definition, first it is not clear what "normal" current and abnormal current are. Normally, the current carried through an ON-state relay is fully dependent on the load and may change over a broad range in accordance with how the load changes. So, again, what is the "normal" current?

Second, what is the relation to the current carried through an ON-state relay? In other words, with the output switching element being in a full conductive condition, is it also the "rated operating voltage" which has been applied to an output circuit of the relay before switching to the ON-state? In fact, after the relay switches ON, any operating voltage cannot be applied to an output circuit of the relay, as in the ON-state there remains only a small voltage drop (shares or units of volts), depending on the current carried on the switching element.

Third, as is known from the theory of the power solid-state devices, the unique criterion, limited to a carried current, is the temperature of its internal semiconductor structure which strongly depends on the presence or absence of a heatsink, and the type of cooling (natural, air compulsory, water, etc.), the frequency of load switching, the power factor, and the duty cycle.

In connection with the above, the definition *rated operational current* must be changed to a more correct form: "the maximal effective value of a current carried through an output switching element of solid-state relay in stationary mode for a long time period, when the temperature of its internal structure does not exceed maximal admissible temperature for the given kind and a material of semiconductor structure, for the specific type of a heatsink and cooling conditions at 40°C ambient temperature."

10.3.2.5 Rated Uninterrupted Current

In connection with the above, it is definitely not clear what the *rated uninterrupted current* of the solid-state relay is. It is defined in Section 3.1.7 of

IEC 62314-1 as the "value of current stated by the manufacturer, which the solid-state relay can carry in uninterrupted duty." If this is so, can't the rated operating current (from Section 3.1.6 of IEC 62314-1) of the relay also carry current in an "uninterrupted duty"? What is the difference between these two currents?

10.3.2.6 Rated Conditional Short-Circuit Current

A new concept is introduced in Section 3.1.11 of IEC 62314-1: the *rated conditional short-circuit current*, which is treated as the "value of [a] prospective current, stated by [the] manufacturer, which the solid-state relay, protected by a short-circuit protective device specified by the manufacturer, can withstand satisfactorily for the operating time of the device under the test conditions specified in the relevant product standard."

We make no further comments; it is simply impossible to understand the definition.

10.3.2.7 Leakage Current

The definition for *leakage current* is presented in Section 3.1.12 of IEC 62314-1 as the "r.m.s. value of maximum current, stated by the manufacturer, which solid-state relay can carry in OFF-state condition."

As the leakage current strongly depends on the voltage applied to the output switching element of the relay, and also on its temperatures, in our opinion it would be necessary to add the words "at the maximum operating voltage and the maximum admissible temperature of a solid-state structure" to this definition.

10.3.2.8 ON-State Voltage Drop

The definition of the term *ON-state voltage drop* is treated in Section 3.1.13 of IEC 62314-1 as the "peak value of voltage, stated by manufacturer, between solid-state relay terminals in the ON-state condition"; this demands additions and refinement.

First, it is not clear why the ON-state voltage drop is to be characterized by the "peak value." As is well known, the ON-state voltage drop characterizes the power dissipation and heating of the solid-state elements in conductive mode. Why must the power dissipation be characterized by a peak value and not by an effective value?

Second, actually the solid-state relay has both terminals of an input control circuit and terminals of output switching elements. In most types of solid-state relays, both input and output circuits contain semiconductor elements: an LED in the input, and a thyristor or transistor in the output. Both of these elements are characterized with ON-state voltage drops. So the question arises: about which "terminals" is the standard talking?

Third, if the standard means the output switching element, a voltage drop on it is determined not only by physical properties of this switching element, but also by the current passing through it. Unfortunately, in the definition given above, the connection of voltage drop with a current for which this voltage drops is missing. All these make the term absolutely indeterminate and unsuitable for practical use.

10.3.2.9 Functional Insulation and Basic Insulation

In Sections 3.2.3 and 3.2.5 of "Terms and Definitions" of IEC 62314-1, two terms, *functional insulation* and *basic insulation*, are defined and further used throughout the standard. According to IEC 61810-1, *functional insulation* is the insulation necessary only for the proper functioning of the relay, and the insulation protected from the electric shock is defined as *basic insulation*. As an explanation of the difference between these two kinds of insulation, in remarks to Tables 10(h) and 11(e) of IEC 62314-1, an instance of "functional" insulation is given as the insulation between contacts of the relay, necessary (as affirmed in the standard) only for the proper functioning of the relay. It is impossible to agree with this assertion, for it is abundantly clear that the same insulation can be "basic" or "functional" depending on the application of the relay. For example, if contacts of the relay make switching in the electric circuits inaccessible to a contact by the person, the insulation between contacts of the relay really is clearly functional, but if contacts of the relay disconnect a voltage source of a part of the electrical installation to which there is an access of a person (direct or mediated, through other electric circuits), this is already basic insulation. On the other hand, the relay is often used for galvanic decoupling circuits with the different potential in the equipment; thus, insulation between the coil and contacts of the relay has no relation to the safety of the person and is clearly functional, whereas in other cases of relay application it is basic. Thus, it can be asked, "How can the insulation marking in the relay be generally defined, that is, without connecting it to the concrete application?" To establish the various demands to electric strength of insulation of the relay only by these definitions determined in advance is impossible. So what is the necessity for defining these terms in general?

10.3.2.10 Solid Insulation

In Section 3.2.4 of IEC 62314-1, the term *solid insulation* is given and defined as "a solid insulating material interposed between two conductive parts."

First, the LED on an input of solid-state relay and photodetective diode, connected in a control circuit of the output switching element, are both encased in plastic (i.e., not conductive). Despite this, sometimes optical clear insulation material is interposed between these two elements enclosed in

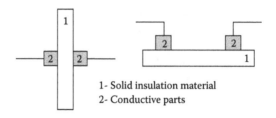

1- Solid insulation material
2- Conductive parts

FIGURE 10.7
Examples of different properties for the same insulation material, which are suitable for the *solid insulation* definition.

dielectric cases to increase the dielectric strength. So, the definition is not suitable for this situation.

Second, both examples shown in Figure 10.7 are suitable for the "solid insulation" definition, but the properties of this insulation for the same insulating materials and same conditions are absolutely different. It is not clear what the necessity was to introduce this term into the standard and to give it a special interpretation. In our opinion, the term and the definition are absolutely unnecessary in a standard on solid-state relays.

10.3.2.11 Conclusions

In conclusion of our review of the "Terms and Definitions" section of IEC 61810-1, we feel it necessary to emphasize that there are many special terms and definitions, including some that are newly invented and not always correct, in which the standard terms currently in wide use in the semiconductor industry and characteristics are not applied. These include such terms as *maximum repetitive OFF-state voltage, maximum nonrepetitive OFF-state voltage, critical rate-of-rise of ON-state current, critical rate-of-rise of OFF-state voltage, surge (nonrepetitive) ON-state current, holding current, peak pulse current, maximum RMS, ON-state current, control current, control voltage, turn-ON time, turn-OFF time,* and *minimal load current.*

There is nothing that justifies the "invention" of new terms while totally ignoring terms that are already widely used. This does not do honor to the authors of this section of the standard, and only creates bewilderment and disappointment for the reader.

10.3.3 Characteristics of the Solid-State Relay

10.3.3.1 Introduction

In this section, rather than repeat what we've already written, we shall address only new terms and the new characteristics of solid-state relays offered in the standard. Thus, in Section 4.4 of IEC 62314-1, an absolutely new system of classes for the electric loads has been introduced. These classes have not existed before and are intended for use only in the given standard.

The existing system of load classification, known widely as the *Utilization Categories* and embracing all possible types of electric loadings, is completely ignored in IEC 62314-1. In our opinion, such an approach, when for each new standard a new system of classifications is invented that completely ignores existing systems, is not justifiable; and consequently the standard, in this part, should be completely rewritten and incorporate the terms currently employed throughout the industry.

10.3.3.2 Overload Current Profile

The term *overload current profile*, introduced in Section 4.3 of IEC 62314-1, is treated in Section 3.1.9 as "current-time coordinates for the controlled overload current," and in Section 4.3 overload current is treated as "a multiple of rated operational current (Table 4) [that] represents the maximum value of operating current under operational overload conditions. Deliberate overcurrents not exceeding ten cycles of the power-line frequency which may exceed the stated values of Table 4 are disregarded for the overload current profile." It is difficult to understand the meaning of these sentences, and all the more difficult to use them in practice. At the end of Table 4, "Minimum requirements for overload capability test conditions" are introduced for load categories that have been specially "invented" (Table 1) for the standard and are not used anywhere else in the technical literature. These categories and terms (as mentioned in this chapter) are filled with ambiguities which have no place in an international standard. For example, the load category *LC A* is described as "resistive or slightly inductive loads." How much is *slightly*? Is the term *slightly inductive* suitable for a standard?

Instead of all these convoluted, unconvincing, and unintelligible word combinations in the engineering specifications on solid-state switching devices, there exist precise, clear parameters that are widely used and usual for technical specialists' parameters:

I_T(RMS): maximum continuous ON-state current (RMS)

I_{TM}: maximum peak repetitive ON-state current (for half cycle)

I_{TSM}: maximum peak nonrepetitive surge ON-state current (for half cycle)

I^2t: maximum surge nonrepetitive ON-state fusing current

In our opinion, only these four parameters should be used in the standard, instead of all the tables and new parameters invented for characterizing the load and overload capacity of solid-state relays.

10.3.3.3 Rated Control Circuit Voltage and Rated Control Supply Voltage

Section 4.5 of IEC 62314-1 introduces such parameters as *rated control circuit voltage* and *rated control supply voltage*. In the clarification of these terms, it is given that the difference between these two voltages can occur owing to the transformer, rectifier, and other elements built into an internal control circuit of the relay, on which an additional voltage drop can be created. The introduction of the two different "voltages" and the explanation given for this result in bewilderment only: who and with what purpose needs to know about voltage differences between internal elements and internal circuits inside the solid-state relay?! In our opinion, this information is absolutely senseless and even harmful as it only confuses the consumers.

10.3.3.4 Switch-ON Voltage and Switch-OFF Voltage

In the same section (4.5 of IEC 62314-1), such terms as *switch-ON voltage* and *switch-OFF voltage* are introduced. In actual specifications of solid-state relays appearing on the market, the "control voltage" characteristic is specified as the interval of control voltages provided proper relay functioning, for example 3–12 V DC. This is simple and absolutely clear. What purpose is served by inventing specifically for this standard two new and unclear terms? For example, what is the switch-OFF voltage? In IEC 61810-1, no definition or explanation for this term appears!

10.3.4 Marking and Documentation

Table 10.2 of the "Marking and Documentation" section of IEC 62314-1 lists the technical parameters required of the producer for indicating the relay directly, or referring to it in the technical specifications. This table contains a set of parameters that we have already criticized and does not contain any important parameters that we have not also addressed. There is, however, one new parameter (No. 3d) that attracts attention, "safety maximum load integral i^2t between 1 ms and 10 ms," which appears in the standard for the first time and has not been included in Section 3 ("Terms and Definitions") or in Section 4 ("Characteristics of Solid-State Relay"). The requirement for inserting the parameter in the technical specifications which has not been mentioned at all in the standard seems to us inadmissible.

10.3.5 Construction Requirements

In Section 7.1, the requirement for the materials used in the relay is formulated as follows: "The maximum permissible temperature of incorporated materials used in solid-state relays shall not exceed their operating limits, which shall be verified by testing according to 7.3." The given formulation, in our opinion, is erroneous. It should be written, "The maximum *operating* temperature of incorporated materials used in solid-state relays shall not exceed their *permissible* temperature, which shall be verified by testing according to 7.3," thus giving the formulation the opposite sense.

10.3.6 Tests

In Section 8.2.1.a.2 of IEC 62314-1, it is emphasized (by underlining) that the overload capacity of the solid-state relay "utilizing a current-controlled cut-out device in addition to an overcurrent protection ... shall be tested with the cut-out device in place." In other words, if the solid-state relay is used in a concrete electrical installation together with the protective automatic circuit breaker, it should be tested for overload capacity together with this circuit breaker. So, what are we testing in this case: the solid-state relay or the circuit breaker? According to standard requirements, it is the circuit breaker.

We already noted in this chapter that standard requirements connected to testing solid-state relays by voltage higher than the rated level do not specify concrete points (circuits) between which the test voltages should be applied. It is incorrect and should be corrected also for Section 8.4, "Insulation Tests." However, this will be only a partial solution of the problem. More complex is the problem of a test withstanding the voltage of the output switching element of the solid-state relay.

The problem arises because many constructions of solid-state relays contain built-in internal structures for overvoltage protection (more often, varistors) which, naturally, have lower pickup thresholds than the installation protected by them (i.e., the solid-state switching element of the solid-state relay). Thus, short-term, step-up pulses of a voltage on the relay terminals at tests will lead to pickups of this protective element. The long duration, for example approximately 1 minute, of the step-up voltage during the tests will lead to the destruction of a protective element. Thus, it is impossible to perform normal withstanding voltage tests for many solid-state relay constructions.

The problem may be solved by changing a principle of connection of a protective element to output relay terminals at which this protective element could be disconnected from the terminals for the period of tests due to the removal of external links (Figure 10.8). Such a change in the design of the solid-state relay would allow solving a problem of tests and should, in our opinion, be introduced into the standard as an obligatory requirement in the construction of solid-state relays.

10.4 The International Standard on Reed Switches (IEC 62246-1)

10.4.1 Scope

The IEC 62246-1, 2nd ed., "Scope" section states that it is the generic specification that "applies to all types of reed switches including dry reed switches and heavy-duty reed switches." This formulation leaves a slight bewilderment: if *all* types of reed switches (i.e., dry low voltage, high-voltage vacuum,

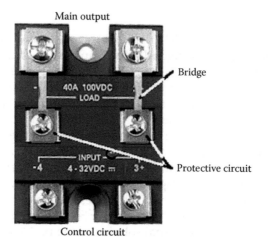

FIGURE 10.8

A solid-state relay with an incorporated protective element which can be disconnected from the main terminals for the period of tests due to the removal of external links (bridge).

heavy duty, mercury wetted, plunger type, and membrane type, among others[1]) are embraced by this standard, why does the standard focus on only two types (dry and heavy duty)? The answer can be seen in the "Terms and Definitions" section of IEC 62246-1, which gives definitions of only these two types of reed switches: dry and heavy duty. From this, we see that the IEC 62246-1 standard, in this case, simply misleads users in its assertion that it relates to all reed switch types.

Further, in the same paragraph it is noted that the standard "applies to reed switches which are operated by applied magnetic field." This is very interesting: in what other field, except magnetic, can the *magnetically operated contact* (which is a Russian name for *reed switch*) scope?

In Note 2 to this section, there is a remark that "for elementary relays with reed switches, this standard is recommended to be used together with the standards IEC 61810-1 and IEC 61811-11." Practically 90% of manufactured reed switches are intended for use in electromagnetic elementary relays. This means that the standard should be coordinated with these overall standards. As will be shown in this chapter, it does not.

10.4.2 Terms and Definitions

10.4.2.1 Introduction

As was noted above, in the "Terms and Definitions" section of IEC 62246-1, definitions are given to only two types of reed switches, dry and heavy duty, and there is no mention at all of reed switches of other types, such as mercury-wetted reed switches. Why?

10.4.2.2 Reed Switch

The definition for the term *reed switch* (p. 3.1.3 of IEC 62246-1) is that it is an "assembly containing contact blades, partly or completely made of magnetic material, hermetically sealed in an envelope." But, first, all these attributes characterize reed switches of any type, not just dry ones. For example, all these attributes are found in mercury-wetted reed switches. Second, use of the term *blade*, strictly speaking, describes only certain contact profiles and does not reflect the complete variety of existing reed switches, among which there are also ball and membrane, plunger type, and the like.[1]

In the development process, the term *dry* is used to designate reed switches, the envelope of which is filled by dry air under atmospheric pressure or special gas (or a mixture of gases) under a heightened pressure, and also to designate vacuum reed switches. Thus this term, as defined in the standard, reflects only a minor case, the type of environment filling a reed switch envelope, and is not an essential reed switch attribute, thus distinguishing it from other types of switches. That is, the definition of the base term *reed switch* in the standard is replaced by a minor definition that suits only some varieties of reed switches. We ask, "Is it logical to give a definition to various kinds of reed switches that does not reflect the given base definition of the reed switch itself?"

10.4.2.3 Heavy-Duty Reed Switch

In the definition of the term *heavy-duty reed switch* presented in Section 3.1.4 of IEC 62246-1, it is given that it is a dry reed switch in which increasing switching ability is provided by means of a special design of the contacts. But why should a reed switch with increased switching ability be necessarily "dry"? Why is the increasing of the switching ability of the reed switches only possible due to a "heavy-duty" contacts design? Is it that filling its envelope with mercury in the so-called wetted reed switches does not allow such switches to increase considerably their switching ability? Unfortunately, the standard only prompts such questions, but does not answer them.

In the definition of the term *heavy-duty reed switch*, a basic mistake is incorporated, in our opinion, in that for such reed switches the design of the magnetic and electric parts is supposed to make them as separate parts. As shown in Reference 1, separation of a magnetic and electric circuit is the major attribute distinguishing the usual electromagnetic relay (and not a reed switch relay) in which these circuits are combined. As a matter of fact, already in the term *magnetically controlled switch* (which, by the way, is another Russian term for *reed switch*), it is certain that just this *electric* contact is controlled by a *magnetic field*. Thus, if these circuits are separated, what purpose is served by the magnetic material for the contacts (see the definition in IEC 62246-1 for the term *dry reed switch*)? If we push this reasoning to the point of irrationality, it is possible to consider that the usual reed switch contains a ferromagnetic material covered by a special electroconductive coating, and in this way electric and magnetic circuits become separated: the electric circuit enclosed in an electroconductive covering, and the magnetic circuit enclosed in ferromagnetic material. On the other hand, the attribute of the presence of a hermetically sealed envelope on contacts is not the main distinctive reed switch attribute, because the very first element working as a reed switch, which was invented in 1922 by Petersburg University Professor V. Kovalenkov, did not have an envelope (Figure 10.10).

In 1936, contact elements were placed in a hermetically sealed envelope by the engineer W. B. Ellwood from Bell Telephone Laboratories. Since then, practically all the types of reed switches manufactured by the industry are supplied with a hermetically sealed envelope. However, today also the contacts of the high-voltage vacuum circuit breakers applied in the electric power industry, and contacts of various electromechanical relays, are placed in a hermetically sealed envelope (Figure 10.9), not being reed switches. Therefore, the presence of an envelope on contacts is necessary, but is not a basic attribute distinguishing reed switches from other devices.

Thus, IEC 62246-1 does not give a precise definition for the very element to which consideration is devoted.

10.4.2.4 Contact Blade

In the definition of the term *contact blade*, given on p. 3.4.1, it is given that it is the metal element "providing the functions of either electric or magnetic circuit or both functions combined as in the case of dry and wetted reed switches." We have already pointed out in this chapter the abnormality of this approach when applied to a reed switch as an element with separate functions of electric and magnetic circuits. The definition as given again includes the separation of the functions to be attributed to all other types of reed switches, except for dry or wetted. Since up to this point in IEC 62246-1, *heavy-duty reed switches* have also been mentioned, the standard implies that the presence of the separation of electric and magnetic circuits is automatically attributed to heavy-duty reed switches. But this is not so in reality. There are many ways to increase the power of reed switches in which the magnetic circuit and an electric circuit are not separated,[1] and there are devices in which they are completely separated (Figure 10.9). So this definition has very little to do with the general attributes of reed switches and refers more to electromagnetic relays with hermetically sealed contacts.

10.4.2.5 Wetted Reed Switches

In Section 3.4.1 of IEC 62246-1, for the first time *wetted reed switches* are mentioned, but a definition of this term is not given.

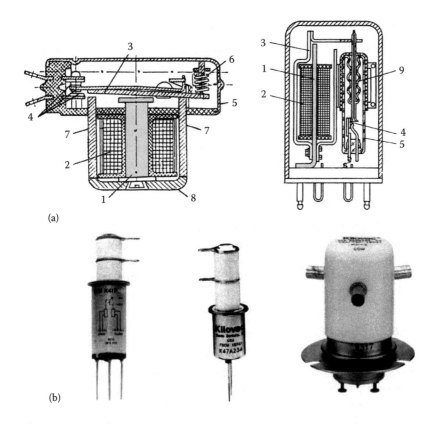

(a)

(b)

FIGURE 10.9
Design of electromagnetic (*not reed switch*) relays with hermetically sealed contacts placed in a gas-filled or vacuumized shell (a) and an external view of reed switch relays with hermetically sealed contacts (b) manufactured by the companies Kilovac, Jennings, and Gigavac. 1: a core sticking out of the hermetic shell; 2: control coli; 3: a stiff ferromagnetic armature; 4: contact straps; 5: a hermetically sealed shell for contacts; 6: a restoring spring; 7: parts of the magnetic core sticking out of the hermetic shell; 8: a removable part of the magnetic core; and 9: a flexible bellow.

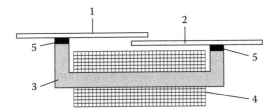

FIGURE 10.10
The first reed switch relay (USSR Author Certificate no. 466, 1922). 1 and 2: contact elements (springs) made of magnetic material; 3: an external ferromagnetic core; 4: a control winding; and 5: dielectric spacers.

10.4.2.6 A Precise Definition of Reed Switches

Due to the absence in IEC 62246-1 of a precise definition for *reed switch*, we shall try to formulate one: "the reed switch is an electromechanical device intended for switching electric circuits by means of moving or deforming at least one of the contact details, sensitive to an external operating magnetic field and placed in a hermetically sealed envelope"; in this definition, a *contact detail* is the internal magnetic sensitive element of the reed switch that carries out the switching of an external electric circuit by means of moving or deforming under the action of the operating magnetic field.

10.4.2.7 Maximum Cycling Frequency

In Section 3.4.8, the *maximum cycling frequency* is defined as the "maximum number of cycles per second at and below which the reed switch still meets the specifications." At the same time, in Section 3.4.20 the *frequency of operation* is defined as the "number of cycles per unit of time." What the "cycling" is and how it differs from an *operation* are not explained in the standard.

10.4.2.8 Maximum Contact Current

In Section 3.4.12, the term *maximum contact current* is defined as the "maximum allowed switches d.c. or peak current in correlation to [a] given number and frequency of operations and load, under specified conditions."

First, the term *maximum contact current* without additional explanations is absolutely unclear as there is also a maximum switched current, a maximum short-term current carried through closed contacts, and a maximum long-term current carried through closed contacts. The term should be specified, in our opinion, as the *maximum switched current*.

Second, as the limits for switching current and also for the current carried through closed contacts are characterized ordinarily by their effective values, it is not clear why for alternating current it is necessary to consider a peak value (as required by the standard) instead of the RMS value of the current.

Third, the full cycle of the switching process includes make and break operations. As is known, the switching abilities of contacts at make and at break (especially on a DC) are essentially different. This has been reflected in IEC 62246-1 in Sections 3.4.23 and 3.4.24, where such concepts as *limiting making capacity* and *limiting breaking capacity* (and now the RMS value for AC, for it is underlined) are defined. These concepts are interpreted as the "greatest value of electric current which an output circuit is capable of making (breaking) under specified conditions." This raises a question: what exactly is the difference between the terms *maximum current* and *greatest current* used in the standard? The standard does not answer this question. There are no illustrated purpose and scope of the term *maximum contact current*, on the one hand, and the terms *limiting making capacity* and *limiting breaking capacity*, on the other. It is intuitively clear that the term *limiting capacity* means a small number of cycles withstood by contacts at making and at breaking of any super-overcurrents specifically stipulated. But, in our opinion, the international standard is not the place for intuitive guesses and fabrications.

Fourth, what is the purpose of using the term *output circuit* in Sections 3.4.23 and 3.4.24 instead of the term *contacts*, as used in all other sections?

10.4.3 Rated Values

10.4.3.1 Section 4.2

In Section 4.2 of IEC 62246-1, one more value is expedient to add: 120 operations per second, as the value implemented by the reed switch at feeding the control coil from a source with a frequency of 60 Hz to the row of recommended frequencies.

10.4.3.2 Section 4.4

In Section 4.4, a row of recommended rated values of *open-circuit voltage across contacts* for AC is given as follows:

0.01; 0.1; 5; 12; 24; 30; 40; 50; 100; 110; 120; 127; 150; 170; 175; 200; 220; 250; 265; 300; 380; 400; 500; 1000; 2000 V (RMS).

And for DC:

0.01; 0.03; 0.05; 0.1; 1; 1.5; 4.5; 5; 6; 6.5; 10; 12; 15; 17; 20; 24; 28; 30; 36; 40; 48; 50; 60; 80; 100; 110; 150; 170; 175; 200; 220; 250; 265; 280; 350; 400; 440; 500; 600; 800; 1000; 1200; 1500 V.

This section seems very strange. First of all, the term *open-circuit voltage across contacts* is not present in the "Terms and Definitions" section. There is the term *maximum contact voltage*; is this the same or not? Second, these voltages rows do not coincide with the rows of the rating voltage values for electromechanical relays (see IEC 61810-1) in spite of the fact that in Section 1 of IEC 62246-1, it is recommended to use both of the standards together in the case of reed switch relays (90% of all reed switches applications). Third, there is the strange condition that for DC the voltage row terminates at 1500 V, whereas for AC it terminates at 2000 V RMS. The last value (2000 V AC RMS) makes much more serious demands on reed switch insulation than the 1500 V DC value and is equivalent not less than 2800 V DC. So why does the voltage row for DC terminate with 1500 V? In mass production, there are dry reed switches that withstand voltages of 5, 10, and even 15 kV DC. Such reed switches are manufactured in large quantities by various firms. What are we to do with these reed switches that are not described in IEC 62246-1? Finally, what is the necessity for a row of values of the voltages differing among themselves from 5 to 7 V (e.g., 100, 110, 120, 127, 150, 170, 175, 200, 220, 250, or 265 V)? What reed switch manufacturer will develop them separately for 100 V, for 110 V, for 120 V, and for 127 V? Why isn't this row given in the form of 100 V, 150 V, 200 V, 250 V, and so on? It is enough to analyze the technical specifications of reed switches from various manufacturers to be convinced that actually just such a division of the voltages is put into practice. And as for some (e.g., 0.01, 0.03, 0.05, 0.1, 1.0, 1.5, 4.5, 5.0, 6.0, or 6.5 V), this is very interesting, for authors of the standard certainly saw reed switches with rated voltages such as these.

It would seem that the person who developed the standard has selected these rows arbitrarily, matching them with neither common sense nor existing practice.

10.4.3.3 Section 4.5

In Section 4.5, a recommended row for rating values of currents is given as follows: "1; 1.25; 1.5; 1.6; 2; 2.5; 3; 3.15; 3.5; 4; 5; 6.3; 7; 7.5; 8 A or the decimal multiples or submultiples of these figures in amps: 1, 2 and 5." We raise the same question: what is the reason to extract 1.5 and 1.6 A separately, and 3 and 3.15 A, and so on? What is the meaning of 1, 2, and 5 at the end of the row? Why doesn't this row coincide with a row of rated switching currents, as recommended in Section 4.11: 1, 10, 15, 30, 50, or 100 mA; or 0.3, 0.5, 1, 2, 3, or 5 A? What is the necessity for representing two these rows separately?

10.4.3.4 Section 4.6

In Section 4.6, there is a recommended row for rated loads, in volt-amperes, that is, for AC loads. But where are the values, in watts, for DC loads? Aren't the reed switches intended for switching a DC load?

10.4.3.5 Section 4.10

In Section 4.10, there are rows of recommended values for *rated operational voltage*. This is yet a third term for designating a voltage on reed switch contacts after "maximum contact voltage" (Section 3.4.14) and "open-circuit voltage across contacts" (Section 4.4). What is the difference between all these terms, and why are there no explanations in the "Terms and Definitions" section for the last two of them? Why does this new voltage row differ from that given earlier in Section 4.4? Why does this new row of a rated operational voltage differ from the rated operational voltage given in Annex E of IEC 62246-1?

10.4.3.6 Section 4.11

All of the above refers to the term *rated switching current* in Section 4.11, which has not been included in the "Terms and Definitions" section, and to values of the current that are not consistent with the values given in Section 4.5.

10.4.3.7 Section 4.12

In Section 4.12, *rated insulation voltage* (i.e., a test voltage, according to the definition given in Section 3.4.25) is given in a row of RMS values of the voltages ending with 600 V for AC. This is amazing, since in Section 4.10, a row of rated operation voltages comes to an end with the value of 1000 V AC (RMS), and in Section 4.4 the "open circuit voltage across contacts" ends with the value of 2000 V AC (RMS). In this section, it says that for the reed switch intended for switching 2000 V AC, the standard recommends testing at 600 V AC!? In what cases and for what are such "tests" necessary? And the boundary value of 1500 V DC also is not true, as the test voltage should always be much above the maximum operating voltage. Besides, for contacts of the relay (and reed switches are used very widely as contacts of the relay) intended for use in the electric power industry, there are special standards in which the test voltage is selected as the double operating voltage plus 1000 V. In our opinion, this section of the standard demands rewriting.

10.4.3.8 Section 4.13

Values for the *rated impulse voltage* that are given in Section 4.13 as 800, 1500, 2500, 3000, and 4000 V do not encompass constructions actually existing in the market; for miniature and measuring reed switches, the voltage 800 V is too high a value, and for dry vacuum reed switches with rated voltages of 5, 10, and 15 kV, the voltage 4000 V is too low a value. This voltage row needs to be corrected.

10.4.3.9 Section 4.14

The utilization categories presented in Section 4.14 in the form of a table, taken apparently from standard IEC 60947-5-1, also contain rather strange formulations for some classes, in particular for classes AC-12 and DC-12, "control of resistive loads and solid state loads with insulation by opto-coupler"; AC-13, "control of solid state loads with transformer isolation"; and DC-14, "control of small electromagnetic loads having economy resistor in circuit." In our opinion, such incorrect and uncertain formulations for load kinds, as "solid state loads with insulation by opto-coupler," "solid state loads with transformer isolation," and the like, are not acceptable for use in the new international standard on reed switches even if they also have been used earlier in another international standard. In our opinion, classes AC-12, DC-12, AC-13, and DC-14 should be removed from the standard, all the more so since these classes are not used anywhere else in the standard (see Tables 5, 6, 7, and E1 of IEC 62246-1).

10.4.3.10 Section 4.16

In Section 4.16, a row of recommended values for *limiting continuous current* is given. It is not clear why the recommended row of values for this current does not coincide with other rows of currents given in the standard. In addition, in the standard the procedure for testing reed switches on conformity to this parameter is not stipulated. This being the case, what is the purpose for the recommended row for these values?

10.4.4 Tests and Measurement Procedures

In Section 7.7.1, the procedure for measurement of the contact circuit resistance, based on the Kelvin four-wire circuit method (a measurement voltage drop at the presented value of the current carried), is given. The measurement is to be made at voltages not exceeding 6 V and a current of 1 A. In our opinion, the specified value of a current (1 A) is too high for miniature low-power reed switches and can lead to their faulting. Besides, because of overlapping in reed switch magnetic and electric circuits, the current, which is carried through closed contacts, creates a magnetic field aimed at breaking them. In low-power miniature reed switches, carrying a current of 1 A can lead to an appreciable decrease of contact pressure and, accordingly, to an increase in resistance of the closed contacts during the measurement process. In our opinion, there must be various values of test currents for reed switches with different switching powers.

In the testing procedure for dielectric tests (Section 7.8.1A), the rapidity with which the value of the test voltage is to be increased should be specified.

In measurement procedures of the operate time and release time (Section 7.10), mention is made about fixing the moments of turn-ON and turn-OFF of a current in the coil of a reed switch for the measurements. But in Section 3.3, it is written that the *operate time* and *release time* are terms for times connected with the application and removal of the operating magnetic field on the reed switch. It is very important to note that the moment of the turning-ON of a current in the coil and the moment when the magnetic field in this coil will reach its rated value are not at all the same moment! They differ by a significant value known as the *time constant* of the coil, which is commensurable with or even exceeds the operate and release times of reed switches. Thus, the test procedure as presented is inapplicable and requires reconstruction in respect to the effect of the coil parameters on the measurement results.

Section 7.11 refers to *contact sticking* and contains two subsections: *thermal sticking* and *magnetostrictive sticking*. For thermal-sticking tests, the unit under test is kept in a closed condition of contacts for 24 hours at the maximum temperature, and for magnetostrictive-sticking tests, it is after 2000 cycles of operation-release, then the release value and break time are measured. Changes of these parameters are used as the criteria for an estimation of the extent of reed switch sticking.

It is necessary to note that in the theory of reed switches, there are two principal causes for sticking: the first is the magnetostrictive effect, which is a grinding of the contact surfaces after repeated wear and remaining in the closed position under an act of molecular forces, and the second is electric erosion of contacts at DC switching in which an acute ledge is formed on one of the contacts, and a crater on the other. The pinching of this ledge in the crater leads to a sticking reed switch. We do not know what *thermal sticking* is, and efforts to find any reference presenting this phenomenon in the technical literature have been to no avail. The other matter, the magnetomotive force of making (ampere-turns to make), depends to a strong extent on the reed switch temperature owing to changes of the magnetic properties of a ferromagnetic material from which the contact elements are manufactured (Figure 10.11). On the market, there are even special, so-called thermal reed switches, operating at certain temperatures and used as temperature sensors.

On the other hand, it is known that the turn-on time of reed switches strongly depends on the rate of the current in the control coil (i.e., at the magnetic field level that is abruptly applied to them; see Figure 10.12). Therefore, it is abundantly clear that a change of a magnetic condition of the contact details under the effect of temperature will also change the turn-ON time of the reed switch. This change, however, has no relation to reed switches sticking. Thus, instead of the analysis of one of the most common reasons why reed switches stick, the electric erosion of contacts, the standard invents some new aspect of reed switch sticking not known in the technical literature.

Test procedures and requirements for vibration, shock, and acceleration (Sections 7.19, 7.20, and 7.21) do not specifically reflect mercury-wetted reed switches. But, as the standard comprises all types of reed switches, these sections should be expanded so that they include mercury-wetted reed switches as well.

The electric loads given in Table 2, Section 7.23.4, of IEC 62246-1 (Table 10.2) as the standard conditions for electrical endurance tests refer only to dry reed switches (why?) and also are restricted to resistive loads by values 1 to 100 mA at voltages of 0.03 to 200 V DC and values of 10 mA to 1 A at voltages of 0.03 to 230 V AC. And for the combined load (Table 3, Section 7.23.4), the maximum current test does not exceed 100 mA. These values are not connected in any way with the actual parameters of electrical endurance of real reed switches, made by real manufacturers; see Table 1 of IEC 62246-1. In addition, they are not connected with the values of rated switching currents and rated switching voltages given in Sections 4.10 and 4.11 of IEC 62246-1. This is certainly puzzling!

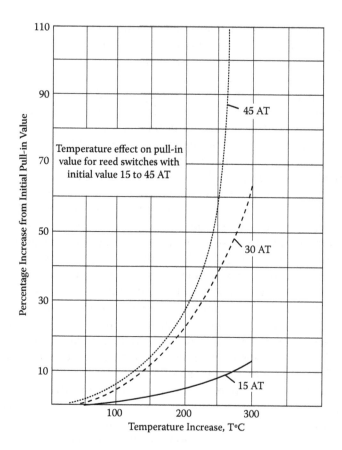

FIGURE 10.11
Impact temperature on a pickup current (pull-in value) for reed switches with various initial
sensitivity (ampere-turns, or ATs).

In Section 7.26, "Voltage Surge Test" (specifically, Subsections 7.26.3 and
7.26.4), we are given the following recommended values:

- V_{PEAK}: 1000 V, 1200 V, and 1500 V
- t_1: 10 μs
- t_2: 700 μs or 1000 μs

Only a specialist in the field of EMC may understand what this list means,
but by no means can a specialist in the field of reed switches be expected to
understand.

Why are there no references in these sections to the matching standard,
IEC 61000-4-5, which presents, in detail, a technique of conducting of such
tests and parameters of test pulses? Why are there no explanations for t_1 and

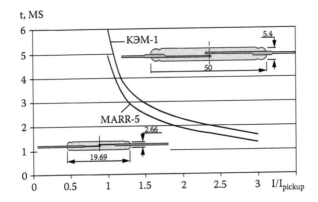

FIGURE 10.12
Experimental curves gained for reed switches of different types and sizes, and manufactured by different companies in different countries. I: The actual current in the control coil; I_{PICKUP}: the pickup (pull-in) current; and t: the switch-ON time.

TABLE 10.2

Electrical Endurance for Some Types of Reed Switches Manufactured by Yaskawa

	Make		Break		Electrical Endurance (× 1000)			
					Reed Switch Type			
Voltage and Load	Current (A)	Power Factor or Time Constant	Current (A)	Power Factor or Time Constant	R24U	R25U	R14U	R15U
~110 V resistive	3	Cos φ = 1.0	3	Cos φ = 1.0	—	—	—	200
	2		2		50	200	200	1000
	1		1		300	500	1000	2000
~240 V inductive	10	Cos φ = 0.7	1	Cos φ = 0.4	—	—	500	800
	5		0.5		300	1000	1000	1500
	2.5		0.25		600	2000	2000	3000
= 115 V inductive	0.5	L/R = 40 ms	0.5	L/R = 40 ms	—	—	1000	1000

t_2, as earlier the same designations were applied in the standard to make and break times (see Figure 2 in the Section "Terms and Definitions" of IEC 62246-1)? Why, as the parameters of a test pulse (in the given context, t_1 and t_2 are the times describing *pulse rising edge* and *pulse decreasing edge* of the high-voltage test pulse), are the parameters of the pulse known as the *tele-communication pulse*? Are such pulses used for testing the radio-electronic and communication equipment, in contrast to a standard pulse of 1.2/50 µs (Figure 10.13), used for testing electrotechnical equipment such as the relays, contactors, switches, buttons, and so on? Can the miniature reed switches, for example ORD213, MITI-3, RI-70, KSK-1A80, HSR-0025, and dozens of other

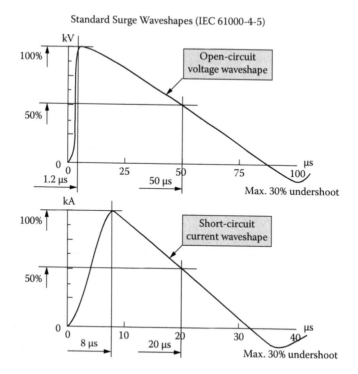

Standard Surge Waveshapes (IEC 61000-4-5)

FIGURE 10.13

Standard waveshapes for testing surge immunity (according to IEC 61000-4-5).

types with breakdown voltages of 100–200 V, withstand an impulse voltage of 1000 V? On the other hand, what test using a 1500 V impulse voltage will verify vacuum reed switches with rated voltages of 5000 or 10,000 V?

Unfortunately, there are more questions than answers, which is absolutely inadmissible for a basic international standard.

In Section 7.28, "Rated Making and Breaking Capacities," the test procedures and requirements are presented in detail. And also in Section 7.32, "Making Current Capacity Test," and in Section 7.33, "Breaking Current Capacity Test," the same procedures and requirements are presented. Why?

As the *procedure* for the making current capacity test (Section 7.32.2), it is written, "Sub-clause 7.23.5 without breaking applies." As the *procedure* for the breaking capacity test (Section 7.33.2), it is written, "Sub-clause 7.23.5 without making applies." However, in Section 7.23.5 the *requirements* (not procedure) for failure categories consideration for the *endurance test* (not for the making or breaking capacity test) are given.

Test procedures and test requirements are not same!

To clarify, making (breaking) capacity and electrical endurance are not the same! According to Section 3.4.22, *electrical endurance* is a "number of cycles until contact failure, with specified electrical loading of the output circuit and under specified operating conditions." According to Section 7.32.1 (7.33.1), "the making (breaking) current capacity test evaluates failures caused by the electrical wear of contact surfaces when making (breaking) current." Are the last definitions sufficiently clear for understanding the difference between *making (breaking) capacity* and *electrical endurance*? In our opinion, the authors of the standard do not understand these differences. There are absences also in all other sections related to the testing of these two parameters. These sections simply are exactly copied in the standard. Why? Truly clear definitions for the terms *breaking capacity* and *making capacity* are given in the standard IEC 60947-1, "Low-Voltage Switch Gear and Control Gear—Part 1: General Rules":

> The rated breaking capacity of all equipment is a value of current, stated by the manufacturer, which the equipment can satisfactorily break, under specified breaking conditions.
> The breaking conditions which shall be specified are:
>
> - the characteristic of the test circuit;
> - the power-frequency recovery voltage.
>
> The rated breaking capacity is stated by reference to the rated operational voltage and rated operational current, according to the relevant product standard. An equipment shall be capable of breaking any value of current up to and including its rated breaking capacity.
>
> NOTE: A switching device may have more than one rated breaking capacity, each corresponding to an operational voltage and a utilization category.

As can be seen, in this definition for breaking capacity (it is analogous also for making capacity), the number of switching cycles is not mentioned at all, in contrast to endurance. That is, the switching capacity is the specified current value, and the electrical endurance is a specified number of switching cycles.

And further, what do the notes about "without breaking applies" and "without making applies" in the procedures mean for the making and breaking current capacity tests (Sections 7.32.2 and 7.33.2)? How is it possible to realize this demand practically, if, according to the standard's claim (Section 3.2.13, Figure 2), the switching process is always accompanied with contact bouncing? In contrast to Figure 2, the figure in Annex G ("Making Current Capacity Test Sequence") and the figure in Annex H ("Breaking Current Capacity Test Sequence") do not include contact bouncing. So, which figure is correct?

It is absolutely clear that all these sections need to be completely rewritten.

For tests on rated conditional short-circuit currents (Section 7.29) in the standard, it is stipulated that the item being tested has to be connected in series with a fuse or an automatic protective switch capable of a fast disconnection

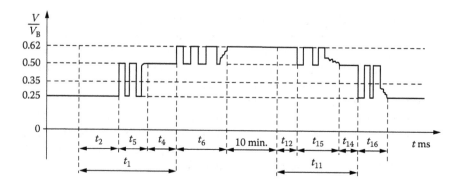

FIGURE 10.14
Oscillograms of switch-ON and switch-OFF processes in the reed switch. t_5, t_6, t_{15}, and t_{16}: Contact bounce times (see Figure 2, IEC 62246-1, 2nd ed.).

of a short-circuit current. However, the fuses on high currents or automatic circuit breakers are not the calibrated devices that have undergone the latest metrological tests and have matching certificates; on the contrary, the time-current characteristic of such devices is usually rather approximate and has a wide dispersion of parameters. Therefore, in our opinion, they are not suitable for tests for conformity to the standard. At the same time, in the IEC 61000-4-5 standard, the technique and parameters of the test pulse intended to test the ability to withstand short-circuit currents (see Figure 10.14) are presented. The special test equipment is manufactured for these tests by many firms. We do not see any reason to ignore the requirements of the IEC 61000-4-5 standard and to use the "homebrew" methods recommended in the IEC 62246-1 standard.

10.5 Conclusion

Many of the mistakes and inaccuracies found in the IEC standards discussed in this chapter are dangerous symptoms indicating a very low quality of international standards and a high hazard in their practical usage, in particular for DPR design and improvement.

The considerable number of mistakes and unintelligible definitions, and the absence of connections with terms currently used in practice, make it impossible, in our opinion, to use the standards IEC 61810-1, IEC 62314, and IEC 62246-1 in their existing form. These standards require complete rewriting in view of the analysis given in this chapter.

References

1. Gurevich, V. *Electric Relays: Principles and Applications.* Boca Raton, FL: CRC Press, 2005.
2. Gurevich, V. *Protection Devices and Systems for HV Applications.* New York: Marcel Dekker, 2003.
3. Girgis, A. A., Nims, J. W., Jacamino, J., Dalton, J. G., and Bishop, A. Effect of Voltage Harmonics on the Operation of Solid-State Relays in Industrial Applications. *IEEE Transactions on Industry Applications,* vol. 28, no. 5, 1992, pp. 1166–1173.
4. Gurevich, V. *Electronic Devices on Discrete Components for Industrial and Power Engineering.* Boca Raton, FL: CRC Press, 2008.

11

New Concepts for DPR Design

Microprocessor digital protective relays (DPRs) are extensively displacing all other types of relay protection devices from the market since all the world leaders in this area have almost completely ceased production of all other types of protection relays for the electric power industry. Since the first introduction of DPRs in the electric power industry more than 20 years ago, these devices have caused considerable enthusiasm among specialists. Quite a number of articles have been published in the technical literature praising the benefits and advantages of DPRs in all aspects. However, as any sensible professional should know, an ideal technical device does not exist, and even the DPR has many technical problems and shortcomings that have already led to a significant reduction in the hardware reliability of relay protection[1-4] and increased its cost. Since there is no alternative to DPRs today, we face a challenge to counterbalance the negative impact that DPRs brought into relay protection. In this chapter, I offer a new concept of DPR design capable, in my view, of solving many of the actual problems of DPRs.

Today, there are hundreds of DPR models made by dozens of manufacturers in the market. Each type of DPR is built in a separate body, which may be totally different from that of any other type of DPR (even of the same brand); see Figure 11.1.

Historically,[5] there is a large number of noninterchangeable and incompatible DPR designs. This means that if a certain module of any DPR installed in the particular substation or power station fails, it can be replaced only by the same one produced by the same manufacturer. Thus, after you have spent a small fortune purchasing a DPR from one manufacturer, you fall into economic dependence on this manufacturer for the next 10–15 years, since after you have chosen one manufacturer it no longer makes any difference if there are other manufacturers in the market, as you cannot use their products. And the only way to get out of this is to pay another small fortune for a DPR from another manufacturer (and, thus, you switch from one bondage to another).

And what does the manufacturer do with an absolute monopoly? Right: the manufacturer increases the price! The price of one spare DPR module can reach almost one-third or even a half of that of the entire DPR! As you have no other choice, you pay that price. The insufficient hardware reliability of DPRs[1-4] results in a serious economic problem since the built-in self-diagnostics feature so much advertised by manufacturers does not help reduce failures and breakdowns.

FIGURE 11.1
Configuration of modern DPRs of different brands.

The popular way of increasing the hardware's reliability as backup is also problematic due to the high cost of DPRs and the shortage of resources to reequip the basic set of protections apart from the backup set. Another way of increasing the reliability of electronic equipment related to the preventive replacement of limited life units, for example power supply units with electrolytic capacitors, is hardly ever used in practice due to the same reasons.

The full incompatibility of DPR software, sometimes even between different versions of the same program (not to mention among software from different manufacturers), results in another problem hampering the use of DPRs. The same energy company may use four to five types of DPRs, and service staff should learn all these essentially different programs; this results in serious problems due to the so-called human factor. And this happens in the background of the ubiquitous sophistication of DPRs and their software. Below is the opinion of a leading Russian protection engineer concerning relay protection devices from one of the world's leading manufacturers of DPRs:

Terminal Siprotec 7SJ642 (Siemens) has unreasonable technical and informational redundancy. User manual (C53000G1140C1476, 2005) declares «simple operation through integrated control board or PC with a systems program DIGSI», which is totally untrue. For example, you should enter nearly 500 parameters (settings), despite of inevitable changes to signal matrix, while each signal has its "properties" influencing the operation of the unit (printed-out DIGSI signals matrix takes about 100 pages of text in English). Since there is a need to compile terminal adjustment tasks, where all set-up protocols should be considered, the amount of documentation becomes huge. Big volumes of data need to be entered making the set-up process very difficult. Informational redundancy increases the probability of so called human factor errors. Technical redundancy requires only high-level specialists to be involved in work with the terminal. Available technical documentation includes thousands of pages, which often do not provide necessary data while containing errors.[6]

The Siemens product mentioned above is not the only device with these kinds of problems. The above is also true for any other brand. Unfortunately, this is a common trend today. The lack of simple and basic design and software standards significantly complicates acceptance and periodic testing of DPRs.[7]

The latest trend of increasing the number of protection functions in one module, "charging" excessive additional nonprotection functions, and implementing so-called nondeterministic (not defined) logic additionally affects the reliability of relay protection and increases the unpredictability of DPR behavior under emergency conditions.[8,9] Is there any way to solve these problems?

The concept of DPR design we propose is based on the following basic principles:

1. The operating assemblies of the DPR should be distinctively well separated; thus, the chaotic layout of the assemblies on the printed boards[10] should be replaced with an ordered, standardized arrangement. For example: operating assemblies such as power source units, input currents and voltage transformers with signal preprocessing units, digital input modules, output relay modules, central processor modules, and so on should be arranged on separate printed boards of a standard size with universal connectors.

2. Separate relay protection devices for power plants and substations should be designed and sold as separate universal printed boards (modules), rather than as a number of devices in enclosures of different sizes and shapes to be used by a customer to configure the required DPR. These boards (modules) should allow simple

installation (by guide rails to connect to cross-board connectors) in metal cabinets with separate sections and separate doors. Metal cabinets should be made with technology aimed to protect the contents from external electromagnetic disturbances.

3. The DPR should perform only relay protection functions. The number of functions integrated in one module should be optimized according to cost and reliability and limited by standard.

4. The DPR's software should include a standard basic framework and set of different applications and libraries compatible with the common basic framework.

5. All modules in the cabinet must be connected to two heavy-duty power sources, which are connected together as main and standby.

6. All these principles must be described in the new standard with a conventional name, such as "Principles of DPR Design: Basic Requirements."

But is it possible to realize the proposed concept from a technical point of view?

As mentioned above, the majority of DPRs available in today's market do not include a set of modules strictly separated by functions and design, and they remain a hodgepodge such as a central processing unit (CPU) and switched power supply located on the same printed boar.[10] We have analyzed a lot of different types of the most up-to-date DPRs of the world's leading brands and finally found devices that ideally correspond to the above design requirements. These are Series 900 DPRs of the well-known Chinese company Nari-Relays with universal modules used in protection devices of different types (Figure 11.2).

These turnkey modules do not require any preliminary preparation except for logic and setting configuration. Moreover, once assembled, the DPR does not require any adjustments. In order to assemble the DPR, you only need to install printed boards as shown in Figure 11.2 (the actual set includes the power source board, which is not shown in the picture since it is not required as part of the set in our concept) into labeled rails of the enclosure (in our case, it is the section of the cabinet). Assembly of such complex protection as distance consist of seven separate modules supplied in cartons, and the entire assembling process doesn't take more than 10–15 minutes, after which you can begin entering the setpoints. It is obvious that an ordinary protection technician who does not have expertise in microprocessor technology won't have any problem with the on-site assembly of protection relays consisting of such universal units.

Basically, nothing today prevents early implementation of the proposed concept in a given country. Even a small company can enter the DPR market offering consumers a new concept of cheap and reliable relay protection, equipped with backup block modules. It should start with purchasing sets of universal Nari-Relays modules (with different algorithms written in

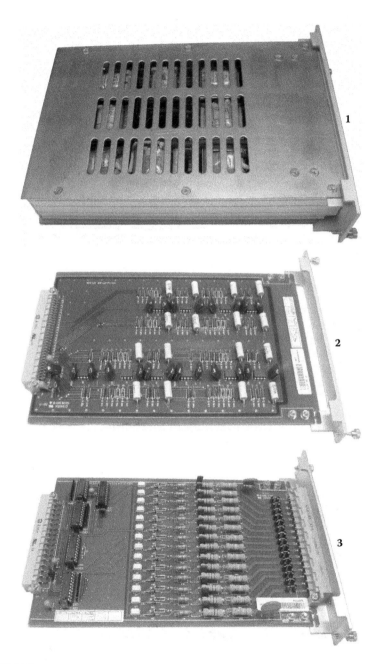

FIGURE 11.2
Set of universal functional modules (220 × 145 mm) arranged on separate printed boards, which different Nari-Relays DPRs consist of: PCS-931 (differential line protection), PSC-902 (distance protection), and so on. 1: input current and voltage transformers module; 2: low-pass filter (anti-aliasing filter); 3: digital input module; 4: output relay module; 5: optical communication module; and 6: CPU module.

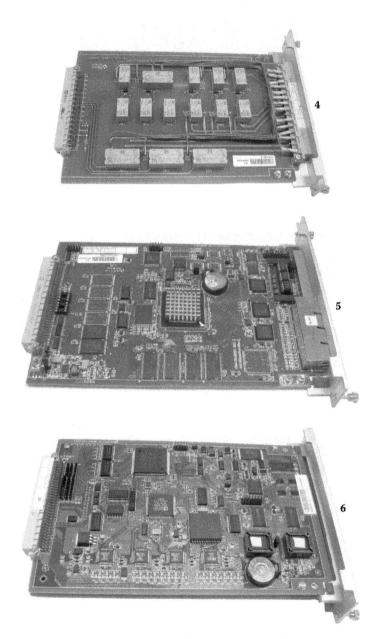

FIGURE 11.2 (continued).

EEPROM and different sets of input transformers) and mastering the production of cabinets.

What are the advantages of the proposed DPR development?

For a Consumer

- Significant reduction in DPR cost value.
- DPRs may be composed of individual modules of different brands that best meet the needs of the operating organization in terms of optimal balance between quality and cost.
- Optimal SPTA set of DPR modules.
- Reduces the urgency of the problem of low reliability of DPRs through quick and easy on-site replacement of failed low-cost modules with standby modules operated automatically; if the basic units are damaged, no need to repair faulty modules.
- Antimonopoly measures since you don't depend upon the DPR manufacturer.
- Enforcing competition between producers thanks to new market players—small- and medium-sized companies specializing in the production of only certain types of modules rather than complete DPRs.
- Simplifying the testing of the DPR and reducing the influence of "human factors."
- Easier software, which allows selecting the most suitable and convenient application (interface) and painless replacement of DPR applications (interfaces).
- Acceleration of technological progress in the field of DPR without complicating the operation and adding any additional problems while upgrading devices to the next generation.
- Reduces the cost of updating the DPR, as the whole DPR will not have to be updated every 10–15 years as is the case today. You will only need to update some individual modules. Moreover, you will be able to update the CPU board more frequently than now, thus speeding up the technological progress in this area.

For a Manufacturer

- No need to produce outdated modules to old DPR installations.
- No lifetime free repairs.
- Significant increase in the sales of individual modules.
- New emerging markets for a basic framework, and a set of different applications and libraries compatible with a common basic framework.

- Specialization in the production of individual and the most profitable types of modules.
- Small- and medium-sized companies that do not have sufficient resources for the development and manufacture of complete DPRs may become competent competitors.
- Competitive advantage for domestic manufacturers, who will be the first to start production of the DPRs as modules in a specific country, compared with foreigners.

The proposed direction of development could open the DPR market to new players: some of them would produce analog input modules equipped with current and voltage transformers, and the others would produce CPU modules, or software. The consumer could assemble the DPR from separate modules of different manufacturers, just as is the case with PCs today, based on the cost and quality of these modules, as well as use the same software for all its DPRs. It would solve many of the issues raised above, and significantly reduce the cost of relay protection. The latter could also allow installing two sets of identical protection types instead of one in order to improve reliability, while the other set would be used as backup, starting automatically upon receipt of the "watchdog" signal coming from the damaged core DPR. In addition, it would eliminate the need to use individual power sources for each DPR; rather, it would enable the use of one double high-capacity power source set of improved reliability for the entire cabinet. And, finally, it would allow installing many service modules, capable of improving the DPR's reliability, in the same cabinet.

Thus, the relay protection maintenance would be simpler as the service staff would not need to read thick folios about different DPRs installed in the facility and study specific characteristics of the software of each DPR type. In addition to easier maintenance and time savings for installation and learning new types of protection, it would significantly reduce the percentage of errors caused by the so-called human factor.

Such design of DPR would solve the problems of testing complex DPR functions. But is the proposed concept economically attractive? According to a report by Newton-Evans Research Co., in 2006 a group of leading world companies, such as ABB, AREVA, SEL, Siemens, and NARI, earned about $950 million selling protection relays, while a second group, which included Basler, General Electric, and Schneider, earned another $500 million. Besides the companies mentioned in the report, large companies such as Beckwith, Cooper Power, Orion Italia, VAMP, Woodward, and others are also participating in the DPR market.

According to the report, it is expected that Western companies would earn as much as $2 billion in 2009 selling DPRs. The Russian market and other former Soviet Union (FSU) countries are represented by both large Western and local manufacturers: these include Bresler, Ekra, RELSiS, Kievpribor,

Meandr Inc., Mechanotronica, Radius Automatika, Energomashvin, ChEAZ, and VNIIR. Since the share of DPRs in relay protection in Russia is as low as 10%, there's still a big sector of the market in which to develop this concept in the near future.

Of course, the share of DPRs in the market is significantly smaller than that of PCs, which earn $52 billion dollars in sales; it is still big enough, however, for successful implementation of the proposed concept.

How should we start to implement this concept? We believe that the starting point should be the development of the above-mentioned national standard by a wide range of professionals, including protection engineers, relay operators, scientists, design engineers, and industry representatives, as well as a contract with the Nari-Relays, as a first step toward the realization of this concept.

References

1. Gurevich, V. Reliability of Microprocessor-Based Relay Protection Devices: Myths and Reality. *Serbian Journal of Electrical Engineering*, vol. 6, no. 1, 2009, pp. 167–186.
2. Gurevich, V. I. Some Performance and Reliability Estimations for Microprocessor-Based Protection Devices [in Russian]. *Electric Power News*, no. 5, 2009, pp. 29–32.
3. Gurevich, V. Reliability of Microprocessor-Based Protective Devices Revisited. *Journal of Electrical Engineering*, vol. 60, no. 5, 2009.
4. Problems of Microprocessor Protective Devices: Specialist's Opinions, Unsolved Problems and Publications. http://digital-relay-problems.tripod.com
5. Gurevich, V. I. Problems of Microprocessor Protective Relays: Who Is Guilty and What to Do? [in Russian]. *Energo-Info*, no. 10, 2009, pp. 63–69.
6. Belyaev, A., V. Shirokov, and A. Yemelianzev. Digital Terminals of the Relay Protection. Experience in Adaptation to Russian Conditions [in Russian]. *Electrical Engineering News*, no. 5, 2009.
7. Gurevich, V. Tests of Microprocessor-Based Relay Protection Devices: Problems and Solutions. *Serbian Journal of Electrical Engineering*, vol. 6, no. 2, 2009, pp. 333–341.
8. Gurevich, V. I. Sensational "Discovery" in Relay Protection [in Russian]. *Power and Industry of Russia*, nos. 23–24, 2009.
9. Gurevich, V. Sophistication of Relay Protection: Good Intentions or the Road to Hell? [in Russian]. *Energize*, 2010, January–February, pp. 44–46.
10. Gurevich, V. Microprocessor Protective Relays: How Are They Constructed? [in Russian]. *Electrical Market*, no. 4, 2009, pp. 46–49; no. 5, 2009, pp. 46–50; no. 6, 2009, pp. 46–50; nos. 1–2, 2010, p. 3.

Index

A

All-Russian Electrotechnical Institute, 275–276
All-Russian Relay Protection Scientific Conference, 260
Alternative protective relays, 287–346
  Bell Laboratories, 8, 298–299
  Feiner, A., 298–299
  HV DC equipment, 325–335
    relay winding, 315, 334–335
  International Electrotechnical Commission, 167–171, 347–380
  memory, 294–301
  overcurrent protective relays, 303–308
  polarized reed switches, 294–301
  power reed switches, 301–302
  Quasitron, multipurpose protection relay, 303
  reed switches, 179, 181, 288–302, 313–317, 320–341
    hermetic magnetically controlled contact, 289
    magnetically controlled contact, 289
    Ryazan Ceramic-Metal Plant, 291
  simple, very-high-speed overcurrent protection relays, 308–325
  solid-state protective relays, 335–346
    current relay with restraint, 342–346
    current relays with independent, dependent time delays, 340–341
    with high release ratio, 338–339
    instantaneous current relay, 337–338
    relay of differential protection, 342
    relay of power direction, 341

  thyristors, 32, 34–36, 39, 176, 189, 220, 334–335, 357
  universal overcurrent protective relays, 287–288
  Yaskawa Company, 301
ALU. *See* Arithmetic logic unit
Amorphous substances in semiconductors, 1
Analog input modules, 58–63
Analog-to-digital converters, 78–83
Arithmetic logic unit, 94
Arsenic, 2
Artificial intelligence, relay protection with, 138
Asynchronous reset-set trigger, 29
Auxiliary seal-in relay with target, 161

B

Bardeen, John, 8
Basic insulation, 41, 257, 347–348, 361
Basic international standard, 347–353
  basic insulation, 347
  documentation, 350–352
  functional insulation, 347
  rated values, currents, voltages, 348–350
  terms, 347–348
  test procedure, 352–353
Bell Laboratories, 8, 298–299
Bipolar transistor modes, 17–24
  active mode, 19, 21, 239
  quiescent point, 21–24, 151
  saturation mode, 19, 23–24, 32
  working points, 21, 23
Brattain, Walter, 8
Brown-out timer, 221
Brownouts, 188. *See also* Sags; Voltage sags
Buffer storage registers, 95

Milton Keynes UK
Ingram Content Group UK Ltd.
UKHW021831071024
449327UK00021B/1477